Coated and laminated textiles

Walter Fung

The Textile Institute

CRC Press
Boca Raton Boston New York Washington, DC

WOODHEAD PUBLISHING LIMITED
Cambridge England

Published by Woodhead Publishing Limited in association with The Textile Institute
Woodhead Publishing Limited
Abington Hall, Granta Park, Great Abington
Cambridge CB21 6AH, England
www.woodheadpublishing.com

Published in North America by CRC Press LLC, 6000 Broken Sound Parkway, NW
Suite 300, Boca Raton, FL 33487, USA

First published 2002, Woodhead Publishing Limited and CRC Press LLC

Reprinted 2007, 2008

British Library Cataloguing in Publication Data
A catalogue record for this book is available from the British Library.

Library of Congress Cataloging in Publication Data
A catalog record for this book is available from the Library of Congress.

Woodhead Publishing ISBN 978-1-85573-576-7 (book)
Woodhead Publishing ISBN 978-1-85573-751-8 (e-book)
CRC Press ISBN 978-0-8493-1448-3
CRC Press order number WP1448

The publishers' policy is to use permanent paper from mills that operate a sustainable forestry policy, and which has been manufactured from pulp which is processed using acid-free and elementary chlorine-free practices. Furthermore, the publishers ensure that the text paper and cover board used have met acceptable environmental accreditation standards.

Printed by TJI Digital, Cornwall, England

Dedicated to all my teachers and to all those who have shared their knowledge with me. 'When you drink water, remember the source.'

Walter Fung
(Feng Qing Xiang)

Contents

Preface

In writing this textbook it has been the author's intention to produce a work of reference for everyone involved in the business of coated and laminated textile products. Products generally begin with the technologist or marketing specialist who initiates the development, progressing to laboratory staff who carry out pre-production preparatory work and the machine operatives who actually produce the goods, through to the technical service specialist and the salesman who meet the customer. Some focus is on why particular starting materials are used and what specialist properties they possess, because from this comes an understanding of how the materials will behave during production and use. This knowledge is important if problems are to be solved as they arise in the plant so that quality coated or laminated fabric can be produced consistently. This understanding is also essential to the investigation of customer complaints or concerns, and finally – but at least equal in importance to all the other considerations – it enables innovation and the design of new or improved products. Existing products are described in some detail, including background information, so that opportunities for improvements and new product innovation may be identified quickly, especially through making use of the new materials and technology which seem to be becoming available almost on a weekly basis.

The book is written in a clear, concise manner – generally free of intimidating (for the layman) chemical formulae and mathematical equations. Little previous knowledge of the industry or subject is assumed, but it is hoped that the underlying scientific principles are explained adequately for readers to understand why compounds contain different ingredients and why it is important to set processing conditions carefully and adhere to them. Included in the book are properties of materials, details of material preparation and actual processing conditions – some from a 'hands on perspective.' Production management and sales and marketing aspects are also discussed. Test methods are presented along with comments, and the scope for research and new product development is reviewed, together with a section devoted to sources of further information for more detailed research and for keeping up-to-date. Suggestions for improvements in existing products and for exploratory developments are made.

Coating and laminating offer methods of improving and modifying the physical properties and appearance of fabrics, and also scope for the development of entirely new products by combining the advantages of fabrics, polymers, foams and films. There are development opportunities in the area of industrial and medical protective clothing, with a view to making it more comfortable and washable without affecting performance. Disposability is likely to be less attractive in the face of ecological waste disposal concerns. There are interesting, novel 'smart' materials such as phase change materials, temperature memory shape polymers and surface modification processes which result in improved adhesion; these are still to be fully exploited by the textile industry. Coating and laminating poses challenges, not only for the technologist, physicist and chemist but also for the mechanical, chemical and production engineer whose expertise is frequently called upon to solve a particular problem as well as to develop more efficient machinery. It is hoped that this book will also be of use in explaining to these specialists the overall picture and the need for accurate control of the production variables, the factors involved and how they all interrelate with each other. Coated or laminated fabrics are handled differently from non-coated fabric, and manufacturing processes also involve materials such as solvent and water-based resins, films, foams and hot melt adhesives in powder, web and film form.

Control and handling of potentially toxic liquids and fumes in an increasingly environmentally conscious world are also important tasks in coating or laminating plant management. The influence of issues such as the environment and global warming on the industry and how they are likely to influence future products and processes is discussed. Coated and laminated fabrics are, by definition, composed of different materials, although the disposal of these materials at the end of their useful life is apparently not an issue at the time of writing – apart from carpets and PVC-based products. Careful thought, however, should be given to the choice of materials for future products in order to facilitate recycling and disposal and to protect the environment. Research work on these aspects and the impact of plastics and textiles in general on the environment has been underway for some time, and these items are also reviewed and discussed.

Coating and lamination cut across virtually all of the groups into which the products of the textile industry can be classified, and thus the scope for development is extremely wide. The processes of coating and laminating are important steps in the production of composites, another important area of technical textiles. The technologist responsible for research and development is working in an exciting area with tremendous potential and opportunities for innovation. Textile analysts draw attention to the fact that in the developed world, technical textiles offer the most potential for innovation and growth and, some say, the main hope for survival of the textile industry in the developed countries. Coating and lamination are two of the processes by which these aims can be realised, but information, imagination, persistence and determination are needed to make the most of these opportunities in an ever-changing world.

Acknowledgements

A book of this nature would be impossible to produce without the assistance and generosity of other people, in proof-reading, checking of information and in the provision of photographs and diagrams, together with permission to reproduce. Thanks are due to the following, in no special order (with apologies to those inadvertently omitted):

David Rigby (DRA Associates), Hugh Anderson (Vander), Bob Wardle (R Jackson Wardle/Werner-Mathis), Dr Harry Fung (EA Technology), Walter Duncan (Synthomer), Peter Thomson (Baxenden), Geoff Formoy and Mark Nagy (Cornelius), Calvin Woodings (Calvin Woodings Consulting), Karen Furneaux, Tony Sager and Stephen Donnelly (BTTG), Astrid Missner (Meyer Machines), Ben van den Berkmortel (Stork Brabant BV, Netherlands), Marcel Mallens (EMS-Chemie), Richard Scott (DCTA), Robert Jackson (WIRA Instrumentation), Tom Govier (SDL International Ltd), Francis Woodruff and Roy Conway (Web Processing (M/C Ltd)), Steve Wallace and Hannah Cameron (Cameron Balloons), Mr DM Repper (James Heal), Dr Volkmar Bartels (Hohenstein Institutes, Germany), Fred Pfister (DuPont, Geneva), John Barnes (DuPont, UK), Corinne Gangloff (Freedonia Group Inc. Cleveland, USA), David Wallwork and Andy Palliser (Clariant, Leeds), Bob Morley, Bill Whitehouse, Tony Longdon and Ted Richards (Acordis), John Stimpson (John Heathcoat & Co), Marc Saubry-Bobet (Bobet, France), John Briggs (Collins & Aikman Automotive Fabrics Ltd), John Williams and Ian Leigh (Noveon Performance Coatings – formerly BF Goodrich), Alan Southall (Beafort Air-Sea Equipment Ltd), John Croyden (Rohm & Haas), Peter Tyers (Bostik), John Wardley, Colin Vowels and Frank Thomas Wolf (Alveo), Mike Hilton and John Skinner (Caligen Foam), Sid Cooper (Joshua Grieves and Company), Angel Massaro-Fain and Jennifer Burkart (The Intertape Polymer Group, USA), CargoLifter Team (Germany), Autoliv, David Farrell (Dritex International Ltd), Paul Pinchess (Bates Textile Machinery Co Ltd), Siubhan Reid-Litherland (JD Power – LMC Automotive Services), Simon Fung (Bombardier), David Convery and Jo McKnight (Penn-Nyla), Roger Bellfield (Carrington Careerwear), David Karsa and Stuart Patrick (Akcross), Peter Cowsill (BASF), Mike Willatts (Stahl), Andy Stelfox, John Russell, Michael Fawcett and Eric Mailler, and to Patricia Morrison and Woodhead Publishing for the opportunity to write this book.

Abbreviations used in references at end of chapters

ATI	America's Textiles International (now America Textile Industries)
BPR	British Plastic and Rubber
EPN	European Plastic News
IMMFC	International Man-Made Fibres Congress, Dornbirn, Austria
JCF	Journal of Coated Fabrics
JSDC	Journal of the Society of Dyers and Colourists
JTN	Japanese Textile News, Monthly
MPI	Modern Plastics International
MRW	Material Recycling Week
PRW	Plastics and Rubber Weekly
TTi	Technical Textiles International
TuT	Technical usage Textiles (France)

1
General survey

1.1 Definitions

There are two definitions of a 'coated fabric' in the Textile Institute's publication, *Textile Terms and Definitions*.[1] The first one is, 'A material composed of two or more layers, at least one of which is a textile fabric and at least one of which is a substantially continuous polymeric layer. The layers are bonded closely together by means of an added adhesive or by the adhesive properties of one or more of the component layers'. The second definition is, 'A textile fabric on which there has been formed *in situ,* on one or both surfaces, a layer or layers of adherent coating material'. For the purposes of this book, this second definition will be regarded as that definition which most closely describes *coated fabrics*; the operative words are, 'formed *in situ*'. The first quoted definition could also be applied to a *laminated* fabric.

Textile Terms and Definitions defines a 'laminated fabric' (or a 'combined fabric') as, 'a material composed of two or more layers, at least one of which is a textile fabric, bonded closely together by means of an added adhesive, or by the adhesive properties of one or more of the component layers'. For the purposes of this book, a 'laminated fabric', also sometimes called a 'bonded fabric', is considered to be different from a coated fabric, in that the layers are already pre-prepared and the second material can be a film, another fabric, or some other material. However, these definitions are of academic interest and there is little benefit in taking a rigid view.

1.2 Historical background

How long ago the coating and laminating of fabrics began, depends upon on what is regarded as being a coated or laminated fabric; some writers see the fabric wrapping the Egyptian mummies as a coated material. For as long as people have been wearing clothes made from fabric, they must have sought ways of making the material less permeable to wind and water, for increased comfort, increased weather protection and, indeed, for survival. Natural oils, such as linseed oil and tar, were applied to the fabric, dried in the sun and then more oil applied in a repeat process. After this had been done several times, water-repellent fabric was obtained, which would afford better protection against the elements. In fact, an 'oil cloth' industry developed in England and Germany during the 1700s, applying a variety of oils to cotton and even to silk, for the manufacture of garments, sails, tents and other coverings.[2]

It is virtually impossible to ascertain when rubber was actually discovered, but it was already being used by the Latin-American Indians as a proofing material for fabric when the Spanish arrived in the mid-1500s. The *New Shell Book of Firsts* attributes the first 'raincoat' to Francois Fresnau, Chief Engineer at Cayenne, French Guyana, who, in 1747, proofed an old overcoat by smearing latex over it. However, many writers regard Charles Macintosh as the founder of the modern coating and lamination industry, but the success of his invention has depended on important contributions from others. He is recorded as having produced a waterproof material by sticking two pieces of cloth together with rubber dissolved in naphtha. Another version is that he produced a film of rubber and sewed it on to fabric with a needle and thread. Producing commercial garments from this rather cumbersome material was difficult until James Syme, a medical student at Edinburgh University, managed to dissolve rubber in a solvent derived from coal tar. Evaporating the solvent away left behind a deposition or film of rubber. Macintosh obtained the patent rights to this process in 1823 and began to produce the waterproof material which has made his name a household word all over the world.[3–5] One of his first customers used the new coated fabric material to equip Sir John Franklin's expedition to the Arctic in 1824.[3]

Macintosh's material was used by travellers riding on the top of stagecoaches, but when the railways arrived and higher standards of comfort and quality were demanded, wearers began to complain about the ungainly appearance of the garments and the offensive smell. In addition, the rubber tended to soften, become sticky and melt in hot weather. These disadvantages were overcome with advances in technology, and especially by the discovery of vulcanisation by Thomas Hancock in 1843. Vulcanisation, which is a crosslinking process, overcame raw rubber's main defect of becoming physically weak with rising temperature. Charles Goodyear in the USA is also credited with discovering vulcanisation quite independently from Hancock in 1841. Hancock, who manufactured rubber goods in Manchester, became an associate of Macintosh in 1830 and together they produced inflatables, including air beds and pillows and pontoons for floating bridges, in addition to garments. It is interesting to note that Hancock patented the use of inflatable bags to raise sunken ships.[5]

Later, in the mid-nineteenth century, nitro-cellulose was discovered and developed as a fabric coating material for cotton. The material became known as 'gun cotton' (Schoenbein), 'pyroxylin' and 'collodion', and was used in many applications, including car upholstery. During the First World War, the British Government invited the Dreyfus Brothers of Switzerland to build a factory at Spondon, Derby to produce cellulose acetate (referred to as 'dope') to be used for coating fabric for aircraft wings. This factory later produced acetate yarns and became British Celanese – later acquired by Courtaulds – and is now the headquarters of Acordis. In the first half of the twentieth century, several other new polymers and synthetic rubbers were invented and, after some development, came to be used as fabric coatings. The most important ones were polyvinyl chloride (PVC), polychloroprene (the best known being Neoprene made by Du Pont), acrylates and polyurethane. These polymers, plus several other specialist materials, are still used in the modern industry, but much research and development has been carried out to improve properties and, more recently, to produce more environmentally friendly materials and methods. This process continues and has assumed greater importance in recent years.

The earliest adhesives used by people were natural materials such as beeswax, tars, gums, derivatives from animal and fish bones, egg white and rice products. Bitumen was used in the construction of the Tower of Babel. Sealing wax was probably the first commonly used 'hot melt' adhesive. The modern lamination industry has at least part of its origins in the garment industry. Cellulose acetate was the first manufactured plastic to be used to bond collars for shirts in the 1930s, but the process used solvents to soften the cellulose acetate and, although millions of shirt collars were produced every year, it was a 'wet' and therefore not a convenient process. Research work was carried out to discover better methods. In 1948 Harold Rose began to experiment with polyvinyl acetate plasticised with dibutyl phthalate; he developed a continuous process for producing a material which could be used in a 'dry' lamination process for garments. No solvents were involved, and the new interlining could be heat activated using a Hoffman press. He was awarded a bulk order for the material and immediately set up a company called Staflex in 1951.[6] Other materials, including hot melt powders, were later developed, and the turnover of Staflex reached £5 million by 1965 and £30 million by 1975. It is interesting to note that, in the early days, polyethylene was used in collars, but it was discontinued because of discomfort in wear caused by lack of air permeability. Hot melt adhesives are now the focus of much attention because they are solventless and also do not require large amounts of energy to dry off water.

1.3 Fabric coating

The basic method of producing a coated fabric is the direct spreading of a polymer, in the form of a thick liquid or paste, using a sharpened length of metal called a doctor blade or knife. The doctor blade is set up over a fabric, which is held flat by the application of tension – both lengthways and, ideally with the use as stenter pins, widthways – in a coating machine. The thickened polymer fluid is placed in front of the doctor blade and is spread over the surface of the fabric by the forward movement of the fabric into a heated oven. The thickened liquid can be either water based or solvent based, and evaporation of the water or solvent leaves behind a polymer deposit on the fabric. In the case of PVC, the thickened liquid is neither water- nor solvent-based, but consists of a mixture of PVC powder, liquid plastisols and other chemicals. The thickened liquid is referred to as the compound or resin, and is a mixture of the polymer and certain other ingredients to produce specialist properties or to assist the spreading process. This process, described in detail in Chapter 3, allows coated fabric to be produced on a much larger scale than could be achieved by hand. It was first used for coating rubber, and is still sometimes referred to as 'the rubber spreading' machine even though it is now used for other polymers. Hancock's original machine probably did not have stenter pins and, in fact, some older machines do not have them, relying entirely on lengthways tension. Compounding or mixing of resin is an important process, both in the actual formulation of the recipe and also in the actual physical mixing, see Chapter 2. In the quest for more comfortable, softer, more aesthetically pleasing products, transfer coating and more technically sophisticated methods, such as rotary screen techniques, have been developed. These methods are discussed in Chapter 3.

Coated fabrics are engineered composite materials, produced by a combination of a textile fabric and a polymer coating applied to the fabric surface. The polymer coating confers new properties on the fabric, such as impermeabilty to dust particles, liquids and gases, and it can also improve existing physical properties, such as fabric abrasion. The fabric component generally determines the tear and tensile strength, elongation and dimensional stability, while the polymer mainly controls the chemical properties, abrasion resistance and resistance to penetration by liquids and gases. Many properties, however, are determined by a combination of both these components, and both base-fabric and polymer must be carefully selected by a thorough consideration of the properties required in the finished article.

The most familiar coated fabric is probably the waterproof jacket worn for protection from the rain. Protective clothing for the police, firemen, postmen and the services also made from coated fabric. Examples of domestic articles made from coated fabrics include shower curtains, mattress tickings, flame retardant (FR) upholstery, wipe-clean table cloths and curtain linings. Aesthetic effects and easy clean features can also be achieved by fabric and polymer combinations, and man-made leathers used for seat covers and apparel are essentially coated fabrics. Coated fabrics in the industrial area include materials for tarpaulins, life jackets, life rafts, aircraft safety chutes, hovercraft skirts, protective coverings, awnings, aircraft fuel tanks and flexible containers. Seat upholstery for automobiles, coaches and trains is likely to be coated to improve abrasion resistance, improve FR properties and reduce dust particle penetration. The product enjoying the largest growth at the time of writing is the airbag, which is coated mainly in order to control porosity, although coating also reduces edge fraying and makes panel cutting and certain other processes easier to carry out.

1.4 Fabric finishing

A process related to fabric coating is 'fabric finishing', where a chemical or a polymer covers only the yarns making up the fabric without closing the gaps in between. In fabric coating, the small gaps in between the individual yarns are covered to varying degrees. In the case of a waterproof coating, the gaps are fully 'bridged' by the coating, the polymer forming a continuous layer on the fabric surface. In other coated fabrics, the gaps may not be fully covered, and these fabrics will be porous to varying degrees. Fabric finishes are usually applied by a 'dip and squeeze' process known as padding or impregnation, which is widely used in textile factories. The need for distinction between 'coating' and 'finishing' is again largely academic – the two merge in certain instances, because repeated impregnation with some polymers can produce a continuous coating on a fabric.

Most apparel fabrics or household textiles have a fabric finish of some description on them, e.g. an easy care finish on cotton, a mothproof finish on wool, a water-repellent finish, or a soft finish to improve the handle or soft touch of the fabric. The technology of fabric finishing covers a very large area with many dozens of different chemicals, and the term also embraces mechanical finishing processes such as embossing. This area is outside the scope of the present book, but certain chemical fabric finishes can also be applied by a coating process involving foamed compounds and a description of this process is included in Chapter 3. The foam processing method of fabric 'finishing' can provide benefits both in terms of properties obtained and also in environmental factors.

1.5 Fabric lamination

Polymer materials, which may not be easily formulated into a resin or a paste for coating, can be combined with a fabric by first preparing a film of the polymer, and then laminating it to the fabric in a separate process. There are various techniques and several different types of adhesive and machinery used in the lamination process. Producing an adhesive bond, which will ensure no delamination or failure in use, requires lamination skills and knowledge of which adhesive to use. It is generally relatively simple to produce a strong enough bond, the challenge is to preserve the original properties of the fabric and to produce a flexible laminate with the required appearance, handle and durability.

Adhesives are available as water-based or solvent-based fluids or as a 'hot melt' material, i.e. a material which is either a solid or a jelly and which melts on the application of heat. Hot melt adhesives are produced in the form of films, webs (rather like a net curtain), granules, powders or as jellies. They can be of several different chemical types, such as polyolefin, polyurethane, polyester, polyamide, or alloys, or blends of different polymers or copolymers. Of course different machinery is required to apply the different forms of adhesive and this must be done under controlled production conditions to produce an acceptable bond without stiffening the fabric or any other undesirable effects.

Which lamination process and which machinery to use are also determined by the physical properties of the fabric being processed and by the performance requirements, such as bond strength and durability of the laminate being produced. The fabric may be a flat woven, which is dimensionally stable and which can be tensioned to produce a flat uniform surface for the lamination process to take place without any difficulty, or it may be a stretchy knit material or a fabric with a pile or raised surface, which is easily distorted or crushed. Machines and methods of lamination are available to process each of these fabrics without altering the physical properties or appearance. The machinery selected should also produce the required performance specifications at a commercial production rate and cost.

As has been mentioned, lamination techniques have been used in the garment industry for collars, waistbands and hems, generally replacing or supplementing sewing. The lamination process shortened production times, reduced costs and allowed more consistent quality. Lamination techniques are also used to replace sewing in other areas, such as the automotive and even sail making industries. Laminated fabrics are used in waterproof and breathable protective clothing, car seat upholstery, footwear, and in numerous other applications. According to the Textile Institute's definition, car headliners, parcel trays and door casings are also laminated fabrics – they are certainly covered by a laminated fabric. As will be seen, environmental considerations also play a significant part in process selection, and they are likely to become more important in the future. Chapter 3 will explain and discuss the main methods.

1.6 Composite materials

Composites are textiles and plastics in a macroscopic combination – alloys are combinations on the microscopic scale. In general, the chemical properties are determined by the plastic component and the physical properties by the fibre. An example is the glass fibre/plastic group of composites; the glass fibre has very high tensile strength, but is very brittle because it is extremely sensitive to cracks and surface defects. If the surface is protected by plastic in a composite, the formation of cracks is prevented and the composite has very high strength. Composites are produced using some of the techniques used in manufacturing coated and laminated fabrics, e.g. surface coating, impregnation and lamination. Coated and laminated fabrics are composite materials, but the term 'composite' has come to be recognised as an engineering material produced from combination of plastics with both fibres and fabrics, frequently used in structural, specialist and high technology applications. The science of composites depends significantly on mechanical engineering, in addition to plastics and textile technology, and is outside the scope of this book.

1.7 The commercial scope of coated and laminated textiles

Fabric coating and lamination are technical processes and the products are, therefore, technical products which form an important sector of the textile industry. Much depends on definitions and how products are categorised, and it is difficult to find figures which specifically apply to coated and laminated fabrics. A recent analysis by David Rigby Associates (DRA) estimates that technical textiles constitute over 40% of the textile industry in the developed countries of the world; and it is particularly important because in these developed countries it is the major sector for growth.[7–9] This figure is expected to grow to 50% within the next few years. However, even in developing counties such as China, where technical textiles represented only about 12% of the industry when the report was published (April 1997), substantial growth, to reach 20% by the year 2000, was forecast. Significant rates of growth are expected in the developing world in general, including Eastern Europe and South East Asia. In 1997, the total global textile fibre consumption was estimated to be approximately 53 million tonnes, of which technical textiles were believed to constitute over 10 million tonnes, i.e. 19% worth at least $54 billion (US dollars) – even before coating and conversion which adds further value. These figures, however, apply to the four product areas of yarns and fibres, fabrics, nonwovens and composites. Coated and laminated fabrics are grouped in the fabrics product area, although some nonwovens are coated and laminated.

Many nonwovens incorporate a polymer binder of some description, which is similar to the materials used for textile coatings. Both nonwovens and composites are growing substantially faster than actual fabrics (see Tables 1.1 and 1.2).

The Techtextil Symposium (for Technical Textiles), held annually at the Messe, Frankfurt, divides the technical textile area into 12 application areas – agricultural, building and construction, clothing (interlinings, sewing threads, waddings and fibre fillings), geotextiles, household textiles,

Table 1.1 World consumption of technical textiles and nonwovens by region ('000 tonnes)

	1985	1995	2005 (forecast)	% compound average growth 1989–95	% compound average growth 1995–2005
West Europe	1674	2367	3111	3.5	2.8
East Europe	407	296	418	–3.1	6.6
North America	2240	3057	3886	3.2	2.4
South America	181	280	428	4.4	4.3
Asia	1262	2696	4510	7.9	5.3
Australasia	83	120	179	3.7	4.1
Rest of world	214	505	1011	8.9	7.2
All world	**6062**	**9321**	**13688**	**4.4**	**3.9**

Source: David Rigby Associates World Technical Textiles Consumption Model and reproduced with kind permission.

Table 1.2 World consumption of technical textiles by product type, 2000–2005

	2000 10^3 tonnes	2005 10^3 tonnes	Growth (% pa)	2000 $ million	2005 $ million	Growth (% pa)
Fabrics	3760	4100	1.7	26710	29870	2.2
Nonwovens	3300	4300	5.4	14640	19250	5.6
Composites	1970	2580	5.5	6960	9160	5.6
Other textiles*	2290	2710	3.4	11950	14060	3.3
All textile products	**11320**	**13690**	**3.9**	**60260**	**72340**	**3.7**

*Includes ropes, twines, thread, fibrefill, etc.
Source: David Rigby Associates/Techtextil and reproduced with kind permission.

industrial textiles, medical uses, transportation textiles, environmental protection, packaging, protective clothing (both industrial and leisure), and sports applications (not clothing). Coated and laminated fabrics appear in most, if not all, of these areas, especially transportation, industrial, clothing, and medical uses. Significant growth is expected in all 12 application areas, especially in the geotextile, environment and protective clothing categories. Transportation textiles, which includes uses in automobiles, other road vehicles, aircraft, marine and rail, is the largest single area in technical textiles. This is shown in Table 1.3, prepared from DRA's World Technical Consumption Model.[7,8]

Information is also available on the volumes and various types of the European protective clothing market with six areas classified.[9] Clothing for purely personal use, e.g. for sports activities, is not included. The largest areas are for dust and particle barrier protection and for gas and chemical protection. Over half of the total volume (1996 figures) of 206 million m^2 is nonwoven fabrics (124 million m^2), the remainder being woven or knitted. A proportion of non-disposable garments is likely to be produced from coated or laminated fabric. With future restrictions on landfill, there may well be a decrease in the amount of nonwoven disposable garments and an increase in washable coated protective garments. Some industrial protective garments are heavy, bulky, movement restricting and uncomfortable, but with development and use of new fibres and materials, there is opportunity for improvement. Leisurewear protective clothing is significantly lighter and more comfortable today compared to, say, the mid-1980s.

A Japanese report (October 1999) put the world clothing market volume at $1062 billion with an annual increase of 2.5% each year until 2010. It stated that sportswear represented 6% of the total in 1996, but predicted this to be 13% in 2010.[10] The report commented that technical clothing for active sports has entered the 'casual' field, that technical fabrics are on the increase, regardless of the age of the purchaser, and that top brands are preferred to cheaper ones.

The production of fabric for protective clothing is expected to increase in developing countries, as industrial workers demand a level of protection and comfort comparable to that available to their fellow workers in the USA and Western Europe. In developed countries, government legislation on health and safety and the threat of employer legal liability for workers' safety has already produced an increase in personal protective equipment (PPE). Protective clothing, whether for the sportsman, the military or the industrial worker, represents a large area in technical textiles. These classifications merge and overlap, and it is difficult to find figures which apply to the individual areas. *Datamonitor* puts the 1999 market value for the UK sports clothing market at £2675 million or $4314 million. Waterproof coats

Table 1.3 World consumption of technical textiles by application, 2000–2005

	2000 10^3 tonnes	2005 10^3 tonnes	Growth (% pa)	2000 $ million	2005 $ million	Growth (% pa)
Transport (auto, train, sea, aero)	2220	2480	2.2	13080	14370	1.9
Industrial products and components	1880	2340	4.5	9290	11560	4.5
Medical and hygiene textiles	1380	1650	3.6	7820	9530	4.0
Home textiles, domestic equipment	1800	2260	4.7	7780	9680	4.5
Clothing components (thread, interlinings)	730	820	2.3	6800	7640	2.4
Agriculture, horticulture and fishing	900	1020	2.5	4260	4940	3.0
Construction – building and roofing	1030	1270	4.3	3390	4320	5.0
Packaging and containment	530	660	4.5	2320	2920	4.7
Sport and leisure (excluding apparel)	310	390	4.7	2030	2510	4.3
Geotextiles, civil engineering	400	570	7.3	1860	2660	7.4
Protective and safety clothing and textiles	160	220	6.6	1640	2230	6.3
Total above	**11340**	**13680**	**3.9**	**60270**	**72360**	**3.7**
Ecological protection textiles*	230	310	6.2	1270	1610	4.9

* Already counted in several categories above.
Source: David Rigby Associates/Techtextil and reproduced with kind permission.

and jackets are estimated at 9.5% of this total, i.e. £254 million or $409.8 million.[11] Waterproof protective clothing has undergone a revolution between 1940 and 1980, in the quest for more thermal comfort, and comfort in general, combined with light weight, fabric softness and high performance in protection from the elements. This process continues with developments in breathability, warmth without bulk, phase change material, and better performance at lower cost. This search for more comfort with high performance has spread to footwear, socks, and even gloves. The process was probably initiated with the introduction of Gore-Tex by W L Gore and Associates in the mid-1970s. Since then the sports world has seen the introduction of a large number of laminated waterproof breathable fabrics and waterproof and breathable coatings.

The European Personal Protective Equipment Directive 89/686/EEC ensures that industrial protective clothing not only provides effective protection against the hazard, but is also well designed and comfortable to wear. Similar regulations are already in place in other developed areas of the world and are likely to be introduced in developing countries, which will present opportunities for future market growth. Even in developed countries, there is still the need for improved performance and more satisfactory test methods as living standards continue to increase. In recent years, additional requirements due to ecological considerations pose more challenges and offer opportunities for research and development.

The biggest single item in the transportation area is the automobile, with about 48 million units made each year. The volume of cars made in the developed countries is now steady, and there is production overcapacity. However, there is still opportunity for substantial growth in the developing world, especially in Eastern Europe and South America over the next five years and in South East Asia over the next ten, see Tables 1.4 to 1.6. Each car contains about 7 m^2 of car seat fabric laminated to polyurethane foam, and most also contain at least one airbag, which is more often than not made from a coated fabric. Airbags have seen quite spectacular growth in recent years – the fastest growth item in the entire textiles industry – but, although the growth is thought to be past its peak, steady growth is still expected for the next few years. However, much research is being applied to reduce injuries in traffic accidents, and the introduction of new or improved products could prolong this growth for longer than many analysts would predict. Acordis Industrial Fabrics estimate that the 1999 figure of 18000 tonnes of yarn (equivalent to 75 million m^2 of fabric) for airbags in Europe will increase to 25000 tonnes (115 million m^2) by 2004. Worldwide, the 1999 figure of 50000 tonnes (200 million m^2) is expected to reach 80000 tonnes (300 million m^2) by 2004.[12]

With the ending and phasing out of import quotas and China's entry to the World Trade Organisation, investment in technical textiles is likely

Table 1.4 World personal vehicle sales ('000 units)

	2000	2001	2002	2003	2004	2005	2006	2011	2016
North America*	18652	17545	17300	17511	17712	17910	17865	18898	19861
% change	*4.1*	***–5.9***	***–1.4***	***1.2***	***1.1***	***1.1***	***–0.3***	***5.8***	***5.1***
Argentina/ Brazil	1385	1535	1681	1864	2034	2181	2303	2708	3253
% change	*9.1*	***10.8***	***9.5***	***10.9***	***9.1***	***7.2***	***5.6***	***17.6***	***20.1***
Western Europe	14941	14649	14806	15143	15587	15844	15997	17373	18027
% change	*–1.6*	***–2.0***	***1.1***	***2.3***	***2.9***	***1.6***	***1.0***	***8.6***	***3.8***
Eastern Europe	2469	2173	2446	2738	3048	3302	3572	4354	4938
% change	*–0.2*	***–12.0***	***12.5***	***12.0***	***11.3***	***8.4***	***8.2***	***21.9***	***13.4***
Japan	4260	4336	4426	4523	4609	4680	5026	5432	5556
% change	*2.5*	***1.8***	***2.1***	***2.2***	***1.9***	***1.5***	***7.4***	***8.1***	***2.3***
Asia/Pacific	3160	3173	3379	3573	3784	4025	4281	6114	8135
% change	*10.8*	***0.4***	***6.5***	***5.8***	***5.9***	***6.4***	***6.4***	***42.8***	***33.1***
Other†	4354	4529	4698	4872	5038	5197	5278	6232	6913
% change	*4.6*	***4.0***	***3.7***	***3.7***	***3.4***	***3.2***	***1.5***	***18.1***	***10.9***
World	**49221**	**47940**	**48735**	**50225**	**51813**	**53140**	**54323**	**61110**	**66682**
% change	*2.5*	***–2.6***	***1.7***	***3.1***	***3.2***	***2.6***	***2.2***	***12.5***	***9.1***

* North America includes USA, Canada and Mexico and the figures refer to cars and light trucks that are used as personal transport vehicles.
† The total for sales in 'Other' also includes a statistical balancing item to compensate for inconsistencies and inadequacies in national data and to equate sales to world production.
N.B. For countries in each heading, please see foot of Table 1.6.
Source: National Sources, JD Power-LMC. Reproduced with kind permission of JD Power-LMC.

to become more essential for the survival of the textile industry in many of the developed nations of the world. Some forecasters predict 75% or more of fibre consumption in the more advanced economies will eventually be in technical applications. These products are less price sensitive in the market place than regular apparel and household goods, and also command better profit margins. It should be noted, however, that, although China is the largest exporter of textiles, it is also, since 1997, the world's third largest importer.[13] There is scope for the export of high tech products to China, both for industrial purposes and also for high quality consumer products for the rapidly growing number of Chinese with large disposable incomes.

Table 1.5 World light commercial vehicle* sales ('000 units)

	2000	2001	2002	2003	2004	2005	2006	2011	2016
North America	1113	1065	1084	1103	1141	1158	1377	1578	1543
% change	*−0.2*	*−4.3*	***1.8***	***1.8***	***3.4***	***1.5***	***19.0***	***14.6***	***−2.2***
Argentina/ Brazil	286	330	395	459	500	537	569	653	779
% change	*8.1*	*15.5*	***19.6***	***16.2***	***9.1***	***7.4***	***5.9***	***14.7***	***19.4***
Western Europe	1785	1682	1587	1602	1674	1739	1778	1765	1859
% change	*5.5*	*−5.8*	***−5.7***	***1.0***	***4.5***	***3.8***	***2.3***	***−0.8***	***5.4***
Eastern Europe	416	344	418	502	594	678	722	880	1031
% change	*17.1*	*−17.3*	***21.5***	***20.1***	***18.3***	***14.0***	***6.5***	***22.0***	***17.2***
Japan	1617	1642	1687	1735	1810	1841	1882	2222	2295
% change	*−0.4*	*1.5*	***2.7***	***2.9***	***4.3***	***1.7***	***2.2***	***18.1***	***3.3***
Asia/Pacific	2169	2207	2279	2381	2494	2661	2856	3648	4282
% change	*23.2*	*1.7*	***3.3***	***4.5***	***4.8***	***6.7***	***7.3***	***27.7***	***17.4***
Other	432	456	482	499	514	525	541	575	650
% change	*4.2*	*5.7*	***5.6***	***3.5***	***3.0***	***2.2***	***2.9***	***6.4***	***13.0***
World	**7818**	**7727**	**7931**	**8281**	**8728**	**9139**	**9726**	**11321**	**12439**
% change	***8.2***	***−1.2***	***2.6***	***4.4***	***5.4***	***4.7***	***6.4***	***16.4***	***9.9***

*Light commercial vehicles are those of less than 6 tonnes gross vehicle weight (GVW).
N.B. For countries in each heading, please see foot of Table 1.6.
Source: National Sources, JD Power-LMC. Reproduced with kind permission of JD Power-LMC.

Demand for coated fabric in the USA is expected to reach 635 million square yards (531 million m^2) by the year 2000 – an annual increase of 3.9% from the 525 million square yards (439 million m^2) in 1998 (see Table 1.7). The Freedonia Group have categorised the US market into eight product areas.[14] The growth areas are believed to be in transportation, both automobile and non-automobile applications, and in specialist niche protective clothing for the increased safety of industrial workers, for example from pesticides and biological agents. Growth is likely for apparel to protect sensitive microelectronics against contamination from the human body; anti-static and lint-free properties are also essential in this application. Specialist sports and leisure industry goods, and apparatus for improving the environment, such as better and more economical filters for gases and

Table 1.6 World heavy commercial vehicle* sales ('000 units)

	2000	2001	2002	2003	2004	2005	2206	2011	2016
North America	532	448	481	547	580	597	605	639	714
% change	***–9.6***	***–15.8***	***7.4***	***13.8***	***6.1***	***2.8***	***1.4***	***5.6***	***11.7***
Argentina/ Brazil	92	93	91	90	99	107	116	139	169
% change	***22.9***	***0.7***	***–2.3***	***–0.9***	***10.1***	***8.6***	***7.8***	***20.1***	***21.6***
Western Europe	377	336	317	328	351	371	387	465	518
% change	***5.3***	***–11.0***	***–5.4***	***3.2***	***7.1***	***5.8***	***4.4***	***20.0***	***11.4***
Eastern Europe	179	178	216	247	271	280	289	325	385
% change	***18.9***	***–0.9***	***21.5***	***14.6***	***9.6***	***3.5***	***3.3***	***12.3***	***18.5***
Japan	81	77	80	98	134	147	153	152	170
% change	***–2.2***	***–5.7***	***4.7***	***22.3***	***36.3***	***9.9***	***3.8***	***–0.2***	***11.5***
Asia/Pacific	674	717	804	898	971	1081	1163	1201	1376
% change	***6.2***	***6.4***	***12.0***	***11.7***	***8.2***	***11.2***	***7.6***	***3.3***	***14.6***
Other	42	45	49	52	55	57	61	80	106
% change	***7.6***	***6.1***	***8.2***	***6.7***	***6.6***	***2.0***	***8.2***	***30.3***	***33.1***
World	**1979**	**1893**	**2037**	**2260**	**2462**	**2640**	**2774**	**3001**	**3438**
% change	***2.5***	***–4.3***	***7.6***	***10.9***	***8.9***	***7.2***	***5.1***	***8.2***	***14.6***

Source: National Sources, JD Power-LMC. Reproduced with kind permission of JD Power-LMC.
*Heavy commercial vehicles include trucks of more than 6 tonnes GVW, and (except in USA/Canada) coaches.
Countries in each grouping are as follows:
North America – USA, Canada, Mexico
Western Europe – Germany, Italy, France, UK, Spain, Portugal, Ireland, Netherlands, Denmark, Bel/Lux, Greece, Sweden, Norway, Finland, Austria, Switzerland
Eastern Europe – Bulgaria, Croatia, Czech Republic, Hungary, Poland, Romania, Russia, Slovakia, Slovenia, Turkey, Ukraine
Asia/Pacific – Korea, China, India, Malaysia, Indonesia, Taiwan, Thailand, Philippines
Other – Australia, Colombia, Venezuela, South Africa plus others

liquids, also present opportunities. The current trend for real leather is expected to moderate and the durability of coated fabrics could be an advantage, especially in contract furniture markets. Another growth area with potential for technical products is the aviation and aerospace industries; air travel is set to grow at about 15% for the foreseeable future, with

Table 1.7 Coated fabric demand in USA (millions of square yards)

	1989	1998	2003	% annual growth (1989/98)	% annual growth (1998–2003)
Coated fabric demand	355	525	635	4.4	3.9
Item					
Motor vehicles	79	137	182	6.3	5.8
Furniture	74	101	115	3.5	2.6
Industrial	46	53	61	16	2.9
Protective clothing	28	47	56	5.9	3.6
Wall coverings	37	45	47	2.2	0.9
Book coverings	27	32	34	1.9	1.2
Non-automotive transportation equipment	12	25	36	8.5	7.6
Awnings, tents, others	52	85	104	5.6	4.1

Source: Freedonia Group Inc. (Cleveland OH) and reproduced with kind permission.

competition between airlines intensifying. Populations in the developed world have higher disposable incomes and more leisure time and will take part more in active sports, such as sailing, windsurfing and ballooning – all of which are likely to require coated fabrics.

The fear of occupational exposure to bloodborne pathogens, such as the human immunodeficiency virus and hepatitis B, led to a growth in use by the medical profession of barrier materials for clothing. These range from micoporous PTFE (polytetrafluoroethylene) film based products to laminated nonwovens. Expensive goods have to be washable and able to withstand sterilisation treatments. Disposable garments, however, incur environmental concerns, such as landfill or incineration. A significant growth in the market for incontinence garments is also expected as the aged populations of the world increase. The US retail and institutional market is expected to reach $2.1 billion in 2005, whilst the worldwide figure is expected to be $5.8 billion.[15] At present the bulk of this market is in disposable items but, in a similar way to medical products, the environmental factors may well produce a need for products which are more durable and washable.

1.8 The technical scope of coated and laminated textiles

The textile coating industry is heavily dependent on the polymer and speciality chemicals industries for polymers, additives, process aids, and many other chemicals necessary to produce the specified customer properties at the quality standard and durability demanded. The manufacturing procedures used in textile coating and lamination are related to similar processes in the film and paper industries; both films and papers are coated with polymers and laminated to other materials, and in fact the dividing line between paper and some nonwoven fabrics is quite diffuse. The nonwovens fabric industry is possibly the biggest single user of many polymer resins as binders. Textile coating uses some polymers used in the paint industry, such as acrylics and polyurethanes, and in fact some analysts now refer to the paint industry as the coating industry. An analogy has been made to painting; the best results are obtained in textile coating by the application of a few thin layers, rather than one thick layer – just like painting.

Coating machines of more complexity than those used in textiles are used in film processing, and the volumes and processing speeds are far higher. Paper is also coated and laminated and, again, the volumes and speeds of processing are higher than those used in textiles. However, films and paper are in general more dimensionally stable than textiles, but there are other problems which arise when producing at such high speeds. There are probably lessons to be learnt and synergies to be gained by comparisons with these related industries.

Other industries related to fabric coating are rubber, where coated fabric has its origins (many rubber materials are still used as textile fabric coatings), and the highly technical and growing area of fabric/polymer composites. The present book, however, will deal only with the more 'traditional' textile aspects, and will only mentions these other industries briefly. Fabric coating is also related to fabric printing; indeed, by some definition a print is a coating of ink, held on by binders, some of which are polymers used in coating, and the printing ink contains thickeners similar to those used in fabric coating. An 'all over' print is indistinguishable from a very lightweight coating, and the rotary screen printing machine is a convenient way of applying such coatings.

Fabric coating and lamination draws on several disciplines, including textile technology, chemistry and polymer chemistry and engineering skills. Textile technology is required for design, construction and preparation of the base fabric, which determines tear and tensile strength and coating suitability. The fabric must also be prepared by the appropriate finishing processes, to ensure satisfactory polymer adhesion and to provide a soft, flexible handle and drape. Polymer chemistry is required to decide polymer type, e.g. PVC, polyurethane, acrylic, and also the knowledge and skills necessary to mix coating recipes. Coating compounds incorporate chemical additives to produce specific properties, for example FR chemicals, plastisols, processing aids, etc. Chemical engineering and mechanical engineering knowledge is required to design coating machine ovens with the correct drying rate and fume and solvent (or water) extraction, and to produce goods with the specified technical properties at commercial production rates. Material handling, especially of coated (and laminated fabrics), is sometimes neglected but is an extremely important factor. It is frequently critical in determining profitability, because soiled, damaged, and creased fabrics are responsible for much downgrading.

There is now more instrumentation, as the 'human element' is reduced and more automatic and also more reliable controls are built into processing machinery. Information technology and computers are playing an important part in work scheduling, and are optimising use of machine time and resources by better communications, stock control and more efficient purchasing. These trends are likely to increase with even further advances in information technology and more use of computers generally.

People-management skills are needed to train and motivate both operatives and staff, to cultivate team spirit and to maintain a culture of continuous improvement. These are necessary for innovation and to produce quality products with minimum second quality material – preferably with 'zero' defects to ensure commercial success. The commercial bulk production of goods is as challenging as the original development in the laboratory – perhaps more so, because the operatives are generally less qualified than laboratory staff. The work can be somewhat repetitive, but requires conscientious attention to detail at all times, especially when producing at speed, because many metres of fabric may be processed incorrectly before a fault is noticed. Regular machine maintenance and meticulous cleaning are also essential requirements.

Environmental considerations are becoming increasingly important, and governments and local authorities strive to improve air quality by the reduction of volatile organic compounds (VOCs) and other process emissions. Good air quality is not only important for health and safety reasons, but also in the prevention of global warming and climate change. There are now restrictions on effluent and on the use of certain potentially harmful chemicals used in compounds for textile coating. Reduction in the quantity of waste going to landfill and the encouragement of recycling are also having their effect on the coating and lamination sector of the textile industry. The recyclability of coated fabrics is under review and could well become a significant factor in choice of materials. It would seem that many coated fabrics, comprising two or more polymers joined together, will be difficult to recycle. With this problem in mind, research work is being carried

Table 1.8 Properties that can be obtained by fabric resin coating or by film lamination

General fabric properties

Improved dimensional stability

Less stretch distortion

Less pattern distortion

Increased tear strength (use of silicone resins)

Reduced tear strength (if desirable)

Anti-fray/reduced unravelling of edges

Stiffening (when desirablez

Improved handling (in processing)

Easier and more accurate panel cutting/sewing

Increased friction, more grip (if required)

Increased slide (less friction if required)

Improved crease resistance

Improved abrasion/pilling/snagging

Reduced tendency to curl

Reduced porosity/permeability to gases/liquids

Full impermeability to gases/liquids

Improved/easier cleanability

Soil resistance

Water/liquid repellence

Wipe clean properties

Thermoformability

Heat sealability

Weldability

Thermal insulation

Acoustic insulation

Increased bulk

Increased thickness

Improved printability

Improved embossability (patterns/grained effect)

Artificial leather/suede

Appearance/aesthetics

Gloss/matt appearance

Smooth continuous surface

Opacity

Full black-out

Coloration (by pigments in coating/film)

Enhanced printability

High visibility, 'dayglow'

Reflective surfaces

Metallic effects

Pearlescent effects

Iridescent effects

Glitter effects

Fluorescent

Photoluminescence (glows in dark)

Hologram effects

'Chameleon' effects (colour change with heat)

Barrier properties to produce

Shower resistance

Waterproofness/water resistance

Chemical resistance

Dustproofness

Windproof

Barrier to gases/liquids

Protection against liquids, chemicals, microbes

Functional processing aid to assist manufacturing processes, e.g. vacuum moulding techniques, '*in situ*' moulding techniques

Filtration

Thermo-shielding

Light shielding

EMI/RFI* shielding

Using coating or film as a matrix/container for functional material

Pigments (various types – see appearance/aesthetics)

Reflective/fluorescent/ photoluminescence material

Table 1.8 (*Continued*)

Using coating or film as a matrix/container for functional material cont.	
Metal dust/flake (e.g. nickel, iron for magnetic strips)	Hollow glass/other material microspheres (for weight saving or thermal insulation)
Conductive material	Blowing agents for soft touch–pore structure
Phase change material	Adhesives (may be microencapsulated)
Antistatic material	Fillers for bulk, weight, economy
Humectants (moisture retention materials)	Heat/light/UV stabilisers
Antimicrobial chemicals	Specialist fillers/chemicals
Fragrances/perfumes	**'High tech' fillers**
Deodorising chemicals	Micro-instruments
† Microcapsules (containing different substances)	Sensors
FR chemicals (high loadings possible)	Probes
Activated carbon (for absorption)	Heating elements
Oil absorbing	Thermoshape memory polymers
UV/IR absorbing chemicals	Nanocomposites
Radiation absorbing material	

Summary: Coatings or film lamination can be used as:

a) coverings or as a barrier for protection, separation or containment
b) for appearance modification for decorative or functional purposes
c) improving dimensional stability, controlling stretch, preventing edges from fraying or curling
d) for control of porosity, e.g. for filtration
e) as a matrix for holding some functional material, chemical, pigment or other agent
f) as a processing aid, for example in '*in situ*' moulding, vacuum technique or thermomoulding
g) combining the specialist properties of polymers with the flexibility, strength, drapeability and covering power of a fabric.

* EMI/RFI = electromagnetic/radio frequency interference
† Micro-encapsulation can be used as a means of improving the durability of chemical finishes or for gradual release of chemicals, e.g fragrances, perfumes, anti-microbial chemicals. Finishes micro-encapsulated and applied to garments may be resistant to many machine washes. 'Post it' note paper is produced from micro-encapsulated, pressure sensitive adhesive. Phase change materials, when micro-encapsulated, confer 'cool when hot' and 'warm when cold' properties to garments.

Table 1.9 Novel properties and exploratory development chart

Material	General property
Fibre or Fabric	Tear and tensile strength Stretch and recovery Covering power Filtering ability Speciality types available: high tenacity FR (moderate to very high) anti-static 'anti-cut' energy absorbing anti-microbial aramids/UHM* polyethylene, high strength to weight ratio and other exceptional properties. Easy handling and storage (rolls) Easy fabrication
Polymer/resin/ compound	Covering power from 100% downwards Chemical resistance Thermoformability Enhanced printability, receptive to artwork Some have weldability See Table 1.8 and 2.1
Filler	Coloration opportunities: pigments pearlescent metallic FR chemicals Encapsulated material UV and other radiation absorbing Heat stabilisers Many other properties possible, virtually all textile chemicals may be used as filler or additive, see Table 1.8
Film	100% covering with smooth surface Barrier properties Printability/receptive to artwork Some have weldability Allows film technology to be combined with fabric technology and fabric properties Speciality films available: Tedlar – PTFE – wide temperature resistance Kapton – polyimide – wide temperature resistance PVDC (exceptional impermeability) Decorative films – e.g. Mearl (Mearl Corporation) Reflective surface (heat, light, other radiation)
Foam	Soft touch Improved aesthetics (deep sew lines, e.g. car seat covers) Thermal insulation Sound insulation

Table 1.9 (Continued)

Material	General property
	Bulkiness (enhanced safety) Impact absorbing Controllable filtering ability
Non-woven/ compound fabric (e.g. tufted fabric, wadding)	Soft touch Thermal insulation/sound insulation Controllable filtering ability Bulkiness (enhanced safety) Impact absorbing Non-woven fabric can be a speciality fibre or an inexpensive base material, i.e. allow a cost saving

* UHM = ultra high modulus

out in Germany to develop water dilutable polyester polymers as coating material for polyester fabric, the reasoning being that a coated fabric composed of the same or a similar polymer will be easier to recycle.

In addition to modifying base-fabric properties, the polymer coating can be used as a matrix or 'container' for speciality chemicals or pigments. For example, high visibility or 'day glow' articles can be produced by mixing the appropriate treatments into a polymer coating and applying it to a fabric. This route is generally more versatile than dyeing the fabric with high visibility dyes – a higher concentration of pigment or dye can be loaded into the coating than can be dyed on to the fabric. Speciality chemicals, such as anti-microbial or anti-static agents, applied as finishes to the fabric, may be easily removed by washing, wiping or abrasion. However, if these chemicals are mixed into the polymer coating, they will be more difficult to remove. Other speciality chemicals which can produce novel effects include photoluminescent chemicals, which emit light in the dark, dyes which change colour with temperature, UV light absorbing agents (against skin cancer), 'phase change material' (which provides heating when cold, cooling when hot), anti-odour agents, chemicals, perfumes and fragrances, and FR chemicals. Micro-encapsulation in millions of tiny capsules, each about 3–5 micrometres in diameter, can prolong the durability, and hence effectiveness of many of the chemicals mentioned. Encapsulation protects the chemical or substance from evaporation, washing off or other means of removal, and releases the substance gradually as a number of the capsules are ruptured. A summary of physical and chemical material properties which may be obtained by the process of coating and lamination of textiles appears in Tables 1.8 and 1.9.

Coating and lamination are versatile techniques and offer the means to innovate and develop new products. However, the coating and lamination technologist must keep up to date with developments and requirements and seek continuous improvements in each of the 12 technical areas (as categorised by Techtextil) in order to identify opportunities immediately they arise. He or she should be well informed about the latest fibres and polymers which, when combined together, could offer unique, novel properties to provide for a particular need, solve a particular problem or offer the opportunity to develop a new product.

1.9 References

1. McIntyre JE and Daniels PN, *Textile Terms and Definitions*, Tenth Edition, The Textile Institute, Manchester 1995, 67.
2. Schwarz EWK and Bennett JA, 'Coated fabrics', Chapter 7 in *Textile Chemicals and Auxiliaries* (Editor Speel HC), Reinhold, New York 1952, 143–5.
3. Robertson P, *The New Shell Book of Firsts*, Hodder Headline, London 1995, 339.
4. Lister WN, 'Waterproof coated fabrics', Chapter 10 in *Water-proofing and Water-repellency* (Editor Moilliet JL), Elsevier, Amsterdam 1963, 297–9.
5. Wootton DB, 'Introduction and historical background', Chapter 1 in *Textile Reinforcement of Elastomers* (Editors Wake WC and Wootton DB), Applied Science Publishers, London 1982, 1–12.
6. Rose H, *Fusible Interlinings, Origins and Technology*, Highgate Publications, London 1979, 292.
7. Byrne C, *Technical Textiles; World Market Prospects to 2005* (based on a study for Techtextil, Messe Frankfurt), David Rigby Associates Report April 1997. Also in Further Reading 2 (below).
8. Rigby D, 'Growth opportunities in technical textiles in Asia', *Texcon Conference*, Ahmedabad 15 Dec 1997. From DRA/Techtextil special report *Technical Textiles; World Market Prospects to 2005.*
9. Davies B (David Rigby Associates), 'Trends in the European clothing market', *IMMFC*, Dornbirn 17–19 Sept 1997, and also in *Technical Textiles*, Vol 41 Feb 1998, E17–E20.
10. Dusi E, 'Sportswear and market trends towards the third Millennium', *JTN Monthly*, Oct 1999, 58.
11. www.marketprofiles@datamonotor.com 6 Jul 2000, Market Capsule, Sports Clothing.
12. Anon, 'Fibres-Western Europe; 18000t airbag yarn', *Chemical Fibers International*, Vol 51 Feb 2001, 6.
13. Anon, *Textile Outlook International*, Jan 2000, 142.
14. Anon, 'Industry News', *JCF*, Vol 29 July 1999, 6–10.
15. Bitz K, 'Soaking up sales in adult incontinence', *Nonwovens Industry*, Mar 2001, 83–6.

1.10 Further reading

1. Adanur S (Editor), *Wellington Sears Handbook of Industrial Textiles*, Technomic, Lancaster PA, USA, 1995.
2. Byrne C (David Rigby Associates), 'Technical textiles – an Overview', Chapter 1 in *Handbook of Technical Textiles* (Editors Horrocks AR and Anand SC), Woodhead, Cambridge 2000, 1–23.
3. Dusi E, 'Sportswear and market trends towards the third Millennium', *JTN Monthly*, Oct 1999, 58.
4. Horrocks AR and Anand SC (Editors), *Handbook of Technical Textiles*, Woodhead, Cambridge 2000.
5. Lennox-Kerr P (Editor), *Flexible Textiles Composites*, Textile Trade Press/WRC Smith, Manchester/Atlanta CA, USA 1973.
6. Moilliet JL, *Water-proofing and Water-repellency*, Elsevier, London 1963.
7. Rose H, *Fusible Interlinings, Origins and Technology*, Highgate Publications, London 1979.
8. Wake WC and Wootton DB (Editors), *Textile Reinforcement of Elastomers*, Applied Science Publishers, London 1982.

2

Materials and their properties

2.1 Introduction

A coated fabric combines the benefits of the base fabric with those of the polymer with which it is coated. The resulting coated fabric will have many properties which cannot be offered by either component individually, and careful consideration is necessary to select both base fabric and coating polymer. The base fabric provides the mechanical strength of the composite material and supports the layer of coating applied to it. For quality coated fabrics, quality base fabrics are essential. This point is made because newcomers to the industry sometimes believe that the coating can cover fabric defects, and so second quality fabrics may be sent for coating. In fact, the defect is frequently made more prominent and the cost of rejected coated fabric, with the added value of coating, will be significantly higher than that of the base fabric alone.

2.2 Fabrics

Polyester and nylon are the main fibres used, because of their strength and general resistance to moisture, oils, micro-organisms and many common chemicals. Generally, polyester is more resistant to light and ultraviolet (UV) degradation than nylon while nylon is more resistant to hydrolysis. Polyester, however, has grown at the expense of nylon because of its better dimensional stability and shrink resistance, lower extensibility and generally lower cost. High tenacity nylon and polyester yarns are used in many coated articles for extra strength, and aramid fibres are used where more specialist properties, such as high strength to weight ratio and resistance to high temperatures, are required. Acrylic fibres are used for some applications where very high UV resistance is necessary, such as for awnings, car roofs and hoods for convertibles. Cotton was the first fabric used in textile coating and it is still used in large quantities. In applications where strength is required, however, it has been replaced by nylon and polyester which have considerably higher strength to weight ratios. Cotton is also vulnerable to wet rotting and microbial attack, but it has certain advantages over synthetic fibres, such as better polymer adhesion. The rougher surface of the cotton and the short fibre length provide more opportunity for mechanical anchoring of the polymer. The smoother, continuous filament synthetic fibres frequently require more specialist means of promoting fibre–polymer adhesion, especially with PVC plastisols and rubber coatings. However, cotton or fabrics produced from spun yarns cannot in general be direct coated to produce lightweight, coated fabrics, especially waterproof materials, because the fibre ends may cause pin holing, or be 'scraped or teased up' causing a raspy surface, usually with poor abrasion and waterproof performance. This, of course, does not apply when the coating thickness is sufficient to cover any fibre ends completely. Fabrics incorporating some continuous filament textured yarns, such as false twist textured, can, in some cases, be direct coated and the yarn texture can improve the coating adhesion by mechanical means.

There are speciality variants of nylon and polyester, such as high tenacity (HT) and low shrinkage (LS), which are used in coating applications, such as tarpaulins and conveyer belts. Glass woven fabrics are use as bases for PTFE coatings for industrial uses, such as calender belts and building structures. These applications make specific use of the glass properties of very high strength with very low elongation and excellent flammability resistance. Glass fibres cannot be used in applications that involve significant flexing, because of their poor flex resistance.

2.2.1 Fabric construction

Only a relatively very small number of fabric constructions are employed for polymer coating, i.e. plain woven, twill and basket constructions. Excessively loose constructions could distort dimensionally, especially on the bias, whilst excessively open structures would lead to resin penetration. If the fabric construction is too open, the coating resin may not be able to 'bridge' the gaps between the yarns to form a continuous layer necessary for performance such as waterproofness. To combine lightness of weight with high tear strength, rip-stop constructions are sometimes used, i.e. a stronger yarn is included in the fabric every 5 mm or so. Woven fabric constructions, especially plain weaves, are generally rigid with little stretch, and any required stretch has to be introduced via the yarns. Yarn mobility in the fabric construction, as well as inherent yarn tensile strength, determines fabric tear strength. If the yarns can move and bunch together, higher tear strength will result compared to a situation where they are broken individually. It can be appreciated that coating penetration between individual threads will tend to reduce tear strength. A modified Oxford weave which enhanced tear strength and allowed direct coating was patented in the 1960s. This was later developed into the Poly-R^R weave which has double the tear strength of a standard weave.[1] Knitted fabrics are used when significant elongation is required but, as will be seen, knitted fabrics cannot generally be direct coated because of their stretchiness and open construction. Fabrics woven from continuous filament yarns are generally relatively stiff compared with fabrics woven from spun yarns, and especially knitted fabrics. When direct coated these woven fabrics become stiffer, and they can be very stiff if excessive resin penetration occurs. The softest handle can be obtained with the minimum of penetration, but this could result in poor resin adhesion. A balance, therefore, has to be reached between coated fabric handle and resin adhesion. As has been stated, fabrics woven from spun yarns – which usually have better drape and softer handles compared to continuous filament woven fabrics – generally cannot be direct coated but, as will be seen later in Chapter 3, they can be coated with crushed foam or transfer coated. Woven fabrics produced from continuous filament textured yarns, or from a blend of yarns, have properties somewhere in between, and may be direct coated in certain circumstances.

Knitted fabrics, which cannot generally be direct coated and which invariably have softer handles with better drape than woven fabrics, are usually transfer coated, see Chapter 3. Coated fabrics with excellent drape and soft handles are made possible by this process. The knitted base fabrics are sometimes slightly raised or napped to improve coating adhesion and to maximise the softness and flexibility of the coated fabric. A notable difference between direct coating and transfer coating is that in direct coating, the resin is on the back of the fabric and is on the inside of the garment. On the other hand, with transfer coated fabrics, the polymer forms the face of the material and is on the outside of the completed garment.

Nonwovens are an industry in their own right and resins, referred to as binders, are extensively used in their manufacture. Nonwovens, with some exceptions, generally have limited tear strength, poor handle and drape and are not used for apparel, apart from disposable protective clothing. Many cannot generally be direct coated because of their rough surface, and because they are not strong enough to be tensioned on a coating machine. Needlepunched nonwoven fabrics, however, are used as floor coverings; they are coated with both water-based resins and hot melt coatings to improve surface abrasion and to confer mouldability properties, see Section 3.3.13.

The main fibre properties which influence choice as a coating substrate are summarised in Tables 2.1 and 2.2. There are also papers published on fabric substrates for coating.[1–8]

Table 2.1 Properties of fibres commonly used in coating and laminating

Fibre	Density (g/cc)	Melting point (°C)	Tenacity (g/den)	Stiffness (flexural rigidity) (g/den)	Limiting oxygen index (% oxygen)	Abrasion resistance	Resistance to sunlight and UV
Acrylic	1.12–1.19	150d	2.0–5.0HT	5.0–8.0	18	Moderate	Excellent
Modacrylic	1.37	150d	2.0–3.5	3.8	27	Moderate	Excellent
Nylon 6	1.13	215	4.3–8.8HT	17–48	20	Very good	Poor – good if stabilised
Nylon 66	1.14	260	4.3–8.8HT	5.0–57	20	Very good	Poor – good if stabilised
Polyester	1.40	260	4.2–7.5HT	10–30	21	Very good	Good – excellent if stabilised
Cotton	1.51	150d	3.2	60–70	18	Moderate	Moderate
Wool	1.15–1.30	132d	1.0–1.7	4.5	25	Moderate	Moderate
Polypropylene	0.9	165	4.0–8.5HT	20–30	18	Good	Poor – good if stabilised
UHM polyethylene	0.97	144	30	1400–2000	19	Good	Good
Aramid	1.38–1.45	427–482d	5.3–22	500–1000	29–33	Moderate	Moderate
Carbon	1.79–1.86	3500d	9.8–19.1+	350–1500	64+	Brittle	Excellent
Glass	2.5–2.7	700	6.3–11.7	310–380	not burn	Brittle	Excellent

Notes:
1. d = does not melt but starts to decompose.
2. HT = high tenacity.
3. UHM = ultra high modulus.
4. Thermoplastics begin to *soften* at temperatures below their melting point and therefore thermoplastic fibres can deform or be damaged by the action of temperature well below their melting point. This is more likely if the heat is combined with mechanical pressure as for example in a lamination or moulding process. Surface pile can be crushed or deformed or glazing can occur.

N.B. Data compiled from several different sources and intended only as a guide.

Table 2.2 Fabric substrates for coating

Fabric	Advantages	Disadvantages
Cotton	Excellent coating adhesion No bonding agents required Low thermal shrinkage	Low strength to weight ratio Absorbs moisture Vulnerable to mildew, rotting, insects
Polyester	Strong with HT, low shrinkage and many other variants available Relatively inexpensive Resistant to mildew, rotting and insects High abrasion resistance Blendable with cotton for garments and other end uses Good overall properties/price	Low moisture absorbency (for garment comfort) Limited resilience (not used in carpets)
Nylon	Strong with HT variants available Good elasticity and resilience High abrasion resistance Resistant to mildew, rotting and insects Good thermal absorbency (used in airbags)	UV resistance low unless protected Fabric may sag due to moisture absorption Relatively expensive compared to polyester
Polyethylene, polypropylene	Lightweight Inexpensive Chemically inert Resistant to mildew, rotting and insects	Low melting point – especially polyethylene Adhesion difficult to some substances
Aramid	Very high tensile strength Excellent strength to weight ratio High melting point Good FR properties	Expensive Degraded by sunlight/UV
Glass	Very high temperature resistance High tenacity Excellent FR properties Good dimensional stability Resistant to mildew, rotting and insects Zero moisture absorbance Excellent UV stability	Adhesion difficulties Relatively heavy Brittle, poor flexing properties

N.B. There are now available many variants of these fibres specially developed for specific applications, including low shrinkage (LS), high tenacity (HT), ultra high modulus (UHM), FR resistance, anti-wicking polyester, etc.

2.3 Preparation of fabrics

2.3.1 Scouring

A fundamental requirement for any coated or laminated fabric is that the coating or material joined to the fabric must be strongly combined to that fabric, and at must not separate throughout the lifetime of the product. One of the most frequent reasons for premature breakdown of a coated or laminated article is delamination. For good adhesion, the fabric must normally be clean and free of any soiling. Coating and lamination are essentially joining operations, and all the principles governing satisfactory adhesion apply, see Sections 2.7.1 to 2.7.3. Any dirt or contaminant on the fabric – or on the machinery from which the contaminant can transfer to the fabric – is likely to affect the adhesion of the coating to the material to which it is being joined. Waxes, oils and silicone-based materials, even in small amounts, are especially likely to reduce bond strength. Cotton fabrics should normally be desized and synthetic fibres such as nylon and polyester should be scoured. Sometimes this is not done to save costs, but there is always the risk of poor bonds being formed. However, a continuous scouring range set up in front of a finishing stenter is relatively inexpensive to run, once the initial costs of installation have been met.

Scouring itself can sometimes cause problems – if the fabric is not properly rinsed, the scour chemicals themselves may become contaminants and can reduce adhesion. A typical scouring liquor will contain a chemical such as sodium carbonate to ensure that the pH is alkali, a wetting agent, and detergent to remove lubricants and oils and hold them in suspension to prevent re-deposition back on to the fabric. In regions with hard water, it is sometimes necessary to include a water softener for effective scouring. The textile chemical companies have developed agents which perform many of these functions. If the material does not wet out readily, a wetting agent may be necessary, but this is an extra chemical which needs to be removed so that coating or lamination adhesion is not jeopardised. There are available wetting agents which thermo-degrade on stentering (e.g. Sandozin NRW by Clariant), and are therefore less likely to influence adhesion. Occasionally frothing can occur during scouring, especially at high speeds, and anti-foaming may be used. Excessive amounts of anti-foam agents, and especially silicone based anti-froth agents, should be avoided, because of possible influence on coating adhesion.

Virtually all fabric finishes and residual scour bath chemicals will reduce coating or lamination adhesion from a hypothetical ideal of completely 'clean' fabric – difficult to achieve in a commercial process. However, fabric, especially synthetic fabric with absolutely no finish or residual lubricant on it, would be very difficult to process, because of the generation of static electricity. The static electricity would cause fabric to cling to rollers and to the sides of containers, giving rise to creasing, and to static electrical shocks which will be a nuisance and pose a risk to the safety of operatives. Even seam slippage agents, fabric softeners and, in the case of nylon, after-treatments to improve dye fastness (syntans), can reduce adhesion substantially – although some syntans reduce adhesion less than others. Sometimes the loss in bond strength can be tolerated, and sometimes a finish such as a fluorocarbon is actually applied beforehand to help control resin penetration during coating. If finishes are necessary, they should be kept to a minimum concentration and their effect on adhesion should be examined before proceeding to bulk production. It is possible to apply finishes to one side of the fabric by a foam processing or rotary screen technique, which will minimise the risk to adhesion. However, it is still advisable to check beforehand – the finish may be penetrating through the fabric to the other side. When adhesion is especially critical, and for the best possible adhesion, the fabric should be 'pure finished', i.e. no special finish applied.

2.3.2 Heat stabilisation

Fabric coating or lamination generally entails a heating process, and thermoplastic base fabrics should be heat set at a temperature higher than the temperature that will be used in the coating or lamination process. If this is not done, the fabric will shrink – or try to shrink – during production and possibly even when the finished article is being used. The general relationship between heat and shrinkage of polyester yarn is shown in Fig. 2.1. Creases may be caused and, even if the fabric is held to width on pins, it may well shrink a little when released and put on to a roll. Ideally the base fabric for coating should be in a stable and generally relaxed state on the coating machine. Even non-thermoplastic fabrics such as cotton can be mechanically stretched under tension and, when this tension is released, fabric creasing may occur or patterns may be distorted. It should be tensioned sufficiently to present a flat and uniform surface for uniform and even application of the coating resin. Excessive tensions, including those induced by shrinkage, could cause creasing. Other fabric faults, such as creases or slack selvedges, will also give rise to problems in coating or lamination. Fabrics must be presented to the coating or lamination head in a stable relaxed state, so that they are flat and at the correct width. If they have not been finished correctly, e.g. pulled out excessively in width, they may physically shrink ('neck-in') at the point of lamination or coating, thus presenting the operatives with major problems. Fabric must be supplied at the required width and then stentered to the correct width. Excessive pulling out of woven fabric can also cause excessive penetration of resin, stiffening of fabric or reduced tear strength, as well as resulting in an under-

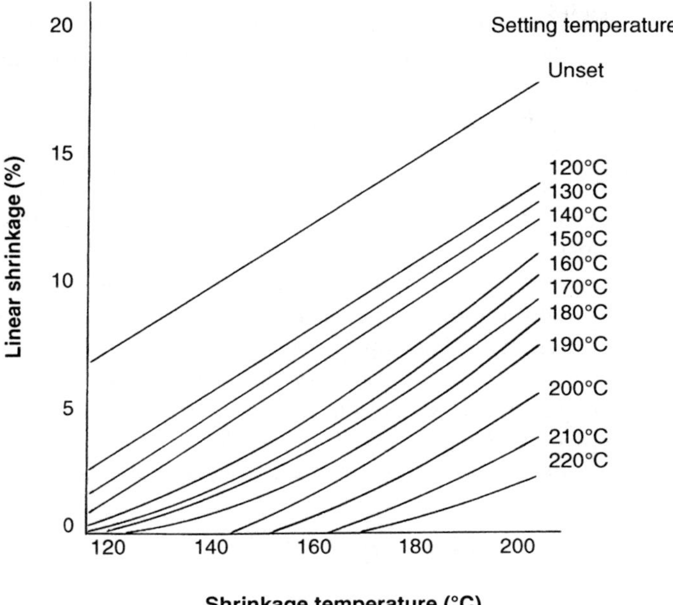

2.1 The effect of setting temperature on the dimensional stability of DuPont polyester fabric. Technical Information from DuPont (UK) Ltd (April 1999). Reproduced with kind permission.

The higher the setting temperature, the more thermal-dimensionally stable the fabric, i.e. no or less shrinkage on the application of heat which usually occurs during a coating or laminating process. However, width loss due to excessive lengthways tension will still occur.

weight and out-of-specification fabric. Priming chemicals (such as RFL – resorcinol formaldehyde latex – see Section 2.6.4) to improve the coating adhesion of PVC and rubber coatings, can be applied by impregnation during the heat setting process. Fabric heat setting should be carried out after scouring, especially when aesthetics are important, because any stains on the fabric can be 'set in' and difficult to remove later. Woven nylon fabric may present problems because, on leaving the stenter, it is likely to be completely dry. The moisture absorbency of nylon is up to 0.4% and, if the material absorbs this moisture during storage on the roll, trellis-like creases may be formed.

2.3.3 Batch preparation

Batches for coating or lamination are normally made up from several pieces sewn together to produce one long continuous length for uninterrupted continuous runs of production. Needless to say, all individual pieces should be of the same width. The joining seams must be carefully sewn and must be strong enough to withstand the longitudinal tensions required for direct coating. Poorly prepared seams can result in creases, streaks in the coating and many metres of unsaleable coated fabric on each side of the seam. An excessively thick seam may burst open if caught under the doctor blade, cause the fabric to rip and could even damage the coating head. Burst seams, ripped fabric or holes in the fabric all necessitate considerable cleaning up, resetting and restarting of the process – all very time consuming and costly in lost production time. In an ideal world, there would be no seams. Fabric batches should be as long and in one continuous length as is practically possible. Even in the modern factory, this is usually what a person can lift manually! Sewing up individual pieces for scouring, heat setting or any other process always presents a risk of creases or potential creases – least handled, is best handled – preferably in one continuous length.

2.4 Materials for coating

2.4.1 Properties of plastic materials

Apart from the natural fibres such as cotton and wool, most other material which may be used in coating and laminating processes, i.e. fabrics, adhesives, films and foams, is likely to be thermoplastic. During the coating or laminating process, both fabric substrate and polymer resin will generally be subjected to a heat treatment, and it is important to know beforehand what is likely to happen. Plastics can be broadly divided into two types; thermoplastics, which soften and eventually melt when heated, and thermoset plastics, which do not soften or melt when heated. All plastics are made from long-chain, linear polymer molecules but, in the case of thermoset plastics, the molecules are crosslinked, which makes the whole structure more rigid and largely prevents them moving when heated. The long-chain molecules in thermoplastics, however, not being crosslinked, are more free to move about when heated above certain temperatures specific to the molecular length and chemical nature of the particular plastic. It is possible – as will be seen – to crosslink plastics and rubbers to various degrees, in order to improve durability and wear properties. Crosslinking also generally makes the material more rigid, less stretchy and harder; and a high degree of crosslinking will usually cause the plastic to become brittle. The person who formulates the resin, the compounder, see Section 2.6.1, can control the degree of crosslinking to produce the coated fabric properties required, by choice of type of crosslinking agent and concentration of the agent in the coating recipe.

Synthetic fibres are made from thermoplastic materials which soften, shrink and melt under the action of heat. As has been seen above, it is necessary to heat stabilise polyester and nylon before coating, to prevent shrinkage during the coating process. Man-made fibres, e.g. cellulosic, are regenerated polymers and are not as heat sensitive as thermoplastic fibres. In some polymer/fibre combinations, such as hoses, belts and car tyres, the dimensional stability at high temperatures of cellulosic fibres, such as viscose, is an advantage. Themoplastic yarn shrinkage under the action of heat is one of the causes of delamination which, in turn, leads to premature breakdown.

Thermoplastic properties of some plastics are useful in that they allow the material to be used as hot melt adhesives and, in certain cases, to be joined by welding techniques. In general terms, for a given chemical type of thermoplastic, the shorter the molecular length, the lower the melting point, and the longer the molecular length, the higher the melting point. Adhesives are generally shorter chain length molecules and melt at relatively low temperatures, for example polyester fibre melts at about 260 °C, but there are polyester-based adhesives which melt as low as 100 °C. The thermoplastic nature influences the ease with which recycling can take place; if thermoplastic, the material can be melted down and reprocessed into the same or another useful article. Thermoset plastics are not as easily recycled, and for this reason thermoplastics may be preferred over thermoset materials – if there is a choice. Thermoset plastics are harder, more rigid and more heat resistant, but the vast majority of plastics are thermoplastic.

The main plastics used for textile coating are polyvinyl acetate, acrylics, PVC, polyurethanes and natural and synthetic rubbers. In addition to the main types, there are variants and copolymers and even terpolymers, i.e. resins made from three starting materials. As will be discussed later, adhesives and hot melt adhesives can be produced from copolymers, giving a wide range of properties. The chemical structures and basic chemistry of these polymers are available in standard chemistry textbooks, but there are available a number of excellent summaries of the chemistries with respect to fabric coating.[9–11] Numerous articles are also available on individual polymer types, their properties and their applications.[12–39] These polymers are generally applied to fabric as solutions or dispersions in water or as solvent in an uncrosslinked state. Crosslinking of the polymer can be effected after deposition on the fabric, which improves the durability of the resultant coating to abrasion and its resistance to solvents and water. However, this crosslinking always causes some stiffening and hardening of the coating. The degree of crosslinking is influenced by the nature of the polymer itself, the type and concentration of crosslinking agents added, the catalyst type and concentration, and by the time and temperature used to effect the crosslinking process. The higher the temperature, the greater the degree of crosslinking – a very rough 'rule of thumb' states that the crosslinking rate doubles for every 10 °C increase in temperature. This may not be strictly true in pure scientific terms, but it does emphasise the importance of temperature control in coating operations involving crosslinking polymers. If not properly crosslinked, coatings can be slightly sticky (tacky) to the touch and have inferior wearing properties. In general the degree of crosslinking, which determines performance, has to be balanced with the aesthetics of handle, drape and softness. The term 'crosslinking' generally has the same meaning as 'curing' which appears to be used more often when surface finishes are applied, and 'vulcanisation' which is used in rubber technology.

It is important to realise that a 'plastic' generally consists of the plastic itself plus several additives. Amongst these are UV radiation and heat stabilisers, antioxidants, fillers to improve the mechanical properties, fillers for economy, FR chemicals, reinforcement fibres (turning the plastic into a 'composite'), pigments and other compounds necessary to confer further special properties or to assist with processing. Sometimes it is necessary to add materials specially to make all the various ingredients compatible with each other. When fibres are added, or when the compound is to be coated on to a fabric, coupling or bonding agents – also called adhesion promoters – may also be required. Plastic compounding or mixing of components is a specialised process and, if not carried out correctly, it can cause production problems and variations in quality. In addition, volatile compounds used in formulations for car interiors can cause 'fogging', see Chapter 5. Furthermore, there are many variants of each chemical type, and the terms 'polyurethane' or 'polyester' in fact refer to families of polymers of related chemical constitution and not just to a single type. During the 1960s and early 1970s there was considerable activity in the development of polyurethanes for both direct and transfer coating, and many research papers were published at that time. Although there have been improvements in properties and developments, especially to make materials and processes more environmentally friendly, most of the general principles of the properties and practical application methods are unchanged and much can be still be learnt from these papers.[25–29]

A summary of polymer resin properties is presented in Table 2.3 (some trade names appear in Table 2.4), but some general comments can be made. Polyurethanes generally have good adhesion to fabric substrates and have good elongation and excellent flexibility, even at very low temperatures. They have excellent surface abrasion properties and, in fact, are used as lacquers for leather. Acrylics are extremely versatile materials with many variants and price ranges. They have good clarity, and good UV resistance. Some low temperature cure acrylic resins are available from certain suppliers, such as Rohm and Haas, which, if suitable for the purpose, will result in energy savings.

Table 2.3 Summary of main polymers used in textile coating

Polymer	Properties/Advantages	Disadvantages	Typical products
PVC Polyvinychloride	Versatile material. Plastisols and water-based available which can be compounded to give wide range of properties Good inherent FR which can be improved Good oil, solvent and abrasion resistance Heat and RF weldable for good watertight seams	Cracks when cold Plasticiser migration Moderate heat and age resistance	Tarpaulins, coverings, large tents and architectural uses Seat upholstery 'Leathercloths' Protective clothing, aprons Leisure products, banners, bunting
PVDC polyvinylidene chloride	Very good FR (blendable with PVC and acrylic latex). Very low gas permeability Heat weldable Clear, high gloss	Hard and brittle	Blends with acrylics to improve FR in coatings, e.g. for roller/louvre blinds (film is used as shrinkwrap) (coating on drinks bottles)
Polyurethane PU	Several grades available in solvent and latex form. Tough, good extensibility, good weathering and abrasion resistance Films available for lamination	Some grades (aliphatic) discolour and have limited hydrolysis resistance FR is only moderate Relatively expensive	Waterproof protective clothing Waterproof/breathable protective clothing Aircraft life jackets Adhesives Used as lacquers for PVC tarpaulins and leather
Acrylic	Large number of variants and co-polymers. Wide range of properties. Blendable with other latices. Good UV resistance and optical clarity, generally inexpensive	FR may be poor unless compounded with FR chemicals	Back coatings for upholstery including auto seats Binders for nonwovens and glass fibres Adhesives Used as lacquers for tarpaulins

Table 2.3 (*Continued*)

Polymer	Properties/Advantages	Disadvantages	Typical products
EVA Ethylene vinyl acetate	Good adhesion to all fibres, good tuft binding properties Flexible at low temperature Good thermoplastic properties and easily thermomouldable Heat sealability Relatively inexpensive	May be some tendency to discolour May be sensitive to water – poor washing resistance	Backings for carpets and upholstery Wallcoverings, exhibition board backing Adhesives
Polyolefins LDPE HDPE Polypropylene	Good resistance to acids, alkalis and chemicals. Easily recycled Lightweight Inexpensive	Low melting point FR limited and limited resistance to ageing	Lightweight coverings, tarpaulins, (alternative to PVC) Sacks FIBCs, bulk bags
Silicone	Odourless, inert, good resistance to many chemicals and micro-organisms. Wide temperature service range from –60 to 200+ °C. High tear resistance and puncture resistance of coated fabric Low toxicity, water repellent Available as fluid (no solvent or water to remove) Blendable with acrylics/polyurethanes Coating has some breathability	Attracts oily soiling, seaming is difficult due to release properties Expensive Difficult to print or apply artwork	Airbags Food, medical applications Gaskets, seals Parachutes Oven curtains

PTFE Polytetrafluoro-ethylene	Excellent resistance to acids, alkalis, chemicals, solvents, oils, oxidation and weathering Non-stick properties Outstanding electrical properties Wide service temperature up to 260 °C	Very expensive	Architectural applications Calender belts Food and medical uses Gaskets, seals
Natural rubber (NR)	Excellent stretch and flexibility, general purpose material, working temperatures to 70 °C, fillers improve mechanical properties Many grades and properties obtainable by compounding and blending	Moderate sunlight and oxidative resistance Moderate solvent and oil resistance Flammable – requires FR agents Unmodified is biodegradable	Carpet backing Tyres Life rafts Conveyer belts Protective clothing Escape chutes
SBR (styrene butadiene rubber)	Generally similar to natural rubber but somewhat better resistance to abrasion, flexing and micro-organisms	Generally similar to natural rubber	Carpet backing As natural rubber
Nitrile rubber (acrylonitrile/ butadiene) NBR	Very good oil resistance which increases with acrylonitrile content. Better resistance to heat and sunlight than natural rubber	Limited FR	Oil resistant clothing Oil seals, especially for hydrocarbons Belts and items handling oily or greasy products
Butyl rubber BR	Very low permeability to gases Better resistance to heat, oxidation and chemicals than natural rubber	Solvent resistance limited FR limited Seaming difficult	Items containing gases, e.g. air cushions, pneumatic springs, bellows Protective clothing – especially for chemicals and acids Lightweight life-jackets Life rafts

Table 2.3 (Continued)

Polymer	Properties/Advantages	Disadvantages	Typical products
Polychloroprene rubber (e.g Neoprene-DuPont) CR	Excellent resistance to oils, chemicals and oxidation Working temperature to 120 °C Good FR properties Versatile material Generally inexpensive	Coloration difficult – generally only in black	Protective clothing Aircraft carpet backing Aircraft slide/rafts Life rafts Life jackets Hovercraft skirts Flexible gangway bellow (trains) Airsprings Radome covers Airbags V-belts
Chlorosulphonated rubber (e.g Hypalon-DuPont) CSM	Excellent oil, chemical and oxidation resistance. Generally similar to Neoprene, but higher temperature to 135 °C (some grades to 170 °C). Can be pigmented. More expensive than Neoprene Accepts higher level of filler than Neoprene		Similar to Neoprene. Used where coloration is necessary and higher temperature resistance is required
Fluroelastomer (Viton-DuPont) FKM	Very good heat, weathering, chemical and solvent resistance 'Elastic PTFE'. Service temperature –20 to 200 °C	Expensive Poor resistance to ketones	Specialist protective clothing High performance applications, e.g. seals and gaskets

FIBC, flexible intermediate bulk containers; RF, radiofrequency.
N.B. All rubbers have limited resistance to aliphatic and aromatic solvents, except for Neoprene and Chloroprene which have moderate resistance and nitrile which has very good resistance to both types of solvent.
For more specialist details, see References 43–48, especially the works by Pruett (47 and 48).

Table 2.4 Some trade names and suppliers

Polymer, compound or resin	Trade name and supplier
Polyurethane	Witocoflex (Baxenden) Permuthane-solvent based (Stahl) Permutex-water based (Stahl) Dicrylan (Ciba)
Acrylic	Appretan (Clariant) Dicrylan (Ciba) Plextol-water based (Rohm-Huls) Plexisol-solvent based (Rohm-Huls) Acronal (BASF) Primal (Rohm & Haas-Cornelius) Vinacryl (Vinamul)
Vinyl acetate	Appretan (Clariant) Vinamul (Vinamul)
PVC plastisols/dispersions	Acrylon Lutofan (Bayer/Huls) European Vinyl Corporation (EVC)
Speciality halogen-free barrier resin	Epotal (BASF)
PVDC	Solvin (BASF/Solvay)
Silicone	Wacker Dicrylan (Ciba)
SBR	Butofan (butadiene rich) (BASF) Styrofan (styrene rich) (BASF)
Natural/synthetic rubbers	Acrlonitrile (Clariant) Vibatex (Ciba) British Vita BF Goodrich DuPont
Fluorocarbon (for water and stain repellent finishes)	Nuva (Clariant) Teflon (DuPont) Ashahiguard (Ashahi Glass)

Resin compounders: Synthomer, Noveon (formerly Mydrin/BG Goodrich), Industrial Latex, Texas Adhesives, Vitabond (adhesives).
General textile chemicals: Clariant, Ciba, CHT-Tubungen, Dr TH Boehme, Huls.
Specialist FR chemicals: Albright & Wilson, Thor, Schill and Seilacher, Great Lakes.
Crosslinking agents: BIP (Beetle Resins), Clariant (Cassurit), Ciba (Knittex and Lyofix), Bayer (bonding agents for rubbers and polyurethane).
In addition: Clariant manufacture a whole range of water-based resins of different chemical constitution under the trade name Appretan. Ciba have ranges under the trade name Dicrylan.

PVC is an extremely well researched material that can be compounded to produce very many different properties.[22–24] It is the most used material in fabric coating, is economical and has a high degree of inherent FR properties. However, it has come under attack from environmental pressure groups and alternatives are being researched. Polyvinylidene chloride (PVDC) resins have excellent FR properties which can be improved by blending with PVC. In addition, they have good impermeability to gases and, in fact, PVDC lacquers are used on polyester bottles to prevent the loss of carbon dioxide which would cause the drinks inside to go flat. Styrene butadiene rubber (SBR) latex, see below, is used in place of and in combination with some of the above when economy is required.

2.4.2 Natural and synthetic rubbers

Rubbers are polymers and are closely related to plastics. They are used in combination with textiles in many end uses, and indeed the fabric coating industry is in some ways an off-shoot of the rubber industry. There are various types to suit different applications, and they are versatile in that they can be blended together and they will accept additives to provide specific properties. The rubber compounding industry began with the discovery of vulcanisation in 1839, and the modern industry is the cumulative effort of countless rubber technologists working on almost countless applications of rubber since that time. As can be expected, rubber compounding is a very specialised process – there is usually a minimum of about eight or nine different ingredients in each mixture. The largest application by far for rubber is the tyre, which accounts for about 50% of all rubber production in the world. However, there is not enough natural rubber available and so this has to be supplemented with synthetic rubbers, especially SBR. During the Second World War, the supply of Malaysian rubber was cut off during 1942 and alternative materials were developed, such as polyvinyl acetal emulsions. There are a number of specialist rubbers including nitrile rubber, butyl rubber, polychloroprene, the best known of which is Neoprene (DuPont) and chlorosulphonated polyethylene, the best known of which is Hypalon (DuPont). SBR rubber latex is an inexpensive and versatile material and has many applications where economy is important, e.g. carpets. These rubbers are used widely in fabric coating applications, which are described in detail in Chapter 4.

No one single rubber has excellent resistance to all chemicals, solvents and fluids. All common rubbers, apart from Neoprene and nitrile rubber, have generally poor resistance to both aliphatic and aromatic hydrocarbon solvents. Conversely, most common rubbers have good resistance to oxygenated solvents (e.g. ketones), apart from nitrile rubber and Neoprene.

Butyl, nitrile and Neoprene rubbers have the best acid resistance. Butyl rubber has the best resistance to weathering and oxidation and also has very low permeability to gases. Viton (DuPont), referred to as 'elastic PTFE' by some chemists, has excellent resistance to virtually everything apart from oxgenated solvents, such as ketones. A summary of the properties of rubbers used in fabric coating appears in Table 2.3 and further detailed information is readily available in the literature.[30,31,40–48]

2.5 Materials for lamination

2.5.1 Films

Films have the advantage over fabrics of 100% cover at light weights; they are generally completely impermeable to liquids and gases and are also excellent barriers to dust and other particles. Film/fabric laminates are used in sail making, hot air balloons and airships. Films can be used to produce reflective surfaces with efficiencies much higher than can be produced with a fabric. However, they are generally easily torn and easily punctured and damaged by flexing, but some of these physical limitations can be overcome by lamination to a fabric, and the two components combined can offer a wide variety of useful properties. Film production is a large volume industry which uses many different variants of a relatively small number of polymers, produced mainly for packaging. Depending on the method of manufacture and starting material, films can be very inexpensive, such as polyethylene, or extremely expensive, such as Kapton (DuPont).[49] Film coating is widely used to confer heat sealability, low gas permeability, printability and other specialist properties. Photographic films (cellulose triacetate) will have several layers of coating. Pre-treatments, such as corona discharge, and plasma treatments are also used to confer wettability and adhesion and to aid other processes. The film industry in the USA generally measures film thickness in 'mils' which is one thousandth of an inch and equivalent to 25 micrometres.

Clear unsupported PVC film made by calendering is sometimes used in automotive door casing construction. It is laminated to car interior trim fabric, in a separate process, as a barrier against wetness and dampness within the door casing, and the PVC film is also a useful processing aid because of its weldability. Coloured unsupported PVC films (sometimes referred to as 'foils'), and films produced from PVC/ABS (acrylonitrile-butadiene-styrene) mixtures, usually embossed, are used extensively as coverstock in automotive interiors, laminated to polyurethane foam, polypropylene foam or a nonwoven fabric to provide a 'soft touch'. Within the last few years, however, the use of polyolefin films in place of PVC-based films has grown for a number of reasons. Unsupported films are made by a variety of techniques, including casting and extrusion as well as calendering. Calendering is used extensively in the rubber industry, and it is possible for the freshly produced film to be combined with a fabric during the calendering procedure. Other films types available include polyurethane, natural and synthetic rubbers, Neoprene and Hypalon. It is possible to produce polymer mixtures, sometimes referred to as plastic 'alloys', such as PVC/ABS and process them by calendering. Calendering is discussed further in Section 3.11.

Speciality films are widely used in protective clothing as waterproof and breathable barriers against the weather. They are also used in footwear, and medical applications are being explored.[50–55] The market leaders are Gore-Tex which is a PTFE film, Sympatex (Acordis) which is made from polyester, and Porelle (Porvair) which is polyurethane based. There are also several other branded waterproof and breathable films made from polyurethane. These films can be broadly classified into two types: microporous and 'solid' – which is sometimes referred to as 'monolithic'. The microporous varieties, such as Gore-Tex and Porelle, have very small pores in their structure which are too small to allow drops of water through, but large enough to allow water vapour molecules through, thus accounting for their breathability. The 'solid' film variety, such as Sympatex, breathe by a process of water molecules migrating from hydrophilic site to hydrophilic site within the polymer network. There are advantages and disadvantages claimed for each type; surfactants are believed to cause leaks in microporous films but not solid films, water swells solid films but not microporous films. Generally, solid films have better tear strength and are said to be better barriers to odours and to some microbes. Sympatex claim that the pores in microporous films become enlarged when the film is stretched at elbows and knees, thus facilitating water entry, whilst their film can be stretched 300% in any direction and still remain 'solid' and impermeable to liquid water. There are now a large number of films available which are both water resistant and breathable, made by a variety of novel techniques, e.g. making a film containing microscopic solid particles which on stretching produces tiny holes around the particles. Many of these films, however, do not have the physical durability or performance of Gore-Tex or Sympatex. These films are discussed further in Chapters 4 and 5.

Some films are offered for sale dot printed with hot melt adhesive powder. This is a convenient form; the customer can reactivate the adhesive to produce a laminated product with relatively simple apparatus and without powder scattering equipment, as and when required. There are a number of films with a very wide range of working temperatures and specialist properties, such as Kapton, a polyimide film by DuPont which has a service temperature range from –270 to 400 °C. Tedlar, a PTFE material also by DuPont, has a working temperature range –250 to 260 °C. Both films may be the basis of novel product development, perhaps combined with aramid fabrics. Film/fabric laminates are used in sail making, hot air balloons and have recently been introduced in some airbags. In addition to functional specialist uses, there are novel decorative films available, such as those produced from multi-extrusion processes, which give novel coloured patterns produced by interference light effects. These films, available from Mearl Corporation/Cornelius, are used mainly in the packaging industry, but if laminated to black or dark coloured fabric, they can produce interesting coloured effects. Decorative films are also available which produce metallic coloured effects by the etching of submicroscopic lines on the surface of aluminised film using a laser. These films are also used in the packaging industry, but could be laminated to fabric for creative decorative effects on garments or drapes. By a process of co-extrusion, films can be made with two different chemicals on either side and thus offer novel properties. Some adhesive films are produced in this way and allow two very different materials to be joined together. Adhesive films are discussed in Section 2.7.7.

2.5.2 Polyurethane foam

Polyurethane foams are available in various qualities, such as different densities, FR properties, different porosity and other specialist chemical and physical properties. The largest outlet for polyurethane foam is in textile laminates, used for automobile seats covers and for other coverings in the car interior, see Fig. 2.2. The usual form is sheet foam with a thickness of anything between 2 and 10 mm or more, laminated to the face fabric on one side, and a lightweight 'scrim' fabric on the other side. This tri-laminate is used by most car companies in the world, and is in this form to produce a material which is soft to the touch and which never creases or 'bags'. In addition, when the material is sewn, deep attractive sew lines are produced. The scrim fabric, which is usually knitted nylon or polyester, helps control stretchability, improves seam strength and acts as a 'slide aid' during sewing and seat making. If 'slide aid' is the only requirement, a lightweight nonwoven fabric is used. Fabric used for door panels and other areas of the car may also be laminated to polyurethane foam for a soft touch, for comfort and to help reduce vibration and noise. In these cases, the scrim is generally not required.

The polyurethane foam may be either polyester polyurethane or polyether polyurethane. The first mentioned is generally more easily laminated by the flame lamination technique, see Chapter 3, but has limited hydrolysis resistance. In hot, humid areas of the world, polyester polyurethane foam could break down, and in these areas polyether polyurethane foam is more

2.2 Production of sheet polyurethane foam for lamination to automotive seat cover fabric. The photograph shows Caligen Foam's 60 m block slitting facility. Photograph supplied by Caligen Foam Ltd and reproduced with kind permission.

likely to be used. Unmodified polyether polyurethane foam cannot be flame laminated and chemicals are added to the foam to make this possible. The automotive industry is extremely competitive and cost is so important that, at present, flame lamination is the most economical way of producing laminated car seat fabric in volume. Over 48 million cars are made in the world every year and all contain at least 7 m^2 of decorative face fabric laminated to polyurethane foam in the seats. Fabric laminated to polyurethane foam is also in other parts of the car interior, such as door casings, headliners and sunvisors. Because of environmental concerns regarding flame lamination and recyclability, alternatives to polyurethane foam have been researched. These include various types of nonwoven fabrics and 'spacer' knitted fabrics, which have a different 'touch' and poorer resilience at temperature. Automotive seat foam is discussed further in Chapter 4 and recycling aspects in Chapter 6.

Polyurethane foam/textile laminates are also used in the footwear industry and in speciality items such as waistbands, baby diapers and personal hygiene products. In the 1960s and 1970s, polyurethane foam/textile laminates were used extensively in apparel and for curtains and 800 million metres of polyurethane foam/textile laminate were produced in the USA alone during 1970.[56] However, this figure declined to 70 million metres by 1979. The product lost popularity because of a preference for apparel fabric with better flexibility and drapability, and also because of delamination during washing and drycleaning caused by, it is believed, poor quality control and lack of understanding of the factors influencing good adhesion.

2.5.3 Polyolefin foam

Polypropylene and polyethylene foams are available in roll form. Polyolefins have excellent microbial, oil, solvent and chemical resistance and can be fabricated by vacuum forming and moulding techniques to produce smooth, well defined contours in articles such as automotive interior components. These foams produce a soft touch to headliners, dashboards, door casings and sunvisors, and they also contribute to heat, noise and vibration insulation. The soft touch is slightly firmer than polyurethane foam and is considered by some in the automotive interior industry to be more 'up market'. Polyolefin foams have certain advantages over polyurethane foams, such as no 'fogging' in automobiles and better chemical resistance and ageing properties, but adhesive lamination requires some careful consideration owing to the inert nature of polyolefins. Both hot melt adhesive powders and adhesive films are available for bonding. In applications where high service temperature stability is required, reactive crosslinking adhesives are used. Information is available from the manufacturers such as Alveo (parent – Sekisui), and corona discharge is used in many cases to improve adhesion. Flame lamination is possible to certain substrates and textiles, and it is a much cleaner process than flame lamination of polyurethanes. The materials are partly crosslinked, and actual recycling by melting down and reforming is limited to 10–15% by weight inclusion into virgin polymer for injection moulding. Other recycling possibilities include shredding and thermally bonding the foam chips into drainage mats for construction applications, or agglomeration to reduce volume and controlled incineration. Polyolefins have a high calorific value when incinerated and can be used as fuel to incinerate other material, but the fumes may need treating and FR varieties of polyolefin foam may contain FR chemicals such as bromine. Polyolefin foams have good impact resistance and are used as inserts and padding in luggage, helmets, footwear and sports equipment such as shin guards and shoulder pads, see Fig. 2.3.

PVC foams are also produced which are easy to laminate and have the advantage of excellent weldability. Their 'touch' is quite similar to polyurethane foam.

2.3 One of the many applications of polyolefin foam, making use of the shock and energy absorbing properties. Polyolefin foam also has thermal insulation properties and a firmer soft touch than polyurethane foam. It is also used in door panels and other parts of the automotive interior. Other applications include padding in body protection clothing, luggage, shoe insoles, and wall and under floor insulation. Diagram supplied by Alveo (Sekisui) and reproduced with kind permission.

2.6 Compounding of polymers (resins)

2.6.1 General comments

The process of compounding of coating polymer or resin has already been mentioned several times. Note that, within this context, the word 'resin' has become almost interchangeable with coating compound or polymer. Resins are supplied as emulsions or as solutions in water or solvent. The 'solids' content – sometimes referred to as 'dry' weight – is always specified, and from this the resin 'add-on' that needs to be coated on to meet the specified coated weight of the end product can be calculated. Solids contents of water-based products are generally in the region of 40–60%, solvent-based products somewhat lower, 20–50% approximately. In general, a better quality film or coating is produced at lower add-ons and low solids contents, but for economic reasons, the coating operator will try to achieve the highest add-on possible in a single layer, commensurate with achieving satisfactory results and coating film properties. High add-on is helped by a high solids content but, as will be seen, the resin mix must be stable and process well. There is always a limit to the solids content that can be achieved with a particular mix before problems of thickening and instability occur.

The overall task of the compounder is to formulate a mixed compound, which will process without problems and which will produce the coated fabric properties required by the customer – at the correct price. The task of the compounder starts with material selection and formulation of the recipe, and also includes the physical mixing of all the ingredients to produce a uniform mix, which will be stable during storage and also during the coating process. Poorly formulated compounds can develop lumps or go thick, sometimes because of premature crosslinking. In addition, solid may separate out during storage or processing giving rise to streaks and drag lines. An important feature of a compounded resin is that it should have as long a 'pot life' as possible, but there is a limit to this when resins are to be crosslinked using an external crosslinking agent. In these circumstances, the crosslinking agent is added by the user just before use, but even in these cases the pot life should be long enough to ensure trouble free coating. Many resins are self crosslinking, i.e. they crosslink (after the water or solvent has been removed by evaporation) under the application of heat without the need or addition of other chemicals during compounding. Attention must be paid to the resin viscosity, because it must be the optimum thickness for the method of coating being employed, and a viscometer must be used for reproducibility. If a spindle viscometer such as a Brookfield is employed, the same spindle at the same speed should be used to check different batches of the same product (see Fig. 2.4). Use of a different spindle or the same spindle at a different speed is likely to give different results with the same resin. Although the resin should not thicken during storage or in transit to the coating works, this is sometimes unavoidable. However, any thickening should not be significant, and the resin should be easily returned to the original thickness just before use by the addition of small amounts of water or solvent. Some mixes may become thinner on the coating machine (i.e. show thixotropic properties) and, if this cannot be controlled on the coating machine, it may be necessary to reformulate the mix by using a different polymer or by using a lower solids content.

All types of coating resin need mixing and compounding even if it is only mixing in a thickener to prepare a single component resin for coating. PVC

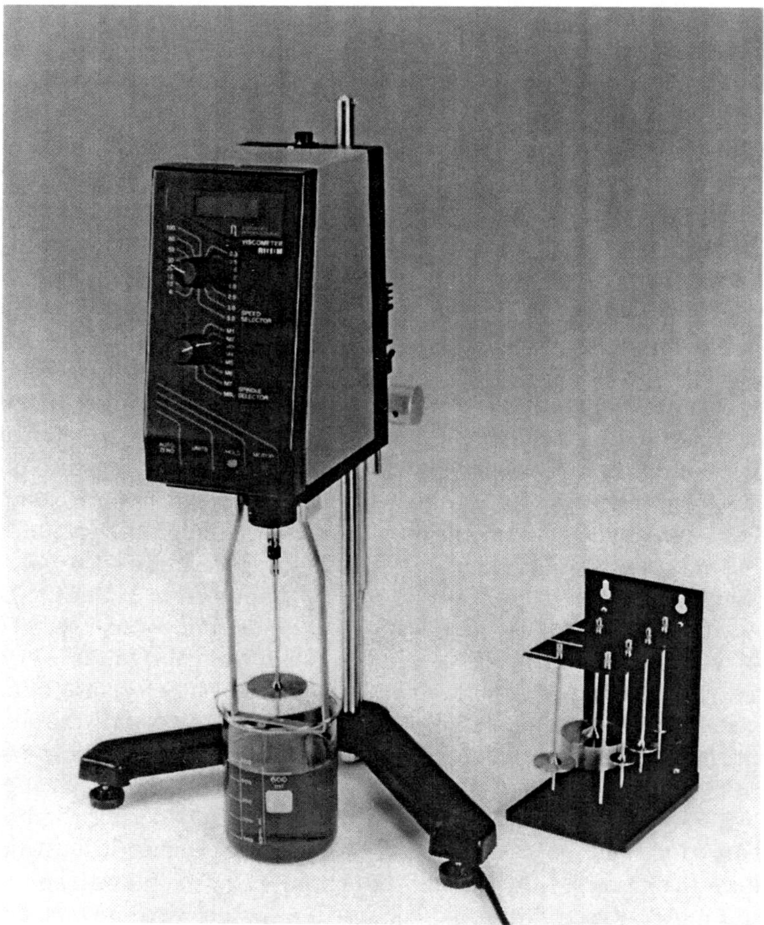

2.4 Spindle viscometer. Measurement and recording of viscosity is a vital part of coating development and quality control to obtain consistent and reproducible first quality products. The spindle and speed used must be noted together with the result. Photograph supplied by SDL International Ltd and reproduced with kind permission.

and especially rubber compounding require much skill and knowledge. Compounding is a specialist operation, the importance of which is frequently underestimated, and many processing faults have their origins in the compounding stage. Unless a competent technician is available to carry out or supervise resin mixing, compounds are generally best bought in ready mixed from a specialist compounder. Even large fabric coating works buy in mixed compounds, especially when pigmented, sometimes as masterbatches which require only dilution. The drawback to buying in ready mixed compounds is that actual precise details of the contents may not be known. In addition, the compounder may buy resins and chemicals from a limited number of suppliers and, indeed, the compounding company may be a subsidiary of a resin manufacturer. In this case, the compounder may have preferences for products which may not be optimum for the particular application, possibly because the profit margin is likely to be less by buying in a competitor's product. The truly independent compounder is able to buy the best products for the job, if necessary from different manufacturers, and use them all in the same recipe. Each coating layer can generally be treated as a separate entity, i.e. the softest resin with the best performance may be used for the base layer and another resin or mixture of resins from an entirely different manufacturer may be optimum for the top or intermediate layer. Having said this, the chemical manufacturers are the source of much useful and essential information, both on chemical properties and also on practical compounding. All are familiar with the major end uses of their products and also the important health and safety aspects. Experimentation in the laboratory followed by thorough testing is of course recommended before any bulk trials.

Every compounding ingredient is likely to have several different manufacturers, but chemical equivalents from different sources may not give identical results in a coating recipe. Apart from the obvious 'solids' content of liquids, there may be purity differences and, in the case of fillers, varying particle size and particle size distribution. The latter factor could give very different results, especially in PVC and rubber compounds. There are numerous auxiliary chemicals and additives, the use of which depends on the properties required, end product and application, but consideration must also be given to safety aspects, such as possible food contact and use by small children. Potentially harmful materials should be avoided and, in the modern environmentally conscious world, disposal at the end of the product's life should also be considered. Compounding of ingredients is likely to require special protective apparatus for safe handling, especially solvents which need precautions against fire risk and good ventilation to minimise breathing them into the body. Powders need care because they may cause skin irritation, dermatitis or some other ailment, especially if handled regularly or breathed in – a face mask may be advisable. The health and safety documents issued by the manufacturer must be read carefully and recommendations noted.

Some indication of the softness can be obtained from the glass transition temperature, T_g, of the resin. A value of say –20 °C is likely to indicate a very soft, stretchy but sticky film which could be useful as a base coat, but not as a top coat, as a value of about 0 °C indicates medium hardness and significantly less tack, whilst a value of about 20 °C or higher is likely to indicate a hard, dry and brittle film. A useful way of examining the physical properties of a resin or resin formulation is to cast a film of the material on to a glass plate or silicone release paper. The casting is dried, crosslinked if necessary and peeled away from the glass or silicone release paper. In this way the tackiness, stretchiness, flexibility and softness of the resin itself can be determined independently of any of the factors associated with coating it on to fabric, such as penetration or the fabric's own individual properties. If an FR coating is being formulated, testing a cast film of the resin formulation on its own could be a way of assessing the effect of the concentration or type of FR chemicals without the interference of all the variables associated with coated fabric, e.g. fabric stiffness, drape, resin penetration, etc. The limiting oxygen index test (LOI) is useful for formulating recipes. However, the actual FR test method specified in the test specification being targeted should be used as early on as possible in the development work. Inclusion of fillers generally reduces film clarity, stretchability and flexibility, and excessive amounts cause the film to crumble.

2.6.2 Compounding of water-based coating polymers

Water-based polymers used for fabric coating include polyvinyl acetates, PVC dispersions, acrylics, polyurethanes, silicones and rubbers. Some trade names of polymers and auxiliary chemicals appear in Table 2.4. The largest outlet for water-based resins is probably in the field of nonwovens as binders. Coating techniques are used extensively in nonwoven technology, but apart from a brief mention these fabrics are outside the scope of this book. In many cases with water-based resins, a single polymer may be used for a particular application or layer of a multi-coat coating, but different grades can frequently be mixed to obtain the required properties. For example, a very soft resin may be mixed with a harder one to soften the harder one or vice versa. Soft resins are invariably sticky (the softest resins are adhesives) and produce a tacky film, while harder resins tend to give films with a 'dry' non-tacky touch. Addition of small amounts of a harder resin to a softer one sometimes reduces the 'tack' of the coating surface.

For direct coating, it is necessary to thicken the polymer to a viscosity appropriate to the fabric construction, resin add-on required, and apparatus to be used, including blade profile. A very broad range of viscosity is used, between say 5000 centipoise and 40 000 centipoise or even higher. It is recommended that the method of viscosity determination, the instrument model and make, e.g. Brookfield, and, for research purposes, the spindle number used and the speed be stated. Some resins are self thickening and require only stirring, but most need the addition of a thickening agent. There are several different chemical types which produce mixes of different consistency, described by terms such as 'stringy' and 'buttery'. Chemical types include aqueous emulsions of acrylic acids, polyacrylate derivatives, vinylpyrrolodone copolymers, urethanes, cellulosics, etc. Addition of small amounts of ammonia is required with some grades of acrylic thickeners. Actual running on a coating head will help decide the best type.

Apart from thickening agents, other additives may be fillers such as chalk or talc for economy, and FR agents if required, see below. Fillers impart opacity, required for curtains, and titanium dioxide may be preferred to cheaper fillers because of its whiteness. Carbon black can be used to produce black-out effects, but other colours are best produced using a pigment of the required colour together with a white base such as titanium dioxide. Fillers also detackify to a certain extent, and matting agents, such as silica powder used especially for top layers in garment coatings, will have some anti-tack properties. There are ranges of different grades of silica or other additives – the chemical manufacturers are able to provide advice on the optimum grade for any particular application. Silicas are quality materials and as such are relatively expensive, while calcium carbonate, clays and talcs are quite inexpensive. Fillers are required to produce bulk, for example in carpets, and can increase toughness and improve abrasion resistance. The densities of some common fillers are (in g/cm^3), clay 2.5 (aluminium silicate), calcium carbonate (talc) 2.8, barium sulphate (barytes) 4.5, carbon black 1.8 and titanium dioxide 3.9 (anatase) and 4.2 (rutile). Barium sulphate is especially useful when mass is important, for example when producing sound insulation effects in automotive carpets. Wetting agents and dispersing agents may also be necessary to form a stable mix, if solids – especially solid pigments – are included in the recipe.

Foamed coatings require foaming agents, with the type depending on the actual foaming operation. For foam finishing (sometimes referred to as 'froth coatings'), relatively unstable foams are used, i.e. foams which collapse during the drying off stage in the stenter. For crushed foam coatings, the foam must be stable and a secondary foaming agent is sometimes required.

Many resins are produced as self crosslinking and can be applied without additional crosslinking agents or catalysts. Some crosslinking is necessary to produce durability in the end product, such as washability, dry cleanability, abrasion resistance and, in the case of carpets and pile fabrics, pile and fibre retention. For non-crosslinking resins and in situations where additional crosslinking is required, ranges of different external crosslinking agents are available. These differ in their reactivity and the properties they confer. The more reactive ones can be crosslinked at lower temperatures and may allow coating processes to be run at faster production speeds. In general, the more crosslinking the better the durability, but stiffening is an invariable side effect, sometimes with discoloration. Some highly reactive crosslinkling agents always cause stiffening of the coating, which may be tolerated in some applications. There is usually a compromise reached during compounding between performance properties required, handle and other aesthetics, and production rate. Resin selection and crosslinking agent selection play an important part, because they all have individual properties such as handle and drape. There are also various catalysts which may be used and the same general considerations apply, i.e. the more reactive the lower the processing temperature which can be used for crosslinking. Catalysts for water-based compounds include citric acid, ammonium chloride and oxalic acid (care – toxic).

In actual mixing of components, it is essential that the thickener, usually diluted with water, is added with careful stirring to the polymer and not the other way around – which will cause lumps to form. A production mixer is shown in Fig. 2.5. Stirring is always started slowly, and speeded up when a stable mix is obtained. Before any bulk mixing of polymers and compounding ingredients is undertaken, it is vital that small scale mixes are first carried out in the laboratory to assess compatibility, stability and any other problems which may arise. There are many different water-based polymers which may be mixed together to produce a particular property. In addition, there are whole ranges of textile chemicals which may also be mixed into the compound, e.g. anti-static agents, humectants (moisturising agents), softening agents, lubricants, etc. Certain waxes, paraffin dispersions or lubricants can also be used as detackifying agents for soft resins. These chemicals may also improve the abrasion properties of coatings but, if used in excessive amounts, can cause some deterioration in other properties. The technical information sheet of each polymer or additive must be studied for pH, whether anionic, cationic or non-ionic. Mixing together an anionic chemical with a cationic may cause precipitation or lumps to form, but it may be possible to use small amounts of them if mixed in separately; however, care is needed and pre-bulk trials in the laboratory are essential. A Werner Mathis oven, shown in Fig. 2.6 and 2.7, is invaluable for trying out compound formulations on small A4 size pieces of fabric before proceeding to full width machine trials which are costly both in time and materials.

A typical recipe for a waterproof coating on woven nylon or polyester using acrylic resins could be as shown in Table 2.5. A typical formulation for crushed foam coating for curtains could be as shown in Table 2.6.

The amount of foaming agent depends very much on the individual resin. Some resins may be self foaming and require little or no foaming agent. Some foaming agents are 'primary' foaming agents, such as sodium lauryl sulphonate; others provide better foam stability, such as ammonium stearate. There are more specialist foaming agents for better foam stability – such as required for carpet production – and reduced side effects.

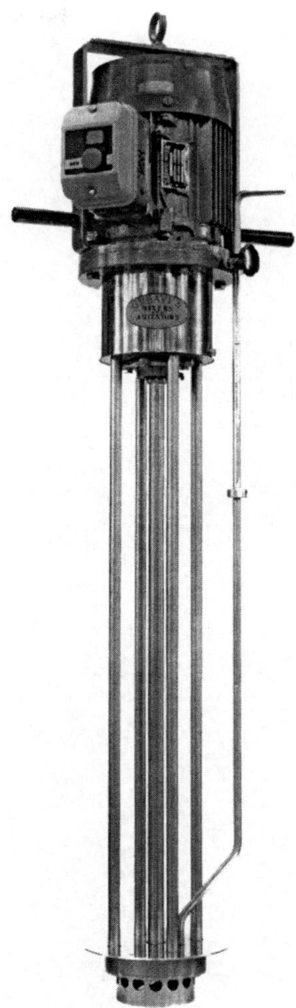

2.5 Greaves GM5 commercial mixer for water-based resins. The mixer should be capable of providing adequate agitation – paddle and disc-type impellers are not likely to be suitable. Mixing should always be started slowly and speed built up gradually. Ingredients should be added in the correct order and thickening only commenced when the mix is perfectly uniform. Photograph by Joshua Greaves & Sons Ltd and reproduced with kind permission.

2.6 Mathis Type LTE-S laboratory coating machine with knife on air, knife on roller and knife on rubber blanket facilities. Air flow can be controlled above and below and can be set to any required temperature up to 200 °C. Photograph supplied by Werner Mathis AG (R Jackson Wardle) and reproduced with kind permission.

2.7 Mathis Oven coating head adjustments. The blade angle can be adjusted and the blade height can be set to an accuracy of 0.01 mm and read off the clock gauges. Photograph supplied by Werner Mathis AG (R Jackson Wardle) and reproduced with kind permission.

Table 2.5 A typical recipe for a waterproof coating on woven nylon or polyester using water-based acrylic resins

	Wet weight	Dry weight
Soft base layer:		
Plextol DV300 (very soft)	100 parts	60 parts
Rohagit SD 15 (thickener), premixed with 3 × own weight of water	9 parts	1.5 parts
Top layer ('dry', non-sticky):		
Plextol BV410 (intermediate soft)	90 parts	45 parts
Plextol BV595 (hard)	10 parts	5 parts
Gasil (silica powder – for detackifying and mattness)	3 parts	3 parts
Rohagit SD 15 (thickener), premixed with 3 × own weight of water	9 parts	1.5 parts

Thickening achieved by very gradual addition of ammonia with stirring viscosity approximately 15 000 centipoise with Brookfield Viscometer.

Excessive amounts of foaming agent may lead to poor abrasion resistance and lowering of other physical properties. Ammonium stearate can migrate to the coated surface and produce 'blooms', a discoloration effect which may be unacceptable in some applications. More effective foaming agents are based on disodium N-octadecyl sulphosuccinamate; these are more

Table 2.6 A typical recipe for producing a crushed foam coating for curtain linings

	Wet weight	Dry weight
Acrylic resin	100 parts	50 parts
Water	30 parts	0 parts
Foaming agent	12 parts	4 parts
Calcium carbonate (filler)	25 parts	25 parts
External crosslinking agent	6 parts	2 parts
Thickener (for required viscosity) approximately	6 parts	2 parts
Ammonia (0.88) approximately	2 parts	0 parts

expensive, but can be used in combination with the less expensive ones to provide a boost. Titanium dioxide could also be used if a higher standard of whiteness is required. Viscosity (before foaming) is usually 60000–70000 centipoise, Brookfield Viscometer model RVT, spindle 6, speed 12, and it is suggested that the resin be foamed to a wet foam density (cup weight) of about 0.2–0.6 g/cm^3). Further information is available from resin manufacturers or compounders or from the literature.[30,31]

2.6.3 Compounding of solvent-based resins

The common solvents used in coating resins are DMF (di-methyl formamide), MEK (methyl ethyl ketone), toluene and iso-butyl alcohol. All mixing apparatus should be of the appropriate type as a precaution against fire, the mixing area should be well ventilated and protective clothing worn. Safety precautions are especially relevant when isocyanates are being handled because they are known to be irritants and respiratory sensitisers. Fire precautions must be taken and resins and mixed compounds stored in a cool and safe area; any residues should be disposed of correctly through a registered disposal contractor. Solvents evaporate quickly and must be stored in sealed containers or a skin will rapidly form on the surface. External crosslinking agents and catalysts are generally supplied in solvents compatible and miscible with the main resin solvent. Polyurethane polymers are formed by the combination of a polyol and a crosslinking chemical of which isocyanates are probably the most used. Because of the potential health risk, isocyanates are supplied in a 'blocked' form and only become reactive when unblocked by heating in the drying oven. There are several sources of information giving details of polyurethane resin properties as determined by their chemical nature.[9–11,26,28] In general, aromatic isocyanates have good chemical resistance but are prone to yellow discol-

Table 2.7 A typical recipe for waterproof coating using polyurethane solvent-based resins

Base layer:	
Polyurethane resin	100 parts
Polyisocyanate crosslinking agent (or melamine formaldehyde resin)	9 parts 5 parts
Acid catalyst	2 parts
Top layer:	
Polyurethane resin	100 parts
Gasil	3 parts

Viscosity 10000 centipoise, Brookfield Viscometer model RVT.

oration; aliphatics have less tendency to yellow but are more costly. Considering the polyol part, of which there are two main classes, ether derived or ester derived, ethers have the greater hydrolysis and anti-microbe resistance of the two, while esters are physically tougher with better chemical resistance. The viscosity of solvent-based resins depends on the solids content and, in compounding, the usual need is to thin the mixture rather than thicken, which is the case with water-based resins. Thinning is easily achieved by the addition of solvent. If thickening is necessary, addition of silica is possible, but this has a matting effect and too much may affect physical properties. Solvents have different boiling points, for example MEK (80 °C), toluene (110 °C) and DMF (153 °C), and they may be blended, if necessary, to produce a steady rate of evaporation to form a good film in a particular drying oven. A too rapid evaporation rate causes blistering, and this may be controlled by increasing the boiling point of the compound by adding a higher boiling solvent to the mixture. Further discussion on solvent coating is given in Section 3.4.5.

A typical formulation for a waterproof coating on woven nylon or polyester could be as shown in Table 2.7.

Polyurethanes, especially those based on polyethers, are susceptible to microbiological attack and the inclusion of a biocide should be considered. Transfer coated fabrics seem to be more vulnerable to microbe attack than direct coated materials.

2.6.4 Compounding of PVC plastisols

The main commercial outlet for plastisols in fabric coating is the production of tarpaulins and other associated covering material, the main physical requirements of which appear in Chapter 4. Raw PVC is generally supplied in powder form and, to convert this to a useful material for coating, it is necessary to prepare a plastisol, which is a complex mixture of PVC, plasticiser and several other specialist chemicals – all required to do a specific job. Great care is needed because the chemicals selected must not have any harmful effects during either mixing or use. For example, an effective heat stabiliser may not have the necessary health and safety approvals. Compounds considered toxic cannot of course be used for materials that will come into contact with food or children's toys, for example. PVC has so many applications, and this has resulted in a very wide choice of chemical additives. In recent years, the added constraints of recyclability and environmental friendliness have also been imposed. In addition articles such as toys, which will be used by children, need much care in plastisol formulation and are subject to both government and EC regulations and scrutiny. For automotive interior applications, plasticisers and other additives should not cause 'fogging', i.e. they should not migrate to the surface and vaporise and deposit themselves on the interior of the windscreen.

There are two separate tasks, formulation of what is included in the plastisol and then the actual mixing process. The first decision is which base PVC polymer or blend of polymer to use; this is followed by choice of plasticiser or mixture of plasticiser. One of the main characteristics of a PVC polymer is the molecular weight – or, more correctly, average molecular weight. PVC polymers are specified by their '*K* value', which is obtained from viscosity measurements and is an indication of the molecular weight and, therefore, of the basic physical properties. A PVC polymer made by the emulsion process with a *K* value between 70 and 80 is generally suitable for flexible coatings. Rigid PVC goods are likely to be made from polymer made by the suspension process with a *K* value around 57 to 68. The manufacturers of PVC polymers supply details of the optimum grade of polymer for a particular application.

Pure PVC is hard and brittle and, for textile coating and lamination purposes, it must be softened by the addition of plasticisers. The PVC is supplied as a powder and this is mixed with the plasticiser which is generally a liquid. The plasticiser is an important component, which determines many of the important properties of the plastisol, such as viscosity and processability. It also influences properties of the resultant coating, including tensile strength, tear resistance, temperature resistance, toxicity and chemical resistance and, therefore, has to be carefully chosen. PVC itself has excellent FR properties, but plasticisers vary in their individual FR nature; different ones require FR chemicals in different formulations. Many of the plasticisers are liquids, and the most common and economical ones tend to migrate very slowly to the surface of coatings and slowly evaporate. This loss of plasticiser from the PVC material causes stiffening and embrittlement of the coating leading to breakdown by cracking. It is this vaporisation of plasticiser which causes 'fogging' on car windscreens. Phthalates, such as dioctyl phthalate, and phosphates are amongst the more common plasticisers which have the basic requirements of good processability and compatibility with the PVC polymer combined with low cost. More specialist plasticisers such as sebacates and adipates are used for good cold temperature flexing; phosphates and chlorinated hydrocarbons give good FR properties. Food contact products require very careful consideration to meet the regulations governing the specific product. There are plasticisers available which meet the American Food and Drug Administration (FDA) requirements for use under specific conditions. Other plasticisers are available for specialist chemical resistance and specialist physical properties, good electrical properties, low odour, etc.

PVC is vulnerable to heat degradation at processing temperatures during manufacture and also during the lifetime of the product. It is also degraded by light and UV radiation, and both heat and light stabilisers must, therefore, be added to the mixture for protection. There are vast ranges of heat stabilisers and stabilisers for other specific purposes. These are now usually barium/zinc compound and calcium/zinc compound mixtures, but some heavy metals have been used and these are a cause for concern for environmentalists, especially at the end of the life of the product. Some fillers are added in order to reduce cost, but others are intended to impart a particular property, such as opacity, and to modify the physical properties. The tensile strength of the coating obtained can be reduced significantly if the wrong filler is used. Fillers, depending on particle size, usually increase the viscosity of the plastisol paste which then has to be readjusted. The filler, in effect, absorbs the plasticiser leaving less for the PVC, thus increasing the viscosity. The tendency of a plasticiser to do this is measured by its 'oil absorption' capability; water-ground mica and fine China clay have high 'oil absorption'. For low oil absorption, the filler particles should be as spherical as possible. Extender polymers, which are low molecular weight PVC materials, are sometimes used to help with processing – making the mix flow better and reducing viscosity if necessary. These are also specific viscosity modifiers which are used to increase viscosity, e.g. aluminium stearate, or to decrease the viscosity, e.g. derivatives of polyethylene glycol. Viscosity can also be lowered by the addition of a solvent, but this increases the fire risk and some processing machines may not be designed to cope with this.

To produce a pore-like structure or a very flexible coated material, a 'blowing agent', such as an azo-dicarbonamide (decomposition range 190–230 °C), is mixed into the plastisol. These agents decompose during processing to produce small voids in the coating giving a sponge-type structure and a soft touch to the surface. The factors to consider in choice of blowing agent are decomposition temperature, particle size distribution and volume of gas liberated, because these variables will determine the pore structure and other properties. Special heat stabilisers are available which can also decrease the decomposition temperature and provide a kicking action to the foaming process.

PVC plastisols require adhesion promoters to improve the adhesion of the base layer to the fabric, particularly with polyester and nylon. Different fabrics need slightly different ingredients and, if the fabric construction allows significant mechanical interlocking, the amount of adhesion promoter may be reduced. Adhesion promoters can be applied to the fabric beforehand by impregnation during the heat setting process or by addition to the plastisol used for the base coating layer.[57,58] For some grades of PVC it is necessary to include an emulsion of a plasticiser into the priming system. As discussed above, adhesion promoting systems are generally not required for cotton and the use of continuous filament textured yarn can improve mechanical adhesion. To produce the required colour, pigments, either in liquid or solid form, are added to the mix. For ease of processing and quality results, the correct viscosity, rheology, flow properties and a completely uniform mixture are essential. Ideally, viscosity should be measured at different rates of shear, so that some idea can be obtained of the performance of the plastisol in actual use on the coating machine, and it can be modified if necessary.

A good stirrer is required for actual mixing, but it is very important that it does not create bubbles or cause air to be drawn in and entrapped. Ideally, mixing is best done under vacuum which would minimise air entrapment. There are air release agents which help, but silicone types may reduce adhesion to the fabric substrate and are generally not recommended. Entrapped air can be minimised by a period of very slow stirring at the end of actual compounding, when all the ingredients have been added. If the viscosity of the paste is low, simply leaving the material to stand for a few hours will remove much of the entrapped air. Entrapped air can cause small pits or blemishes in the surface of the coating. Mixing in of pigments requires special care, especially if they are in solid form. Some compounders first produce a pre-mix by grinding and mixing the solid in a small amount of liquid plastisol, and then mix this pre-dispersion into the main mixture. Generally, pastes should be stored at room temperatures, but preferably below about 23 °C; however, they should be protected against frost which is likely to cause them to go solid and become unuseable even when brought back to room temperature. Poor mixes may well separate out at room temperature within the expected shelf life and, even if only partially separated out, are likely to give very unsatisfactory results if used.

On heating, a PVC resin does not strictly speaking 'crosslink', but passes from the fluid phase to a solid material by a fusing or gelling process, the precise mechanism of which has not yet been elucidated. The transition temperature is about 160 °C, but to achieve optimum physical properties requires heating to at least 170–180 °C for a minimum of 1–1½ minutes.

Table 2.8 A typical formulation for a PVC tarpaulin

PVC plastisol	100 parts
Plasticiser	70 parts
Filler	25 parts
Heat stabiliser	2.5 parts
Light stabiliser (if necessary)	1.0 part
Pigments	10 parts
Dispersing agent	1.0 part

Gelling at lower temperatures produces inferior physical properties. Unlike the crosslinking process, the material must be heated for this length of time for the gelation process to proceed properly. Simply hitting 180 °C for an instant does not 'kick off' crosslinking in the case of PVC, because there is no crosslinking process to take place. PVC is subject to environmental scrutiny at the present time and this is influencing the choice of which additives and plasicisers to use. This aspect is discussed in Chapter 6. Guidance on resin selection and compounding is available from PVC manufacturers, suppliers of compound and the literature.[22–24]

A typical formulation for tarpaulin material could be as shown in Table 2.8.

2.6.5 Compounding of rubber compounds

This is a very large and specialised subject area, because there are so many different types of rubber and additives used in many thousands of applications. There are over a dozen different grades of Neoprene (DuPont polychloroprene rubber) alone, for example, and there are very many different ingredients in a recipe, each ingredient being necessary to do a specific job or to provide some specialist property. The rubber compounding industry began with the discovery of vulcanisation in 1839, and the industry's expertise today is the cumulative result of the work of thousands of rubber technologists working on countless applications since that time. Rubber is processed both from water-based lattices and from solvent-based compounds. Dipping for the manufacture of items such as gloves is generally carried out from lattices. As well as specialist additives, there are certain ingredients included to make the various chemicals compatible with each other. In addition, the many different types of rubber can themselves can be blended to optimise a particular property requirement. The compounder's job is to produce a rubber compound to do a specific job at the lowest possible cost.

The adhesion principles discussed above also apply to rubber coatings, cotton not requiring any adhesion promoting system. There are well over a dozen different classes of rubber, each with its own special properties, and each type of rubber itself has numerous variants. The essential constituents of a rubber compound are: the rubber itself, vulcanising agents, fillers, oils, antioxidants and plasticisers. Frequently, inclusion of one particular chemical improves one property at the expense of another, and a balance must be reached. Finally, the mix must be processable on the apparatus available, produce the specified properties and be at the correct price to be commercially viable.

The compounder's first task is to select the rubber type (see Table 2.3) and, if necessary, consult manufacturers and specialist papers on the subject. [30,31,40–45] The next essential component is the vulcanising agent, or activator, the most common one of which is zinc oxide. In practice, this universal vulcanising agent is usually used together with an additional agent to 'tailor' the properties to meet the requirement. Vulcanisation is the term used to describe the crosslinking process which converts raw rubber into a more useful state, one having elastic properties and more durability. The accelerator is used in conjunction with the vulcanising agent to increase the rate of cure and to produce other improvements. There are many different chemicals used as accelerators, and it is possible to vulcanise rubber under almost any conditions of time and temperature. A compound containing only these basic ingredients may be difficult to process, and fillers and processing aids are also required. These also reinforce the rubber.

The fillers are powders and the processing aids, which may also be plasticisers, are needed to disperse the solid particles and to help control the viscosity. Fillers are sometimes referred to as black or white, the black type being invariably carbon black. Some specialist grades of carbon black can confer significant anti-static properties to the article being coated, e.g. carpets. White fillers can be calcium carbonate, clay, silicates or titanium dioxides, etc. The size and shape of the filler particles can influence the properties of the rubber, and careful consideration is needed in the choice of chemical type. For example, if the rubber article is to come into contact with acids, calcium carbonate would not be used as a filler. There are dozens of different grades of carbon black used, each one producing slightly different properties in the rubber.

Antioxidants are necessary to protect the rubber from atmospheric oxygen and to improve the ageing properties. Here again, careful selection is important in order to prevent discoloration and other adverse effects. In addition to the essential ingredients, there are FR agents, pigments, specialist heat stabilisers, blowing agents for 'soft touch' and bonding agents to join the rubber to the textile and other materials. There are also dispersing agents, thickeners, wetting agents, emulsifying agents, gelling agents, deodorants and other additives to impart special physical and chemical properties.

A typical basic formulation would include:

- rubber polymers
- vulcanising agent/sulphur
- zinc oxide
- accelerator (catalyst)
- fillers
- antioxidant/stabiliser
- softener/lubricant

There are numerous other 'standard' formulations depending on the type of rubber and the end use properties required. Some recipes may contain 20 or more different ingredients.

2.7 Adhesives

2.7.1 Mechanisms of adhesion

There are four basic mechanisms by which adhesion is believed to take place.[59] These are: mechanical interlocking of the materials; diffusion of polymer molecules across the interface; electrostatic forces; and, lastly, interatomic and intermolecular attractions between the atoms and molecules of the materials being joined, i.e. the adhesive and one of the substrates. The adhesive acts as a 'go between' between the two substrates being joined together. The last named group includes chemical bonding, which generally produces a strong and durable bond. Mechanical interlocking and electrostatic forces are physical bonding mechanisms. All types of bond require clean surfaces, free from dirt, grease and other contaminants, especially silicones. Increasing the surface roughness generally improves the bond strength. There are general accounts of adhesives and their properties.[59–62] When bonding problems with films and plastics are experienced, surface treatment of the film or plastic with corona discharge or a plasma process sometimes help, see Section 2.7.3.[63–66] Chemical cleaning or pretreatments with flame are also reported to be helpful.

Mechanical interlocking, especially with rough, natural short filament fibres such as cotton, is an important means of adhesion and, as discussed above, textured yarns also provide facilities for mechanical bonding. Much development has been necessary to improve the bonding of rubbers and plastics to smooth continuous filament synthetic fibres by the use of adhesion promoters and priming materials. The application of adhesive must not affect the appearance, colour or surface texture of the fabric being joined, and should have minimum effect on the handle of materials, especially those which are to be used for further processing. Thus, adhesive must be controlled; it should first 'wet' and flow to cover the surfaces of the materials being joined, and then penetrate to a certain extent, but not so much that it will cause stiffening or penetrate through to the fabric or material face. 'Wetting' depends upon surface energy of the materials; the solid surface being 'wetted' should have a higher surface energy than the wetting agent for it to spread over the surface. Once applied, the second material substrate must be introduced, and the two substrates should be held together in intimate contact until a strong bond is developed. If there is any delay, the adhesive could lose some of its effectiveness, by cooling, premature crosslinking or evaporation of the carrying liquid. If the freshly laminated materials are disturbed before a strong bond is formed – for example by a sudden change in material tension – the final bond strength may be lower than expected. Further processing factors influencing adhesion are discussed in Chapter 3.

2.7.2 Surface tension

'Wetting' is used in a broad sense and does not refer just to water 'wetting' – the liquid can be a water-based adhesive, a solvent-based adhesive or a liquid, hot melt adhesive. Surface tension is the physical property which determines the ease with which the 'wetting' will occur. If the surface tension of the liquid is greater than the surface tension of the solid, the liquid will not 'wet' the surface, but instead will form beads, i.e. small spherical droplets on the surface. If the surface tension of the solid is greater than that of the liquid, the liquid will wet out and spread over the surface of the solid. When fabric is involved, the situation is slightly more complicated because of the 'holes' in the surface, but similar considerations apply. In this case, the surface tension of the fibre – or more correctly that of the finish on the fibre – affects the wetting. All fabrics and fibres will have a finish of some description on them. Even 'pure' finished fabric will have some residual dyebath chemical on it, or the yarn will have residual spin lubricant or finish applied to it during yarn manufacture. As previously mentioned, fabric or yarn with absolutely nothing on it would be almost impossible to process, because of the significant build up of static which would cause it to cling to rollers and other surfaces.

The same surface tension determines whether a fabric will have water or shower repellency. Fluorocarbon finishes on fabrics impart a very low surface tension to the fabric surface, thus conferring shower or water repellency to the material. Certain wax finishes and silicone finishes also have low surface tensions and impart water repellency, but these finishes also attract oil-based soiling, because the oils have an even lower surface tension.

2.7.3 Modification of surface properties

The first fabric used for coating was cotton, which has a naturally rough surface and is made from short staple lengths twisted together to make a continuous yarn. There were few problems with resin–polymer adhesion, partly because of the mechanical locking made possible by the rough fibre surface, but, more importantly, because of the very large number of short staple fibre ends embedded in the resin. When viscose was first used for car tyres, intensive research work needed to be carried out to find a chemical means of improving rubber adhesion to the smooth continuous filament viscose yarn. Most modern coated fabrics are produced from nylon and polyester yarns, frequently high tenacity varieties, and are smooth continuous filament. In addition, for specialist end uses, aramids are used which are also generally smooth continuous filament. Work has been done to try to improve the adhesion of polymers and rubbers to industrial synthetic yarns, using surface treatments such as plasma and corona discharge. In recent work in the USA the surface tension of plastics and fibres has been increased by plasma treatments in the laboratory to 70 dynes/cm (erg/cm^3) which would allow better 'wetting' out of adhesives.[64] Some researchers believe that, in addition to lowering of surface energy, some surface roughening and cleaning results from these treatments. Hence there is more than one reason why adhesion is improved.

The breakdown of the fibre–polymer bond is believed to be the cause of premature failure of many rubber coated articles, including tyres and hoses. Initial strong bonds are essential if tarpaulins and garments are to withstand the flexing encountered during their lifetime and, in the case of the garment, multiple washings. In some cases the fabric substrate surface is buffed or slightly raised before coating in efforts to improve coating adhesion. Sometimes this is done to obtain a softer and more flexible handle, especially with transfer coating.

2.7.4 Adhesive type

Adhesives are available as solutions in solvent or water, or as dispersions in water, or as solids, which melt under the action of heat, see Table 2.9. A variety of different chemical types are available, each with their own properties such as suitability for specific substrates and durability to water, humidity, solvents, heat, etc. The chemical nature generally determines the property, for example polyurethane adhesives are flexible and stretchy, but some may discolour. Polyvinyacetate adhesives are relatively inexpensive, but generally have limited water or washing resistance. There are mixtures of different chemical types that produce specific properties, for example a hot melt adhesive may consist of a blend of copolyesters and modified

Table 2.9 Summary of adhesive types

	Water based	Solvent based	Hot melt
Form supplied	Solution or dispersion in water	Solution in solvent	Powder (various particle size) Granules Gel Web Film
Advantages	Non-flammable Generally safe to use Easy clean up Easy storage Fewer health and safety problems	Generally good tack/grab Quick dry off Good water resistance 'Wets' surfaces easily	Clean No dry off necessary No fumes Instant bond in many cases Storage generally easy
Disadvantages	High energy required to dry off water (latent heat of evaporation is 539 calories per gram) Process may be slow Generally low solids content Limited durability to washing and moisture 'Wetting' of surfaces and spreading sometimes not easy	Fumes potentially toxic Extraction/ emission treatment necessary VOCs environmentally unfriendly Legislation requirements Careful storage is necessary Fire risk Health and safety requirements	Initial plant may be expensive Heat necessary to activate the adhesive which may damage substrates (e.g. pile crush, glazing, stiffening, discoloration) Short 'open time' and loss of tack on cooling Certain operations require high operative skill
Cost	Inexpensive to moderate	Moderate to expensive	Granules generally inexpensive Powders vary from inexpensive to moderate Webs vary from moderate to expensive Films vary from expensive to very expensive

Table 2.9 (Continued)

	Water based	Solvent based	Hot melt
			Gels vary from expensive to very expensive – but may be cost effective if optimised

Flame lamination: Expensive initial plant, operating costs high for small volumes, low to very low cost for high volumes. Regular foams are inexpensive but specialities may be very expensive (the melted foam, 'burn off' is the adhesive). Fumes are potentially toxic and abatement is essential. Alternative methods being sought for environmental reasons.
Notes: Polyester, polyamide, PVC and polyurethane materials are generally relatively easy to laminate but adhesive selection is necessary for the required level of durability (mechanical flexing, wash/dry clean resistance, heat and heat ageing), handle, ease of processing, cost and other specialist requirements. Plasticisier resistance needs to be considered for PVC. Polyolefins are generally relatively inert and require more careful adhesive selection, and substrate pretreatment may be necessary. Bicomponent hot melt adhesive films are available for specialist applications.

polyethylenes. All adhesives must have some affinity for both the materials being joined. As just mentioned above, they must first of all 'wet', cover and penetrate the surfaces to be joined, and then solidify by evaporation of the carrier liquid to form the permanent bond by the mechanisms already mentioned. In the case of a 'hot melt' adhesive, the bond is formed on cooling. Hot melt adhesives are available in several forms; as a 'web' (resembles a net curtain), as a continuous film, or in powder or granular form. Some adhesives are also available as a liquid or jelly; these are 100% (or nearly) active material which does not contain any solvents or water. The earliest use of a hot melt adhesive is probably sealing wax for envelopes.

2.7.5 Solvent-based and water-based adhesives

Solvent-based adhesives are generally environmentally unfriendly, see Chapter 6, and, in addition, many are flammable and their fumes can be harmful to health if breathed in. Solvents are also more expensive than water. In general, however solvent adhesives 'wet' the surfaces to be joined better than water-based adhesives, have more 'grab', and they also dry off faster. Solvent-based adhesives are reported to have a better shelf life than water-based ones, because the organic ingredients are more stable as solutions in solvent than they are as dispersions in water. Water-based adhesives are safer to use and pose less of a problem to the environment, but drying off water (latent heat of evaporation is 539 calories/g) is expensive both in terms of energy and time. Because of the hydrophilic nature of water-based adhesives, many are not, in general, as resistant to water or moisture as are solvent-based types.

2.7.6 Hot melt adhesives

Because of the factors stated above, hot melt adhesives are gaining in popularity, but they need to be carefully selected. For good durability their softening and melting point must be well above the temperature to which they will be exposed during use, e.g. inside a car interior. Other performance factors which must be considered, as with all types of adhesive, are bond strength, resistance to moisture, humidity, heat ageing, light and UV degradation, and any effect on fabric colour. The nature of the materials to be joined, where they will be used within the car, and their physical form, all need to be taken into consideration when deciding which adhesive and which lamination machine to use.

Choice of hot melt adhesive affects the handle for two reasons. The first is the physical property of the adhesive material itself, i.e. if it is hard or soft. The second is the degree to which it sinks into the fabric. The melting characteristics, flow properties and viscosity of hot melt adhesives are important considerations. If too much heat is applied they can flow away from the surfaces to be joined, hence producing a poor bond. This excessive flow may also cause stiffening of the laminate and penetration to the face of the fabric being laminated. The adhesive manufacturers offer advice on which chemical class of adhesive to use, and the best one for the job in hand. They should also have information on adhesive properties, including recommended temperature of bonding, heat resistance, resistance to water and solvents, etc. Manufacturers should also be able to advise on resistance to PVC plasticiser migration, when PVC is being joined, and should have information on adhesive viscosity at the recommended bonding temperature. Of course it is always sensible to get at least a second opinion, and to carry out impartial trials. The manufacturer being consulted may not actually sell the best product or even the best chemical type for the job. They may judge that it is not in their best interest to recommend a competitor's product! But this applies to all products in all industries!

Chemical types include polyethylene, polypropylene (the two chemical types merge and are often referred to as polyolefin), polyamide, polyester and polyurethane, see Table 2.10. There are copolymer varieties of each chemical type, allowing a wide range of properties including melting points, durability to washing and dry cleaning and heat resistance to be obtained. Polyolefins tend to be the most economical, but tend to have lower

Table 2.10 Hot melt adhesive chemical types

Chemical type	Comments
LDPE (low density polyethylene)	Low cost, inert material. Used for fusible interlings, automotive carpets
HDPE (high density polyethylene)	Generally higher melting point that LDPE with better temperature resistance. Low cost. Used in carpets and for fusible interlings, shirt collars and cuffs
Ethylene vinyl acetate (EVA)	Good adhesive properties (tack). Low cost. Used in footwear and to bond leather and paper. Good flexible properties
Polyamide (and copolymers)	Wide range of products and temperature melting points, generally higher than PE and EVA with better heat resistance. Used in garments and applications where solvent resistance is required (dry cleaning). More expensive than PE and EVA. Some reactive types available with high temperature resistance and high durability to chemicals, solvents and water
Polyester (and copolymers)	Wide range of properties, high temperature resistance and good durability. Some suitable for automobile interior trim. Higher cost than PE and EVA. Some reactive types available with high temperature resistance and durability
Polyurethane	Good adhesive properties and durability. Expensive. Moisture curing gels have excellent durability and high resistance to temperature. Used in automotive interior trim, laminates for protective clothing. Good flexible properties with some elongation

Powder suppliers include: Dritex, EMS-Chemie AG, Bostik
Film suppliers include: Sarna Xiro (Guttacol), Cornelius, Bostik, Protechnic, Bemis, Dow
General adhesive suppliers include: Bostik Findlay, National Starch, Henkel
N.B. All adhesive types are generally available in powder form and also as webs and films. Many variants are available from a variety of manufacturers.

durability. The polyurethanes tend to be amongst the most expensive, but they are capable of giving softer, more flexible and stretchy laminates.

2.7.7 Adhesive films and webs

Adhesives in film or web form are generally significantly more expensive than the corresponding adhesive powders. Adhesive webs are discontinuous films, rather like net curtains, and are supplied in roll form made from various chemical types with various properties and melting points. They produce laminates which are flexible and which are porous and breathable. Continuous film adhesives cause stiffening, which may not be a problem in the case of, say, demonstration or exhibition boards or car headliners. A novel product are the slit films made by Xiro^R. These films reticulate, i.e. the small slits open up and present a network appearance on heating, and therefore produce laminates which are porous and permeable to both air and water vapour. This is important for laminated fabric that is to be used in garments. Under the action of heat, these slit films do not shrink inwards and lose width in a similar way to some other adhesive films, and the discontinuous form produces fabric laminates which are softer and more flexible than those produced from continuous films. Film adhesives are also available with small pin holes in them; this helps to prevent the entrapment of air bubbles during lamination and would also be expected to give some measure of water permeability. Bi-component or double sided film adhesives are also available, which allow almost any material to be joined to any other material.

A practical problem which may arise with film adhesives is in being able to feed the film to the lamination machine without loss of width through excessive tension. Films need supporting and unrolling again without too much tension developing, which may cause loss of width and even creasing. Some adhesive films are supplied supported on, or interleaved with, a non-adhesive carrier film. The carrier film is generally there to prevent the adhesive film from sticking to itself, and is presumably used as a processing aid in the adhesive film manufacturing process. This carrier film must, of course, be removed before use, and an additional driven take-up roller may be needed on the lamination machine to do this.

Some mention must be made of pressure sensitive adhesives which are widely used on tapes. They are available on release paper for application to fabric to be used on panels, for example in automotive interiors. Pressure sensitive adhesives are based on a number of polymer types, including acrylic, and are available in different variant types for various applications and degrees of durability and heat resistance. They are relatively expensive but are clean and require no special apparatus to apply.

2.7.8 Hot melt adhesive powders

Adhesive powders are available in most chemical types and also in particle sizes ranging from very small up to about 500 micrometres or so in diameter.[67–72] The choice of size used depends on a number of factors, including the machinery available, the surface nature of the substrates and the handle and properties required. There are a number of processes by which hot melt adhesive powders can be applied, each with its own merits, which are described in Chapter 3. Sometimes adhesive powders are supplied together with, or treated with, a UV fluorescent agent and a UV lamp is positioned near the powder application point. The reason for this is so that the small powder particles treated with UV can be more easily seen, which will allow better control of application.

The smallest particle sizes, 0–80 micrometres, can be compounded into pastes which can be direct coated, usually knife over roll, or dot coated with a high degree of control, using rotary screen print application machines. These precision print machines are capable of excellent results, but require a high level of technical competence and skill to set them up and run them. They have the important advantage of being 'frictionless', because the screen *places* the coating on top of the material. Further details are given in Chapter 3. Dot printing of adhesive can produce breathable laminates with excellent softness and drape which are used in the lamination of waterproof breathable films to fabric.

2.7.9 Moisture curing polyurethanes

These adhesives allow strong adhesive bonds with relatively low add-on, because they crosslink in the presence of moisture. There is generally sufficient moisture in the goods themselves being processed to trigger off the chemical reaction. They are available in hot melt form as a clear jelly in sealed moisture tight drums. They have been applied by spaying, slot die extruding and also by gravure roller. Significant improvements have been made in recent years to make them more manageable and to improve initial 'grab' and 'open time', i.e. the time during which the adhesive is available for bonding once it has been exposed to the air. The 'open time' is believed to vary from manufacturer to manufacturer. Because they actually chemically crosslink and can form strong bonds at relatively low add-ons, durable and flexible laminates with good drape have been found to be possible. For this reason, they are used in washable garments, protective clothing and car seat covers. However, these adhesives are expensive and control of application and penetration into substrates is critical. Moisture cure polyurethane adhesives have been available for a number of years, but it is only within the last few years that significant improvements in open time and initial tack have been achieved which have increased their usefulness.[73,74] Special apparatus is required to handle these adhesives as will be seen in Chapter 3.

2.8 Flame retardancy

2.8.1 Introduction

Statistics show that a significant proportion of household fires start by ignition of textile materials and that a cigarette was the most common ignition source.[75] In the UK all domestic upholstery materials must be FR to a small ignition source, but fires have also been the cause of significant loss of life in public buildings and on trains and on aeroplanes. Flame retardancy standards for public safety are generally controlled or influenced by government departments, such as the Home Office and the Civil Aviation Authority for aircraft. The International Maritime Organisation is responsible for safety at sea. These standards are aimed at reducing ignition, flame propagation speed and the heat and smoke generated, and allowing sufficient time for people to reach areas of safety. Much literature is available on flame retardancy and test methods.[76–83] Further comments on FR testing appear in Chapter 5.

Flame retardancy is required in many coated or laminated products, and a polymer coating can hold a larger amount of FR chemical than a simple finish on a fabric. Of course the fabric will be transformed into a heavier, thicker and less flexible material, but these properties are sometimes less critical and can be tolerated to a certain extent. Examples of FR coated fabrics in the home are furniture upholstery, mattress tickings and roller blinds. Most textiles used for public buildings, such as hotels or offices, generally require quite a high standard of FR properties. Some medical textiles require FR properties because they may come into contact with lasers or are used in areas where there is a high concentration of oxygen. The FR properties of coated fabric used in any transportation application are extremely important, because of the limited means of escape within a vehicle or in the air or at sea. The highest safety requirements must be attached to any application where the general public are involved, and in transportation situations large numbers of people usually occupy a confined space, with the added hazard of panic. When many plastics burn, they give off dense smoke and large amounts of heat which are as hazardous and in many situations, more hazardous than actual flames. These properties must be taken into consideration when FR chemicals are being included in coating recipes.

The common base fabrics used in fabric coating and lamination, nylon, polyester and cotton, have no special FR properties and will generally ignite and burn unless treated with an FR chemical. There are inherently FR varieties of polyester and there are a whole variety of specialist FR fabrics, but these are expensive and most are not generally used in domestic furnishing applications. However, certain specialist fibres are used as fireblockers in transportation seating applications and specialist protective clothing, i.e. Panox (LUCF – Lantor Universal Carbon Fibres), Nomex (DuPont), Inidex (Acordis), etc. Of the resins used in fabric coating, some have relatively poor FR properties, i.e. acrylics, polyurethanes and natural and SBR rubber. Certain resins have good FR properties, i.e. PVC, PVDC/acrylic blends and polychloroprene (Neoprene – DuPont). Addition of FR chemicals can generally improve the FR properties of these resins, especially where some 'synergy' can be exploited, e.g. using the chlorine in PVC or PVDC. The fluorine-based polymers, PTFE and Viton, are extremely FR in their own right. In development work it is sometimes profitable to determine the FR properties of the coating being formulated (first cast it on to a glass plate), using the LOI test which gives results in numerical terms. The LOI is the concentration of oxygen which just sustains a candle-like flame.

A coated or a laminated fabric comprises at least two polymer types, and the FR properties of the combination cannot be predicted using knowledge of the FR behaviour of the individual constituent polymers. The coated or laminated fabric generally has properties of its own which must be determined by testing. The FR properties of both the fabric and the plastic will be changed when they are joined together, depending on the amount of penetration which occurs, and on other variables, among them resin viscosity and add on. A fabric which individually may shrink away, or drip away, from the applied flame and hence not be ignited, may be held to length by the coating and thus will be ignited. Another sample of the same material, but from a different batch, which has been either dyed or stenter finished in a slightly different way, may not shrink back as readily and thus will become ignited and therefore display very different FR characteristics. One researcher has pointed out that the interface of fabric and polymer is, in some ways, an entirely different material to both the fabric and the polymer coating or laminate. If an adhesive has been used, this may be a third, different chemical type which could also display different FR properties and produce further modification of the behaviour of the other two components. There will be, in this last mentioned case, two interfaces between dissimilar substances. These factors add further to the complexity of FR testing, which both research and quality control staff know only too well. The FR properties of a coated fabric will be influenced by coating technique, resin add-on, penetration and the many factors and inter-relating factors which depend on these variables. A study has been made of the effect of coating parameters on the FR properties of a cotton fabric coated with a phosphorus FR compound.[82]

2.8.2 The burning process

The stages of combustion are ignition, growth, propagation and finally decay, but every fire in real life situations is unique because the circumstances and conditions are never exactly the same. The way fabrics burn depends upon a variety of factors and combinations of factors, including fabric stiffness, drape, contact with or proximity to other materials, supply of air, draughts, etc. Smoke results from the incomplete burning of materials and is a dispersion of solid or sometimes liquid particles, together with gases, some formed by the combustion process. Fire is a hazard, not only because of the danger of contact with flames, but also because of suffocation by toxic fumes, injury from heat levels and heat stress, plus all the dangers associated with panic and the inability to escape because routes are obscured by dense smoke. Individual test methods have been devised to take all these factors into consideration, some of them after lessons learnt in actual disasters. We are still learning about causes and methods of prevention – too frequently too late – despite precautions following extensive research. Flame retardancy testing of coated fabric is discussed in Chapter 5.

2.8.3 Basic mechanisms

This section attempts to explain the main principles governing flammability and its control. Many articles have been published on these aspects and also on the relevant test methods[76–83] (see also Section 5.7). Burning depends on three factors, a source of ignition to provide the initial heat energy, fuel or materials capable of burning and a supply of air which contains oxygen (or an oxidising agent), the gas on which combustion depends. Anything which reduces these factors reduces combustion, which is essentially a chemical oxidation process. Heat energy first causes molecules of the fuel to break down into smaller parts called 'free radicals' which are unstable and therefore highly reactive. Burning proceeds by the formation of these free radicals and their subsequent reaction with oxygen. Certain FR agents, under the action of heat, break down producing their own free radicals. The fuel free radicals then react preferentially with the FR free radicals instead of with oxygen, thus inhibiting combustion. An uncontrolled fire with a supply of fuel is self propagating, because heat from the burning material heats up the surrounding air and the material not yet burning.

2.8.4 Mechanisms of FR chemicals

Hydrated chemicals contain significant amounts of water and, when heated, this water is released and cools the flame. The water vapour formed dilutes the oxygen in the air. An example is aluminium trihydrate, which contains 35% of its weight of water. Chemicals such as aluminium hydrates and some boron compounds take in energy (endothermic) on decomposition, and the flame is cooled by this process. Materials which decompose to release non-flammable gases such as carbonates have some FR properties. Some chemicals will function by more than one mechanism.

The most effective mechanisms, however, are inhibition of the free radicals and reducing the availability of fuel by formation of a barrier or protective char. Chlorine and bromine (halogen) compounds have been found to have good FR properties, especially in combination with antimony trioxide, and the 'antimony/halogen synergy' is the basis of many FR formulations; antimony trioxide alone has no FR properties. The halogen compound releases free radicals, which react in the gaseous phase with the free radicals produced by the burning polymer. Thus, reaction with oxygen is inhibited and burning retarded. This method is very effective and is widely used in plastics, but it has the disadvantage of producing potentially toxic fumes.

Some chemicals or combinations of chemicals prevent afterglow and re-ignition; others, on combustion, form a char or barrier which effectively reduces the amount of air reaching the burning material. An example is zinc borate, which forms a glass-like coating and is claimed to suppress smoke significantly. Chemists have developed this concept and produced 'intumescent' coatings which form at relatively low temperatures, i.e. at an early stage of combustion, to produce a voluminous insulating char foam-like substance.[83] This acts as a barrier and inhibits flame spread by restraining the escape of gas formed by burning and also limiting the access of oxygen to the flame. Intumescent coatings can be effective at low concentrations – intumescent paints are available – and continue to be developed. They contain a source of carbon, a 'blowing agent' to increase the volume and fillers and other chemicals.

Phosphorus FR chemicals work by encouraging the formation of char and are reported to suppress glowing, which produces carbon monoxide and carbon dioxide and also carries the risk of re-ignition. Phosphorus FR agents, in combination with certain nitrogen compounds, produce an FR synergy for cellulosic materials. Certain polymers, such as PVC and especially PVDC, already contain high levels of FR chemical species such as chlorine and have inherent FR properties. These polymers have been used as FR compounds themselves, for example the addition of PVDC to SBR in back coatings on automotive carpets.

2.8.5 Disadvantage of FR chemicals and recent developments

The main disadvantages of FR compounds are cost, problems associated with compounding, toxicity of fumes from burning and, more recently, environmental considerations. Products containing FR chemicals need special care when being disposed of at the end of their useful life, especially those incorporating chlorine, bromine and heavy metals. The most efficient FR chemical synergy for plastics, antimony trioxide and organic bromine compounds, are not cheap and they have to be compounded into a polymer system or coating recipe. When compounding with water-based systems, care is needed because the FR chemicals are solids and need surfactants and thickening agents to produce a uniform compound with reasonable shelf life. There is the danger of the solids separating out during storage or transportation or during the actual coating process, giving rise to an unsightly appearance and irregular test results.

Many chlorine and bromine chemicals are believed to be potentially toxic and are subject to control or prohibition. Both antimony trioxide and bromine FR chemicals have been under environmental scrutiny for several years, and this seems to have intensified recently. The search has been on for alternative chemicals systems for some time, but so far there has been only limited success. However, Bayer recently announced a £3 million project to develop non-halogen FR agents. Not everyone is convinced that the materials are hazardous; the bromine industry is calling for independent reports. Others believe that the risk from being burned in an accident is greater than the risk from bromine as an environmental pollutant, see Chapter 6.

Amongst the alternative chemicals put forward are zinc hydroxystannate, zinc stannate and systems based on zinc borate. However, there is a quite widely held view that a higher concentration of the alternative materials is required to produce the same performance that is obtained with a lower level of an antimony/bromine system. In certain applications, it is possible simply to increase the loading of FR filler and chemicals, but this is not easy with fabric coatings, where increased amounts of additives and add-on cause fabric stiffening. Also, there is a limit to the amount of FR chemical which can be mixed into water- or solvent-based resins for fabric coating before the resin becomes too thick, unstable and unsuitable for coating.

2.8.6 Fireblocker materials

Fireblockers were originally developed as barriers to prevent fire ignition on aircraft seats, but are being increasingly used in other areas of public transport such as trains and coaches.[84–87] Seats are generally composed of a fabric covering a cushion of polyurethane foam which, unless highly modified, can burn and emit highly toxic fumes. Combustion modified foams are more expensive and may not have the same comfort properties and physical durability as more regular foams, hence the need for a barrier material between the face fabric and the foam. A variety of specialist highly FR fabrics may be used and these include, Nomex, Kevlar (both DuPont), fibreglass, Inidex (Acordis), Panox (preoxidised acrylic fibre from LUCF), PBI (polybenzimidazole, by Hoechst), Visil (Sateri), etc. and blends of these with other textiles. Coated or laminated fireblockers include Neoprene rubber coated or laminated to another textile (Vonar from DuPont). Fireblockers have also been produced from PVC coatings on to fibreglass. Choice of which type to use depends on various factors, including, cost, weight (important especially on aircraft), comfort considerations, the quality specification and test method to be used.

2.9 References

1. Venkataraman B (Seaman Corp.), 'Industrial coated fabrics', *JCF*, 19 Apr 1990, 241–51.
2. Smith WC (Industrial Textile Associates), 'The importance of proper fabric selection', *6th International Conference on Textile Coating and Laminating*, Dusseldorf 4–5 Nov 1996, Technomic, Lancaster PA, USA.
3. Ford JE (BTTG), 'Fibre and fabric substrates for coating', *Progress in Textile Coating and Laminating*, BTTG Conference, Chester 2–3 July 1990, BTTG, Manchester.
4. Mayer PW, 'The advantages of polyester for use in coated fabrics', *JCF*, Vol 3 Oct 1973, 76–84.
5. Wilkinson M, 'A review of industrial coated fabric substrates', *JCF*, Vol 26 Oct 1996, 87–106.
6. Dartman T and Shishoo R, 'Predictions of performance characteristics of coated fabrics', *Technical Textiles*, Vol 38 Nov 1995.
7. Thomas E, 'Coated materials for specialised end uses', *Progress in Textile Coating and Laminating,* BTTG Conference, Chester 2–3 July 1990, BTTG, Manchester.
8. Haddad RH and Black JD, 'Parameters of woven fabric constructions influencing coating adhesion', *JCF*, Vol 14 Apr 1985, 272–81.
9. Van Parys M, *Coating*, Eurotex (EEC Comett Program), Gent, Belgium 1994, 14–86.
10. Hall ME, 'Coating of technical textiles', in *Handbook of Technical Textiles* (Editors Horrocks AR and Anand SC), Woodhead, Cambridge 2000.
11. Scott RA, 'Coated and laminated fabrics', in *Chemistry of the Textiles Industry* (Editor Carr CM), London, Blackie-Academic 1995.
12. Lomax GR, 'Coating of Fabrics', *Textiles,* Issue No 2, 1992, 18–23.
13. Lomax GR, 'Coated Fabrics: Part 1 – Lightweight breathable fabrics', *Textiles*, Vol 14 No 1 Spring 1985, 2–8 and Part 2 – Industrial uses', *Textiles,* Vol 14 No 2 Summer 1985, 47–56.
14. Loy W, 'Polymers for coating textile substrates', *JCF*, Vol 1 Jan 1972, 150–61.
15. Mansfield RG, 'High-performance coatings and finishes expand uses', *Textile World*, May 1985, 58–60.
16. Harris D, 'Nonapparel coating', *JCF*, Vol 24 Jan 1995, 210–29.
17. Johnson L and Samms J, 'Thermoplastic polyurethane technologies for the textile industry' *7th International Conference on Textile Coating and Laminating*, 13–14 Nov 1995, Charlotte NC, USA, Technomic, Lancaster PA, USA.
18. Anon, *Polymer Emulsions for the Textile and Paper Industries*, Vinyl Products Technical Information Brochure 1985.
19. Kanter GC (Rohm & Haas), 'Using acrylic resins in fabric backcoating', *Textile World*, Aug 1982, 89–94.
20. Huang C (BF Goodrich), 'Thermodynamics of curing acrylic/latex binders', *INDA Conference,* Washington DC, USA 10–11 Mar 1977.

21. Poletti RA (BF Goodrich), 'High performance polymer emulsions; selection, properties and uses', *6th International Conference on Textile Coating and Laminating*, Dusseldorf 4–5 Nov 1996, Technomic, Lancaster PA, USA.
22. Matthews G, *PVC Production, Properties and Uses*, The Institute of Materials, London 1996, 245–50.
23. Anon, *Breon PVC Paste Resin*, BP Chemicals Technical Manual No 2, BP Chemicals, London 1969, 23–36.
24. Mewes H, 'Current world status of PVC coated fabric for architectural structures and related textile development', *JCF*, Vol 22 Jan 1993, 188–212.
25. Lerner A (Millmaster-Onyx), 'Polyesters in urethane fabric coatings', *AATCC Symposium, Coated Fabrics Up-date*, Newton MA, USA 31 Mar–1 Apr 1976, AATCC, Research Triangle Park NC, USA.
26. Koch H-J (Bayer), 'The structure and properties of polyurethane textile coatings', *JCFM (JCF)*, Vol 1 Oct 1971, 118–28.
27. Boron FT, 'Thermoplastic polyurethanes – A versatile material for industrial coated fabric applications', *AATCC Symposium, Coated Fabric Up-date*, Newton MA, USA 31 Mar–1 Apr 1976, AATCC, Research Triangle Park NC, USA.
28. Damewood JR, 'The structure-property relationships of polyurethanes designed for coated fabrics', *JCF*, Vol 10 Oct 1980, 136–50.
29. Davies WD (Lankro), 'Urethane coated textiles', *JCF*, Vol 4 Apr 1975, 212–13.
30. Mausser RF, 'Latex and foam rubber', in *Rubber Technology*, Third Edition (Editor Morton M), Van Nostrand, New York 1987, 518–60.
31. Bloomfield, GF, van der Bie GJ, Rellage JM and Vervloet C, 'Raw polymeric materials', in *Rubber Technology and Manufacture* (Editor Blow CM), Newnes-Butterworths (Plastics and Rubber Institute), London 1971, 73–146.
32. Davids J (Rohm & Haaas), 'Crushed foam coating', *Textile Manufacturer*, Dec 1974, 47–9.
33. Gunnell D (Rohm & Haas), 'Some aspects of acrylic polymers', *Developments in Coating and Laminating*, Shirley Institute Conference, Manchester 27 Mar 1981, Shirley Institute Publication S41, BTTG, Manchester.
34. Karsa DR and Forrester PH, 'Improvements in foamed aqueous polymer dispersion coatings', *Polymer Latex Conference*, London 31 Oct–2 Nov 1978, Plastics & Rubber Institute.
35. Swihart TJ and Awe RW (Dow), 'Silicones for fabric coating', *JCF*, Vol 16 Jul 1986, 11–24.
36. Schwark J and Muller J, 'High performance silicone coated textiles; development and application', *JCF*, Vol 26 Jul 26, 65–77.
37. Breunig S *et al* and Vergelati C, 'Polyfunctionalization of polysiloxanes; new industrial opportunities', *JCF*, Vol 27 Apr 1998, 309–25.
38. Budden G, 'High temperature properties of silicone elastomers', *JCF*, Vol 27 Apr 1998, 294–308.
39. Boddeker B (Dow), 'Silicone coated fabrics; development, applications, outlook', *6th International Conference on Textile Coating and Laminating*, Dusseldorf 4–6 Nov 1996, Technomic, Lancaster PA, USA.
40. Werner AF, 'Variables affecting SBR compounding for rubber coated fabrics', *JCF*, Vol 18 Apr 1989, 220–33.
41. Jacques JC, 'The principles of compounding', in *Rubber Technology and*

Manufacture (Editor Blow CM), Newnes-Butterworths (Plastics and Rubber Institute), London 1971.
42. Evans CW, *Practical Rubber Compounding and Processing*, Applied Science Publishers, London 1981, 19–33.
43. Bament JC, *Neoprene Compounding and Processing Guide Plus Formulary*, DuPont (Hemel Hempstead and Geneva), Technical Information Brochure E-15378.
44. Smith LP, *The Language of Rubber*, Butterworth Heinemann (with DuPont), Oxford 1993, 9.
45. Studebaker ML and Beatty JR, 'The rubber compound and its composition', in *Science & Technology of Rubber* (Editor Eirich FR (American Chem. Soc)), Academic Press, New York 1978, 370–1.
46. Scott RA, 'Coated and laminated fabrics', in *Chemistry of the Textiles Industry* (Editor Carr CM), Blackie-Academic, London 1995, 245–7.
47. Pruett KM, *Chemical Resistance Guide for Elastomers*, Compass Publications, Le Mesa CA USA 1994.
48. Pruett KM, *Chemical Resistance Guide for Elastomers 11*, Rubber World Publications, Akron OH, USA 2000.
49. Mackenzie KJ, 'Film and sheeting materials', *Kirk Othmer, Encyclopaedia of Chemical Technology*, Fourth Edition, Vol 10, John Wiley, New York 1993, 761–87.
50. Spijkers JCW, 'Update Sympatex', *World Sports Activewear*, Jul 1995, 40–3.
51. Painter CJ, 'Waterproof, breathable fabric laminates: a perspective from film to market place', *JCF*, Vol 26 Oct 1996, 107–30.
52. Conroy C, 'The Sympatex story', *The Great Outdoors*, Oct 1986, 63–4.
53. Anon, 'Waterproof breathable membrane system (Permatex), *Textile Month*, May 1995, 48.
54. Anon, 'Magic membrane for all weather clothing', *Textile Month*, Jan 1995, 20–2.
55. Schledjewski R, Schulze D and Imbach K-P, 'Protective clothing with hydrophilic thermoplastic elastomer membranes (Wolff-Walsrode), *JCF*, Vol 27 Oct 1997, 105–14.
56. Editors of American Fabrics and Fashion Magazine, 'Bonded and laminated fabrics', in *New Encyclopaedia of Textiles*, Prentice Hall, Englewood Cliffs NJ, USA 1980, 387.
57. Mewes H, 'Adhesion and tear resistance of coated fabrics from polyester and nylon', *JCF*, Vol 19 Oct 1989, 112–8.
58. Wootton DB, 'Alternative adhesion systems to isocyanates', *Developments in Coating and Laminating*, Shirley Institute Conference, Mancherster 27 Mar 1981, Shirley Institute Publication S41, BTTG, Manchester.
59. Kinloch AJ, *Adhesion and Adhesives*, Chapman & Hall, London 1987, 56–100 and 101–39.
60. Pocius AV, 'Adhesives', *Kirk Othmer, Encyclopaedia of Chemical Technology*, Fourth Edition, Vol 1, John Wiley, New York 1991, 445–66.
61. Shields J, *Adhesives Handbook*, Butterworth, London 1984, 30–79.
62. Fries JA, 'Adhesives for fabric lamination', *JCF*, Vol 12 Jan 1983, 174–86.
63. Shishoo RL, 'Plasma treatment. Industrial applications and its impact on the coating and laminating industry', *6th International Conference on Coating and Laminating*, Dusseldorf 4–6 Nov 1996, Technomic, Lancaster PA, USA.

64. Carr AK and Roth JR, 'Investigation of changes in surface energy of polymers after exposure to a one atmosphere uniform glow discharge plasma (OAUGDP)', *7th International Conference on Textile Coating and Laminating*, Charlotte NC, USA 17–18 Nov 1997. Technomic, Lancaster PA, USA.
65. Herbert PAF and Bourdin E, 'New generation atmospheric plasma technology for industrial on line processing', *JCF*, Vol 28 Jan 1999, 170–82.
66. Plasma Ireland Ltd, Briefing Note, 1999. www.plasma-ireland.com
67. Halbmaier J (Bostik), 'Overview of hot melt adhesives; application equipment for coating and laminating full width fabrics', *JCF*, Vol 21 Apr 1992, 301–10.
68. Miles DC (Dritex), 'Dry powder bonding adhesives in automotive trim laminates', *JCF*, Vol 20 Apr 1991, 229–39.
69. Holme I, 'Bonding and coating', *Textile Horizons*, Jun/Jul 1997, 35–8.
70. Deubelbeiss J, 'Hot melt – reliable route for technical textiles', *Textile Month*, Mar 2001, 22–4.
71. Schnerring K (Lacom), 'Hot melt horizons', *Textile Month*, Mar 2001, 17–20, 48.
72. David FE (Reichold), 'Hot melts for textile coating and laminating applications', *AATCC Symposium, Coating and Laminating Fabrics 2000 and Beyond*, Braintree MA, USA 7–8 Apr 1998, AATCC, Research Triangle Park NC, USA.
73. Green D (National Starch), 'A new beginning in fabric laminating industry; reactive PUR adhesives', *JCF*, Vol 28 Oct 1998, 116–25.
74. Stepanski H and Nebe V, 'Polyurethane adhesives for textiles', *JCF*, Vol 27 Jul 1997, 27–38.
75. Bajaj P and Sengupta AK, 'Protective clothing', *Textile Progress*, Vol 22 Nos 2/3/4, The Textile Institute, Manchester 1992, 2–5.
76. Horrocks AR, 'Flame retardant finishing of textiles', *Review Progress Coloration*, Vol 16 1986, 62–101.
77. Eaton PM, 'Flame retardancy test methods for textiles', *Review Progress Coloration*, Vol 30 2000, 51–62.
78. Mitchell T, 'Flame retardant selection for textile coatings; a composite approach', *JCF*, Vol 23 Apr 1994, 298–310.
79. Skinner GA, 'Flame retardancy; the approaches available', in *Plastics Additives; An A–Z Reference* (Editor Pritchard G), Chapman & Hall, London 1998, 260–7.
80. Oulton DP, 'Flame retardent textiles', in *Chemistry of the Textile Industry* (Editor CM Carr), Blackie-Academic, London 1995.
81. Bajaj P, 'Heat and flame protection', in *Handbook of Technical Textiles* (Editors AR Horrocks and SC Anand), Woodhead, Cambridge 2000, 223–63.
82. Davies PJ and Horrocks AR, 'Optimisation of FR textile coating parameters by factorial analysis', *JCF*, Vol 29 Oct 1999, 118–37.
83. Conway R, 'The development of FR and intumescent systems in textile and coating processes', *6th International Conference on Coating and Laminating*, Dusseldorf 4–6 Nov 1996, Technomic, Lancaster PA, USA.
84. Damant GH, 'Use of barriers and fire blocking layers to comply with full-scale fire tests for furnishings', *JCF*, Vol 29 Jul 1999, 37–57.
85. Keil G, 'Fire-blockers – a protection for passengers', *TuT*, Vol 4 No 2 1991, 46–7.
86. Saville N and Squires M, 'Latest developments in fire resistant textiles', *Textile Month*, May 1990, 47–52.
87. Paul KT, 'Flame retardant polyurethane foams furniture testing and specifications', *Review Progress Coloration*, Vol 20 1990, 53–69.

2.10 Further reading

1. Blow CM (Editor), *Rubber Technology and Manufacture,* Newnes-Butterworths (Plastics and Rubber Institute), London 1971.
2. BP Chemicals, *Breon PVC Paste Resin*, BP Chemicals Technical Manual No 2, BP Chemicals, London 1969.
3. Calvert KO (Editor), *Polymer Lattices and their Application*, Applied Science Publishers, London 1982.
4. Ciesielski A, *An Introduction to Rubber Technology*, RAPRA Technology, Shrewsbury 1999.
5. Evans CW, *Practical Rubber Compounding and Processing*, Applied Science Publishers, London 1981.
6. *Fire Statistics UK 1998*, Home Office Bulletin Issue 15/99, UK Government Statistics Service.
7. Gachter R and Muller H (Editors), *Plastics Additives Handbook*, Fourth *Edition*, Hanser, Munich 1993.
8. Gandhi KL and Pearson JS, *Industrial, Technical and High Performance Textiles*, World Textile Congress, Huddersfield University 15–16 Jul 1998.
9. Gordon Cook J, *Handbook of Textile Fibres*, Vol 1 – Natural fibres and Vol 2 – Man-made fibres, both Fifth Edition, Merrow, Shilden Co. Durham 1984.
10. Hoechst Plastics, *Hostalit,* Fabwerke Hoechst, Frankfurt 1971.
11. Horrocks AR, Price D and Smith J, *Textile Flammability*; *Current and Future Issues Conference*, The Textile Institute, Manchester 30–31 Mar 1999.
12. Karsa DR, Goode JM and Donnelly PJ (Editors), *Surfactants Applications Directory*, Blackie, Glasgow and London 1991.
13. Lewis PA, 'Incorporation of pigments into coatings', *JCF*, Vol 16 Apr 1987, 216–24.
14. Mathews G, *PVC Production, Properties and Uses,* Institute of Materials, London 1996.
15. Morton M (Editor), *Introduction to Rubber Technology*, Van Nostrand, New York 1987.
16. Oertel G (Bayer) (Editor), *Polyurethane Handbook*, Hanser Publications, Munich 1985.
17. Penn WS, *PVC Technology*, Second Edition, Maclaren & Sons, London 1966.
18. Pritchard G (Editor), *Plastics Additives – an A–Z Reference*, Chapman and Hall, London 1998.
19. Pruett KM, *Chemical Resistance Guide for Elastomers*, Compass Publications, Le Mesa CA USA 1994.
20. Pruett KM, *Chemical Resistance Guide for Elastomers II*, Rubber World Publishers, Akron, OH, USA 2000.
21. Roff WJ, Scott JB and Pacitti J, '*Fibres, Films, Plastics and Rubbers* (*A Handbook of Common Polymers*), Butterworths, London 1971.
22. Rubin II, *Handbook of Plastics Materials*, Interscience, New York 1990.
23. *Developments in Coating and Laminating*, Shirley Institute Conference, Manchester 27 Mar 1981, Shirley Institute Publication S41, BTTG, Manchester.
24. Sodano CS, *Water and Soil Repellent Finishes for Textiles*, Chemical Technology Review No 134, Noyes Data Corporation, Park Ridge NJ, USA 1979.
25. Troitzsch J, *International Plastics Flammability Handbook*, Second Edition, Hanser Publications, Munich 1990.

26. Van Parys M, *Coating*, Eurotex (EEC Comett Program), Gent, Belgium 1994.
27. Wake WC and Wootton DB, *Textile Reinforcement of Elastomers*, Applied Science Publishers, London 1982.
28. Woodruff FA, *Coating, Lamination, Bonding and Prepregging* (unpublished book chapter), Web Processing, Whaley Bridge, UK 1994.
29. Woods G, *The ICI Polyurethanes Book*, ICI Polyurethanes and John Wiley, Chichester 1987.

3
Production methods

3.1 Introduction

Quality products can be consistently produced only from quality starting materials using a well maintained processing machine which is specially designed for the particular job. Chapter 2 emphasised the need for and explained the reasons why well prepared fabric and carefully compounded coating resin is essential. This chapter deals with actual production methods and the type of machinery required. However, too frequently the technical emphasis is applied to the coating or lamination head – the place where the actual joining process takes place – and not enough attention is given to fabric and material handling at both the feeding in and taking off parts of the machine. Machinery makers may make claims of high processing speeds, sometimes over 100 m per minute, but in practice this is quite meaningless if rolls of upholstery fabric or foam, no longer than say 50 m or less, cannot be joined on without slowing or stopping the process. The use of larger batches obviously helps with continuous running. Fabric, or any other material, moving at any speed is liable to crease, stretch, ripple or otherwise distort if not handled correctly. The faster the production speed, the more likely this is to happen and, more important, the more second quality fabric will be produced, before the operative will see what is happening or be able to take remedial action. The fabric, once coated or laminated, will have quite different properties from the starting material. It may produce more friction on rollers – because of coating penetration – and may have more tendency to crease or curl – because the two sides of the material now have entirely different properties.

Preparing large batches of material by sewing or joining smaller rolls together does allow a coating or lamination machine to run for longer periods of time without stopping, but each join necessitates a seam which can cause problems. It is better to try to obtain the longest continuous lengths of material from the supplier. In addition, sewing up of smaller lengths beforehand is an extra process and, even if the extra cost may be absorbed, each time a fabric is rolled, there is an opportunity for creases or some other fault to appear. A poor seam can result in two or three metres or more of fabric either side of the join being lost, whilst a burst seam can cause a major delay, especially in spread coating. Least handled is best handled.

Good housekeeping is essential in fabric coating, especially in the case where several layers of coating are applied to the same piece of fabric. Dust or 'fly' (small pieces of fibre) from a dusty atmosphere can land on the surface of the coating, producing soiling which could downgrade that particular area of fabric. Even worse, the contaminant may get trapped under the doctor blade and produce 'drag lines' which may extend for several metres before it is seen and remedial action taken. Bits of dried resin that has not been properly cleaned away may break off the sides of the doctor blade or other parts of the coating machine and cause similar problems. In certain cases, blade profile design can minimise deposits which 'creep' under the blade and build up on the back of it; more detail of this is presented in Section 3.4.2. If the coating machine has stenter pins, the fabric will be pushed down on to them by brushes, and these may touch the wet resin and become soiled. When this resin dries, it may break off and contaminate the fabric being coated. Floors should be kept clean and dust free and, if necessary, be sealed regularly to stop dust particles breaking away from the concrete. The more layers being applied to a coating, the longer is the period of processing time and the greater the chance that the fabric will be downgraded because of dust. In addition, drying ovens and stenters may themselves create draughts, and moving fabric and film can generate static electricity that can attract dust particles. The mention of housekeeping aspects may seem trivial in a technical book, but dust, fly and resin deposits can make all the difference between success and failure, profit and loss.

Health and safety aspects to consider in the coating process include all precautions associated with the handling and storage of volatile and flammable solvents. The problem of static electricity that may be generated whenever synthetic fibres are handled especially requires consideration in the presence of resins in flammable solvents. Fires ignited by static electricity sparks in coating factories are by no means unknown.

3.2 General principles of fabric coating

Coating techniques simply try to control resin add-on and preserve fabric aesthetics and properties as much as possible. This is not easy, and fabric handling is always a challenge. Coated fabric handling may be especially difficult because, even after one layer has been applied, the total fabric weight has increased and the surface nature of one side of the material has changed significantly. Control of resin add-on is one of the fundamental principles of fabric coating, and the various methods achieve this in different ways. Resin solids content or foam density and viscosity are properties of the polymer that are used to control add-on, but machine characteristics and geometry are also relevant. Moreover, control must be achieved without adversely affecting fabric properties. The direct method uses blades of different profile, blade angle and gap separation of blade over roller, or gap separation of blade over table, as means to control add-on. Roller methods use gap separation between rollers, speed of rotation and contact angle; in addition, doctor blades applied against the rollers to scrape off excess resin are sometimes used for more accurate control. The most effective ways of controlling very low add-ons are the gravure roller, the slot die extruder, the rotary screen and foam processing. The slot die extruder and rotary screen have other novel advantages as will be seen.

The prime aim must be to provide the customer with satisfaction by producing a product which consistently passes all quality tests. The technical considerations dictate what the add-on should be to achieve this, and any resin add-on more than this amount is excessive cost. This same principle also applies to lamination processes where excess adhesive, producing bond strengths far greater than the standard required, generates excess cost. These aspects are further reasons why precise control of resin or adhesive is crucial to the success of the business. A fundamental requirement of any coating is the ability to stay adhered to the base fabric throughout the life of the product. Coating is a joining process, and all the rules governing adhesion discussed in Section 2.7.1 on factors influencing lamination adhesion also apply to coating.

3.3 Laboratory and pilot coating and laminating

Evaluation of materials and techniques on small pieces in the laboratory is essential for the success of any coating or lamination factory. Production machine time is expensive and should be used only for actual production or for essential trials that must be carried out in bulk. In addition, trials on full size machines are expensive in terms of the costs of both fabric and chemicals. Ideally a laboratory small sample coating machine should be used, which can reproduce as far as possible the conditions encountered in actual production. The conditions should include airflow as well as temperature control; airflow is especially critical in crushed foam coatings. Conditions which can be optimised by coating many small samples and evaluating the results include time and temperature for optimum running speeds, and different concentrations of various additives, as well as choice of actual resins or combinations of resins. This prebulk trial work can save much production machine time and, equally important, it should produce background information which could be invaluable should problems arise.

Having said this, laboratory work provides information on the resin properties, but full width trials are still essential to establish how the resin will behave and how the fabric will run under commercial conditions.

The best known small sample laboratory coating apparatus is the Werner Mathis oven; this has facilities for knife on air, knife on blanket and knife on roller as well as airflow control, both top and bottom, and temperature control up to 200 °C. There are other laboratory and semi-bulk coating rigs available on the market to simulate all varieties of coating and lamination. These machines can be quite expensive, but it is possible to obtain very useful results using much simpler apparatus provided the limitations are appreciated. A simple metal bar – such as a metal ruler – used as a doctor blade and drawn over a piece of fabric secured with bulldog clips over a piece of glass is perhaps the simplest form of 'experimental coating rig'. Alternatively, a pin-frame can be made from stenter pins and used with blades of different profile. More accurate laboratory work on resin add-on can be achieved with K bars which are precision instruments originally used in the printing industry. K bars are stainless steel metal rods which have been closely wound with stainless steel wire of known diameters. When used to draw resin over a piece of fabric secured over a smooth surface, they will produce a coating of uniform thickness (or a resin film of uniform thickness if drawn directly over the glass plate).

All coatings produced from simple laboratory coating rigs can be dried and crosslinked in a laboratory oven of suitable size, which should preferably be dedicated for this purpose. If solvent-based resins are being evaluated, the oven must be flameproof and arrangements made so that the solvent fumes are removed from the laboratory atmosphere. The Werner Mathis oven is suitable for all laboratory coatings, including solvent-based resins and PVC plastisols. It is not recommended that laboratory work be carried out on PVC plastisols without a suitable oven, because copious fumes are produced when PVC gelling takes place.

Because the application method and machinery used can have such a significant effect on the results obtained, a number of 'universal' coating machines have been developed since around 1990. These machines, primarily designed for development and laboratory use, provide the facilities for a wide variety of coating techniques in a single machine. The techniques include all the doctor blade methods, roller methods, hot melt techniques as well as impregnation and sometimes screen methods. Manufacturers include Web Processing, Coatema and Lacom. 'Multi-pupose', full width coating plants are also offered to respond quickly to changing market requirements.

Lamination can be simulated in the laboratory using a simple flat iron or Hoffman press. It is, of course, simple enough to put samples through a full size calendar – and many small samples may be put through at the same time. The latter procedure is to be preferred because the processing conditions on the calendar can be recorded. When laboratory trials are being done, the use of heat sensitive paper is recommended to ascertain the temperature actually attained at the 'glue line'. The optimum combination of the conditions of time, temperature and pressure can be determined by putting a number of small samples through the machine at different speeds and temperatures. The actual 'glue line' temperatures attained will be known from the heat sensitive paper and the actual peel bond strengths produced by known production parameters can be measured and recorded to build up a data base.

3.4 Direct coating

3.4.1 General comments

Coating is essentially spreading a polymer in the form of a thickened aqueous dispersion or an aqueous or solvent solution on to a fabric to form a continuous layer. It is necessary to thicken the liquid so that it does not sink into or through the fabric. Pre-prepared compounds should always be stirred and the viscosity checked and recorded before applying. The simplest coating procedure is the direct method, sometimes called the 'floating knife' or knife over air technique, where the fabric is stretched flat to form an even uniform surface and is transported under a stationary doctor blade. As the fabric moves forward, it is scraped by the knife and the polymer resin or compound is spread evenly over the surface, see Fig. 3.1. Figure 3.2 shows the knife or blade over roller method of direct coating. The amount of polymer applied, the 'add-on' depends on the concentration of the dispersion or solution – this is the so-called 'solids content'; other parameters also affect resin add-on as will be seen. Detailed accounts of the basic principles are available.[1–5]

Direct coating is mainly used for fabric produced from smooth continuous filament yarns such as nylon or polyester. If fabrics made from spun yarns are direct coated, a raspy rough surface is generally obtained because the fibre ends protrude from the surface. The blade is likely to pull them out a little more and make them stand proud whilst coating them with resin (see Section 3.6). For articles requiring very thick layers of polymer, such as belts, printers' blankets and inflatables, this is not important because several layers of thick coating fully cover the fibre ends. Cotton is still used in some of these industrial coating applications.

3.4.2 Blade profile

The add-on is also influenced by the blade profile and blade angle, and by the fabric tension, which determines the intimacy of contact with the fabric,

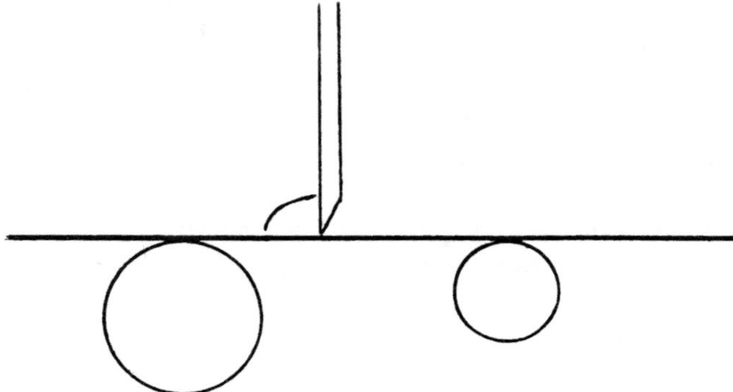

3.1 Knife or blade on air coating. The blade touches the fabric surface, which must be kept flat and uniform by the application of fabric tension. This method is used to apply the first layer of a coating to 'seal' the fabric surface. The second layer is applied either by using a thicker blade or by a knife over roller method to build up the resin add-on.

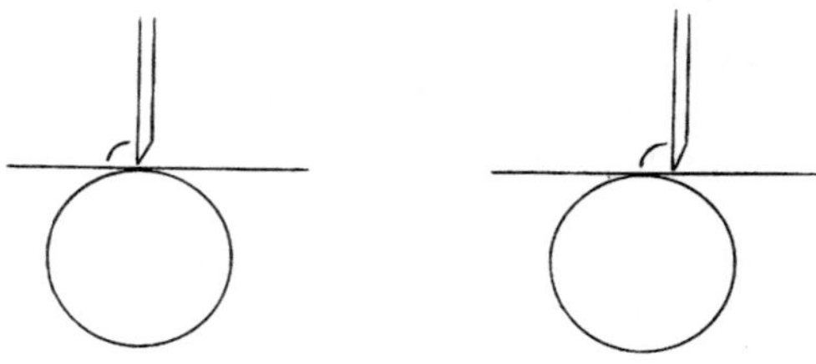

3.2 Knife or blade over roller. In this method a gap is set between the blade and the surface of the fabric to apply a measured amount of resin. It is also used if the weight of the resin being applied is too heavy for the fabric to support, e.g. PVC plastisols.

The position of the blade in relation to the crown of the roller is important, because if it is not directly above it, the situation in effect is similar to the knife on air method. This 'off crown' position is in fact useful for coating thick fabrics which have a raised surface or pile. For a detailed discussion on blade positions and blade profiles, see the paper by Woodruff.[1]

see Fig. 3.3 and 3.4. A thick profile blade produces a higher add-on than a thin, sharp one and a blade angled forwards will tend to increase add-on compared to a perpendicular blade. The forwards angled blade forms a wedge between itself and the fabric and will tend to drive polymer into, and possibly through, the fabric, which will cause fabric stiffening and loss of

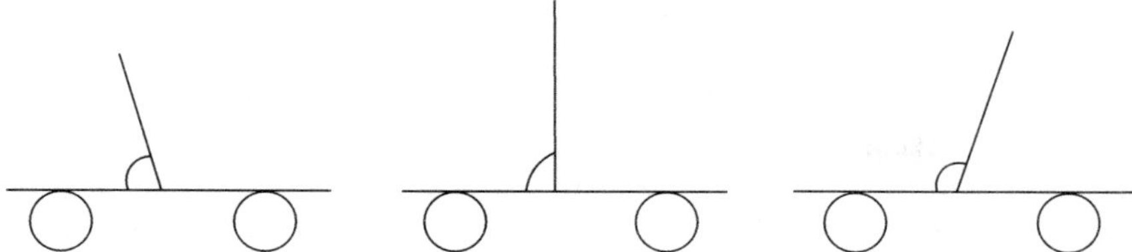

3.3 Resin add-on can be 'fine tuned' by subtle angling of the blade. A blade angled forward produces a wedge with the fabric, and as the fabric moves forward, the resin is driven into it. Angling the blade backwards tends to reduce the resin add-on.

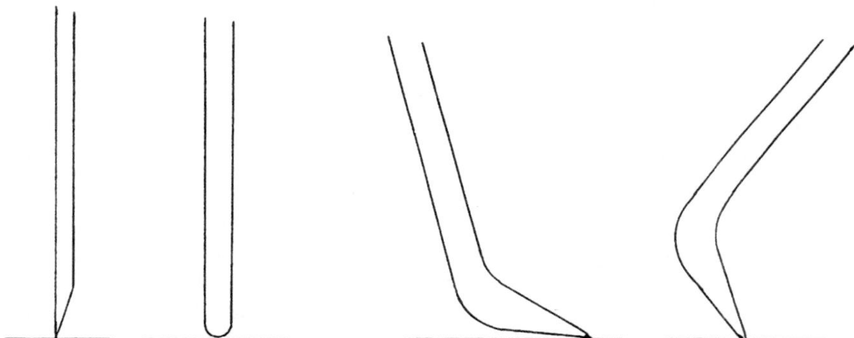

3.4 Examples of blade profiles (schematic). A sharp blade will produce a relatively low add-on. A rounded blade will result in slightly higher add-on. A 'shoe' blade is a versatile piece of apparatus, because the broader the shoe, the higher the add-on, but if this blade is angled forward, it approximates to a sharp blade. The 'shoe' design is to reduce the tendency for PVC plastisols and some polyurethane resins to 'creep' up the back of the blade to accumulate deposits, which when large enough, fall on to the surface of the coating as a contaminate.

tear strength, and will also possibly spoil the fabric appearance. Excessive penetration can be determined by lightly touching the underside of the fabric with the tips of the fingers just after the coating process has started. If the fabric feels wet, excessive penetration may be occurring and appropriate action should be considered. The coated fabric should be examined for stiffness and resin penetration as soon as it emerges from the drying oven. Another indication of excessive penetration is when resin is seen to accumulate on support rollers or bars. This should, of course, be cleaned up as soon as possible before it contaminates the fabric.

How the resin 'wets' the fabric surface also influences add-on and, if the fabric surface is hydrophilic, water-based resin add-on may be higher than expected. Some polyurethane resins and PVC plastisols may work themselves under the blade and appear in the form of small deposits on the back of a sharp or rounded blade during the course of a coating operation. These deposits will gradually increase in size and, when large enough, may break off and fall on to the face of the coated fabric, causing an unsatisfactory appearance and downgraded fabric. In the trade this phenomenon is referred to as 'blobbing' or 'creeps' and is less likely to occur with a shoe profiled blade. Shoe blades are versatile and it is possible to cover a range of coating requirements using a particular shoe blade at different angles.[1,2] Whatever the blade profile, the blade must be robust enough to withstand the movement of fabric beneath and to hold the bank of resin without the slightest 'chatter' or vibration. The blade is usually mounted on a very substantial metal girder mounted over the width of the machine. For quick changeover, sometimes two or three blades of different profile are mounted on the same girder, which is rotated through 180° of arc in the case of two blades, or 120° in the case of three blades, to change the blade profile.

For many products it is usual to apply the base layer using a sharp blade to 'seal' the fabric surface; a thicker blade may then be used to achieve build up of resin. Resin add-on is also influenced by fabric speed, surface geometry and construction. A smooth surface of a closely woven fabric will result in a relatively low add-on, a more open or rough, uneven surface will result in a higher add-on. The first layer which 'fills in the holes in the fabric construction' is generally the heaviest, and it is an especially important one because it determines polymer – fabric adhesion and has a significant effect on coated fabric handle. When the second layer is applied, the surface will be smoother and therefore the resin add-on will be less. For example, in the case of 68 g/m^2 (2 oz) fabric, the first layer add-on could be, say, 25 g/m^2 and the second layer, say, 7 g/m^2. If a third layer is applied, it may only be 3–4 g/m^2 add-on, because the fabric surface has got progressively smoother. If a higher add-on is required, it will be necessary to use the knife over roller or knife over table technique and actually set a gap, using a feeler gauge if there is no instrumentation on the machine. In these cases, the blade does not actually touch the fabric. When calender or roller coating is carried out, it is usual to apply the first and most important layer by the direct method and then build up the resin add-on by roller.

The first layer of coating should be soft and flexible, ideally at least as flexible as the fabric to which it is attached. If it is too hard or not sufficiently flexible, it may eventually, by flexing during garment wear, become mechanically detached from the fabric, even though the fabric–coating bond is quite strong. Soft resins are generally sticky – the softer it is the stickier it normally is. However, for a first, base layer, some stickiness can be tolerated – provided no penetration occurs – because it will eventually be covered by another layer. As already stated, any resin penetration should

3.5 Knife over air and knife over roller commercial coating machine head. Photograph supplied by Web Processing (M/C) Ltd and reproduced with kind permission.

be checked as soon as coating starts by feeling the underside of the coated fabric with the finger tips. If penetration is occurring, the process must be stopped and either the resin thickened or the blade or blade profile changed. It should be mentioned that an excessively sticky layer may present some problems in handling, for example, sticking to itself on the roll, sticking to rollers, being more likely to pick up dust or 'fly' or some other soiling, etc. The top layer, which should be smooth and 'dry' (i.e. non-sticky), should combine flexibility with hardness as far as possible. These factors are important for a coating with good abrasion properties and adhesion durability. Figures 3.5 and 3.6 show commercial coating equipment.

3.4.3 Coating add-on

Establishing the coating add-on is not as easy as it would first appear, especially when a relatively low add-on of say 30 g/m^2 is concerned. Weighing fabric before and after coating sometimes produces evidence that the fabric after coating is *lighter* than it was before! The reason for this is that base fabric weight can vary depending on the widthways and lengthways tension applied. If the base fabric is knitted, and especially if has been raised or

3.6 Commercial coating of high performance products. Photograph supplied by Bobet Coated Fabric Company and reproduced with kind permission.

cropped, the weight difference across the width can be surprisingly quite substantial. Weight samples taken within a few inches or centimetres of each other may be up to 40 g/m^2 different in a 300 g/m^2 base fabric. For coating to take place, the fabric must be held flat and must be under some tension, and this may be sufficient to alter the base fabric weight. The difference can be calculated; for example, even a dimensionally stable woven fabric, 150 cm wide and weighing 300 g/m^2 in a relaxed state, when pulled out 2 cm widthways and 2 cm in 100 cm lengthways (2% underfeed) will weigh only 290 g/m^2. Application of more tension by underfeeding the fabric by 5% and pulling out 2 cm will reduce the base fabric weight to 282 g/m^2, 18 g lighter than may be thought! Thus the combined effects of widthways tension and lengthways tension can cause quite substantial differences to base fabric weight. Another factor which may cause some problems in weight differences – or width differences – is ambiguous instructions to the stenter operatives. Does 'set to 160 cm width' mean 160 cm in between the stenter pins or 160 cm overall width, and does this include the 'fringe' on the fabric edge? This factor could also explain the complaint of underwidth fabric; new operatives may have their own idea if a procedure is not clearly explained or specified.

Modern plants will have automatic computer-assisted sensing and metering devices which will record the amount of resin being applied to the fabric at any given moment. However, the computers may not be aware of resin escaping around the edges of the end plates and, in any case, knowing the base weight of the uncoated fabric is a fundamental requirement in ascertaining add-on. Fabrics may need to be stentered in a particular way to achieve the required amount of residual stretch during making up and, if coating is the final finishing operation, the situation requires careful thought.

3.4.4 Resin viscosity

Viscosity of the compound is another essential variable influencing add-on. A low viscosity resin will flow more easily and thus cause excessive penetration, a high add-on, fabric stiffening and reduced tear strength. A viscosity which is too high, i.e. a thick mix, could cause uneven add-on, a poor appearance and reduced performance, such as low water resistance. It is the job of the compounder to produce a satisfactory mix for the job in hand, but viscosity can change during processing. The flow properties of the compound are referred to as the rheology and, under the shearing action of the doctor blade, the viscosity can become thinner which may cause penetration into the fabric. A resin which behaves in this way, i.e. becomes thinner under a constant force, is termed thixotropic; the converse is rheopectic behaviour. Resins which have constant viscosity under shear are said to exhibit Newtonian properties. Resin viscosity can also change as a result of ambient temperature changes and evaporation of solvent or water. It is, therefore, essential that the bank of resin in front of the doctor blade (the 'sausage') rotates so that the resin does not have a chance to 'age'. Periodic monitoring of the viscosity of the resin is recommended, using a Brookfield Viscometer (with the appropriate spindle and speed) or Ford cups. Ambient room temperature can have a significant effect on resin viscosity. A variation of, say, 10 °C can cause a change of as much as 10000 centipoise, see Fig. 3.7. The workplace temperature could be as low as, say, 10–15 °C first thing on a Monday morning, but could well be over 20 °C by noon.

3.4.5 Drying and crosslinking of resin

The limiting factor governing the amount of compound which can be applied in one layer is usually the drying off process in the coating machine oven. This is an important process, especially in the case of inflateables where a continuous, pin hole-free and surface defect-free coating is essential for impermeability and waterproofness. The first layer is generally applied at a low processing temperature, without curing or with only partial curing, so that the next coating layer will adhere well to it. Full crosslinking of all

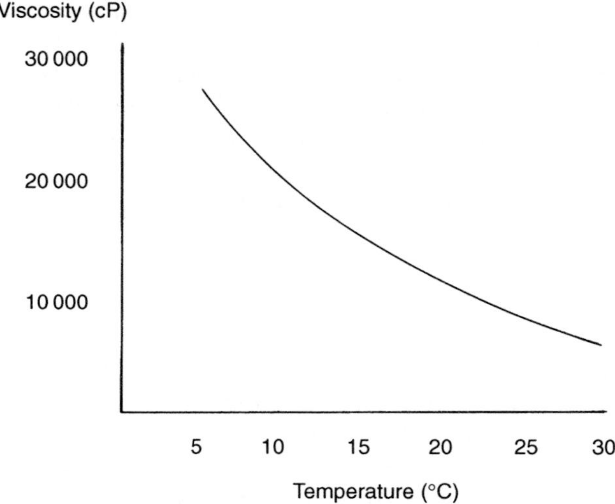

3.7 The effect of temperature variation on the viscosity of a water-based polyurethane resin. The ambient temperature of the coating factory may vary by perhaps up to 10 °C or more. It could be 10–15 °C first thing in the morning rising to maybe 20–25 °C or even higher at mid-afternoon.

layers is carried out only after the top layer has been applied, so that all layers are crosslinked together. If a melamine-formaldehyde (MF) crosslinking agent is used, slightly higher temperatures may be needed compared to an isocyanate crosslinking agent. If crosslinking can be attained at lower temperatures, especially using low temperature crosslinking agents such as polyaziridene, the temperature need only be set to ensure that all solvent is evaporated. Coated fabrics, especially apparel, will be flexed during use and the coating adhesion needs to be of a high standard. In the case of apparel, the fabric–coating bond must also be resistant to washing or dry cleaning and effective crosslinking is normally essential.

The length and design of the drying oven are important because they determine coating speed and quality. The best results are generally obtained when the evaporating process is controlled and gradual, i.e. setting the first chamber of a multichamber oven at a temperature below the boiling point of the carrier liquid, and the other chambers at gradually higher temperatures to achieve resin crosslinking. The actual temperatures will depend on the oven design, air flow and number of chambers available in which the temperature can be independently controlled. Water-based resins may be dried off in a 'profiled' oven for gradual evaporation of water, for example in a four bay oven, with the temperature set at, say, 90 °C in the first bay, 120 °C in the second, 140 °C in the third and 160 °C in the fourth. For a resin based in MEK which has a boiling point of about 80 °C, being processed in a two bay oven, the temperatures may be 70 °C and 120 °C. Exact settings depend on the resin and solvent system being used, add-on, oven design and length and air flow and coating speed; the chemical manufacturers will advise. Too rapid drying can cause blistering or bubble holes and a generally poor result. If this happens, increasing the boiling point of the mixture by addition of a 'high boiling' solvent may be effective, e.g. toluene, boiling point 110 °C or DMF, boiling point 153 °C. Fabric coating has been compared to painting – the best results are obtained by several thin layers of low add-on rather than one or two thick layers. However, machine and operator time is expensive and so a compromise is reached depending on the level of performance and quality required.

PVC does not strictly speaking 'crosslink', but sufficient heat is necessary to fuse or gel the material together properly, otherwise poor results will be obtained, such as interlayer delamination and poor abrasion. An oven dwell time of at least 1–1½ minutes at about 170–180 °C is generally necessary. Whether PVC gelling has been satisfactorily achieved can be assessed by the ease with which acetone attacks a small sample. Acetone will cause even properly gelled PVC to disintegrate, but will do this much more readily if the sample has not been heated sufficiently to allow gelling to take place completely.

In recent years, solvent-based resins have been used less frequently, but certain high performance properties are still not obtainable using the alternative water-based or higher solids varieties; development by the chemical companies continues. Standard tests for all coated fabrics include polymer adhesion (peel bond), tear strength, blocking (tackiness of the top layer), resistance to delamination by flexing and, where applicable, waterproofness. The general test methods for coated fabrics are the BS 3424 series, see Chapter 5.

3.4.6 Products from direct coating

In general, only fairly tightly woven fabrics capable of being pulled flat and uniform can be coated by the direct method. Knitted and stretchy fabrics are normally processed by transfer coating or, in some cases, such as lightweight nonwovens, by rotary screen techniques. Waterproof protective clothing fabric, automotive car seat fabrics, tarpaulins and lightweight material for inflatables are produced by the direct method. Other materials used for inflatables, conveyer belts and printers' blankets, are produced by applying many layers – sometimes 40 or more – of specialist rubber polymers by the direct method, and knife over roller (blanket or table) may be necessary to support the weight of the compound. A gap is set between the blade and the roller before coating commences, and this is done before each layer is applied. For maximum control and precision, the blade profile is ground to match the profile of the roller over which it is used. This is a skilful operation and needs to be done by a highly competent coating engineer or fitter. With all polymer coatings, sufficient heat is essential to crosslink the polymer fully for optimum performance. In the case of rubber, vulcanisation is carried out off the coating machine as a separate operation in an autoclave. The length of coated rubber is wound on to a roll, interleaved with release paper, wrapped with a waterproof wrapper and heated with superheated steam at about 40 psi (2.8 kg/cm^2) pressure (about 140 °C) for about an hour. This vulcanises all the separate layers together and develops optimum properties in the rubber.

3.4.7 Automatic gauging of coating add-on and general control

Most modern coating works will have on-line gauges which measure mass and from this resin add-on can be determined.[6] The most common are beta gauges but IR, UV, X-ray, fluorescence and laser types are also in use. Beta gauges function in pairs by measuring the fabric mass before and after coating and subtracting the two values. On-line measurement of add-on is essential for quality control and allows information to be collected for statistical process control quality systems. Documentation is necessary for ISO 9000 quality standards, and is increasingly becoming a requirement for suppliers to major customers. On-line gauging allows more precise control of add-on, which not only reduces the production of poor quality material caused by insufficient add-on, but also helps avoid the occurrence of excessive add-on which is a waste of resin and leads to excessive cost. On-line gauging of resin add-on, which is one of the fundamental measurements of a coated fabric, is the starting point for automatic control of all the parameters which are associated with it and which control it. These include fabric tension, blade angle, resin solids content and delivery and line speed. The additional variables of oven temperature and air flow rate for optimum drying and resin crosslinking rate are also relevant factors influencing performance. All of these parameters can, in theory, be automated in order to produce consistent quality. Such a control system is likely to be extremely costly but, if production volume is sufficient, it should pay for itself within reasonable period of time. Manufacturers of such control equipment recommend starting with a basic system and adding to it in stages as and when economically justified.

3.5 Foam finishing

Related to direct coating is 'foam finishing', which involves preparation of foam using a solution or a water dispersion of the textile chemical to be applied. The foam is direct coated in the same way as described above for compounded polymers. Foam processing was developed in the USA during the late 1970s as a more environmentally friendly alternative to impregnation or padding with a pad mangle.[7–10] Instead of fully immersing the fabric in an aqueous bath of the chemical finish and then squeezing out the excess water, the foamed chemical, at the appropriate concentration, is direct coated on to one side of the fabric. Thus, there is less water to dry off and less waste because there is no residual liquor left in the pad bath at the end of the production run. The foam does not sink into the fabric, but sits on the surface. The foam collapses on drying and is not actually visible as a separate layer when dry. The actual add-on of the chemical finish is usually of the order of 2–3% or less, on dry weight of goods. This technique can also be used as a method of applying low add-ons of polymers to fabrics. Fabrics produced from spun yarns can sometimes be coated in this way, because the 'solids content' of the foam is very low and the blade need not scrape the fabric surface as much as in a normal direct coating. Some foam processing blades are rounded and the scraping action is gentler than with a sharp knife. However, it may still be necessary to calender some fabrics made from spun yarns for a good smooth surface. Foam processing is also possible using rotary screens, such as Stork. The advantages of rotary screens are discussed in Section 3.12.

Coating on top of a fabric finish may lead to poor polymer adhesion; the foam processing method is especially useful for applying finishes to only one side of the fabric – the non-coated side. However, it is always advisable to check for effect on adhesion because the finish could strike through the fabric and still be having some effect. Non-rewettable processing agents are available which help 'wet out' the fabric substrate during processing and then thermo-degrade once they have performed their function.

Foam processing may be a way of applying colour to one side of the fabric. One advantage may be that it allows the colour in fashion to be applied at the very last minute, removing the need to hold stocks of dyed yarn or fabric, which may become redundant as the fashion or season changes. However, technical competence is essential for reproducible, especially batch-to-batch, colour shade matching.

Foam processing is extremely useful for very heavy weight materials such as carpets, which are usually treated with anti-soil finishes such as fluorocarbons. The fluorocarbon is applied from a very 'dry' foam, i.e. very low density containing very little water, to the face side of the carpet. If the carpet were completely immersed in water it would be quite impossible to dry off all the water at a commercial speed. Fluorocarbons can be applied to carpets made from all fibres, including polypropylene, in this manner. 'Stainblocker' chemicals (e.g. made by Clariant) can be applied at the same time as the fluorocarbon to carpets containing either nylon or wool, to produce excellent anti-soiling properties which are due to the combined action of the two types of finish.

3.6 Foamed and crushed foam coating

3.6.1 Procedures

This method can be used to apply polymer to woven fabrics and knitted fabrics, and also to fabric produced from spun yarns or fabrics of a general open construction which cannot generally be direct coated.[11–16] This is possible because the foam, which is rather like shaving cream, sits on top of the surface of the fabric without sinking into the fabric structure. Foam coating is related to foam processing already described, but involves much higher solids content of polymer in the foam and more stable foams, i.e. foams that do not collapse readily. As has been explained, crushed foam coating increases the number of fabrics which can be coated; it also greatly reduces penetration of resin into the fabric, which allows the production of much softer handles and better drape than can generally be produced by direct coating. In addition, the crushed foam coatings have a degree of permeability and breathability and, weight for weight, they are more opaque than direct coated material. They are also versatile, in that many different varieties of fillers including pigments can be added to the coating formulation to produce different novel properties.

A typical formulation may contain:

- acrylic resin;
- water;
- foaming agent (ammonium stearate or di-sodium N-octadecyl sulphosuccinamate);
- thickening agent;
- filler, e.g. calcium carbonate;
- FR chemical (if necessary);
- crosslinking agents (if necessary).

More detailed recipes and application procedures may be obtained from the chemical manufacturers, e.g. Rohm & Haas (Cornelius in the UK), BF Goodrich and Ciba, or from resin compounders, e.g. Synthomer and Noveon Performance Coatings Europe (formerly BF Goodrich).

The foaming agent di-sodium N-octadecyl sulphosuccinamate is claimed to give better abrasion properties in the coated material, to avoid 'blooming' and also to give better foam life. Flame retardant fillers and anti-static agents can also be added at this stage. Too much filler has a negative effect on durability to washing and other physical properties, but too little may not produce the degree of opacity that may be needed in, say, a curtain fabric.

3.8 Mechanical foam generator for foam finishing and for crushed foam coating. Photograph supplied by Web Processing (M/C) Ltd and reproduced with kind permission.

The basic principle is to prepare foam from the polymer using a mechanical foam-generating machine, see Fig. 3.8 and 3.9. Quality and type of foaming machines vary. Those with square pins (which cut foam bubbles) in the foaming head are claimed to produce foam of more consistent quality than those with round pins (which squash and burst bubbles). The three basic controls that determine foam density are: air input rate, pump speed and rotor or mixer speed. Basic settings depend on the machine and information should be available from the machine or resin manufacturer. The airflow should not fluctuate and is preferably controlled using a flow meter. The rotor or mixer speed can be varied – the faster the speed, the finer the foam produced. The foam should not be used until a significant back-pressure has been built up because, until this happens, the foam density will not be constant. When this has been attained, the foam density is

3.9 A crushed foam coating trial being set up.

determined, usually by taking a sample in a paper cup and weighing. The weight is compared to the weight of the cup full of water and thus the density (the 'cup' weight) can be calculated. This is adjusted by increasing or decreasing the amount of air being introduced and, when the density is correct, the delivery pipe is attached to the traverse in front of the doctor blade set over either a roller or table. The table or roller is necessary to support the weight of the foam, which is too heavy for the knife on air technique to be used. A gap between the blade and the fabric supported by the table or roller is preset by a feeler gauge, if there is no instrumentation on the machine itself. The density of the foam is decided by the properties required and the substrate being coated – generally the more open the structure the lower the density. It can be anything between approximately 0.10 and 0.8 g/m^2. Delivery of the foam to the bank in front of the doctor blade should be at a rate such as to keep the bank as small as possible and ideally it should rotate so that the foam does not 'age' or dry out. Streaks may be caused by old foam interacting with the new foam being run into the resin bank, producing zigzag trail marks which some operators call the 'Zorro' effect. Some coaters feed the new foam in across the whole width of the bank to avoid this.

The layer of polymer foam coating is then dried by passing the foam coated material into the stenter oven. This is a critical process and needs careful consideration. The oven must be 'profiled', i.e. heating must be done in stages, so that the foam is dried gradually. If drying takes place too fast, a skin may form on the surface with undried resin underneath and bubbles formed as moisture escapes through the skin. Overdrying may cause 'mud cracking'. In a four bay stenter for example, the settings for each bay may be 90 °C for the first bay, 110 °C for the second, and 130 °C for the third and fourth. If surface defects do occur, it may be necessary to reduce still further the temperatures on the first or second bays. With some drying ovens, it may be necessary to reduce the airflow if it is too strong or impinges directly on to the foam, which is quite delicate until dry. Drying should not be overdone, however, because premature crosslinking may occur.

After drying, the foam should be examined by pressing a small area with the fingers – if it is not dry enough, the top skin will be ruptured revealing wet undried polymer underneath; if it is too dry, the foam will spring back. The correct amount of drying, believed to be 5–10% water retention, has been obtained if the foam stays crushed and does not spring back. The foam is then crushed between rollers which should preferably be stainless steel/rubber coated. After this process, the material is cured by passing it down the stenter again at, say, 160 °C with a dwell time in the stenter of 60 seconds, or 170 °C at a higher speed to give a dwell time of, say, 40 seconds. The more thorough the curing and the crushing, the better the durability to washing, dry cleaning and to mechanical abrasion of a particular resin formulation. Coatings of higher durability may be achieved by appropriate resin selection and by the inclusion of extra external crosslinking agents, although some softness and drape may be sacrificed.

3.6.2 Products from crushed foam coatings

The polymer generally applied by this process, or variations on it, is acrylic, but water-based rubbers and polyurethane could also be used. Goods produced in this way include apparel, floor coverings, wall coverings, black-out curtains and curtain linings and filter materials. The handle and drape of crushed foam fabric are generally better than those of direct coatings but probably not as good as transfer coated fabrics. The degree and nature of crushing determines the properties of the coated material, such as abrasion resistance, breathability and durability to washing and dry cleaning. Crushing the foam under very high pressures and temperatures results in a relatively clear film with low breathability, but very good waterproofness and high durability. Filter fabrics of varying porosity can be produced by the crushed foam method using mild crushing conditions. Courtaulds Research and others, in the 1970s, produced a waterproof and breathable coating by crushing the foam under moderate pressure and high temperature. This technique has appeared in the technical literature, but much technical competence, perseverance and care is necessary. The Tubicoat MP system (CHT – Chemische fabrik Tubingen) for production of micropous polyurethane coatings is believed to involve foamable water-based resins.

Curtain linings and curtains produced from crushed foams are seen mainly in hotels. Excellent black-out effects can be produced with two or three layers of crushed foam coatings, one of the layers being pigmented black. The other layers are necessary to prevent dulling of the shade of the backing curtaining fabric if it is a bright colour, and to cover the black layer if its appearance is not acceptable. Flame retardant chemicals are usually incorporated into the foam, and flock can be applied to the top layer to produce an attractive surface finish. Further novel effects may be possible by embossing. Crushed foam coating is generally more expensive than direct coating but not as expensive as transfer coating. Only water-based resins can be used, which limits the durability of the coating obtained; crushed foam coatings generally have limited washing and abrasion resistance properties – unless the crushing is carried out under high temperature and high pressure. However, this latter process does not appear to be done in practice, possibly because of plant availability and technical complexity.

3.7 Transfer coating

3.7.1 Transfer coating process

Transfer coating was first developed commercially in the 1960s when polyurethane polymers were being exploited as coating resins for textiles. This 'transfer' technique is used for knitted fabrics which, compared to woven fabrics, are open and stretchy, and cannot be coated by the direct method because they would distort under the tension applied to obtain a flat surface.[17–19] In addition, the resin would sink into a knitted fabric, and probably penetrate through to the face, and fabric stiffening would occur accompanied by a significant loss in tear strength. In addition, fabrics produced from spun yarns such as cotton, which when direct coated generally produce a rough 'raspy' handle, can be readily transfer coated.

The principle of transfer coating is first to spread the polymer on to release paper to form a film and then to laminate this film to the fabric, see Fig. 3.10. In this way, the polymer does not come into contact with the fabric until it is actually in the form of a film. Transfer coating is carried out in a number of steps, as follows. The top layer is applied first to the release paper by a doctor blade and is dried – but not crosslinked – in an oven. The base layer is then applied over this top layer, using a second doctor blade, and

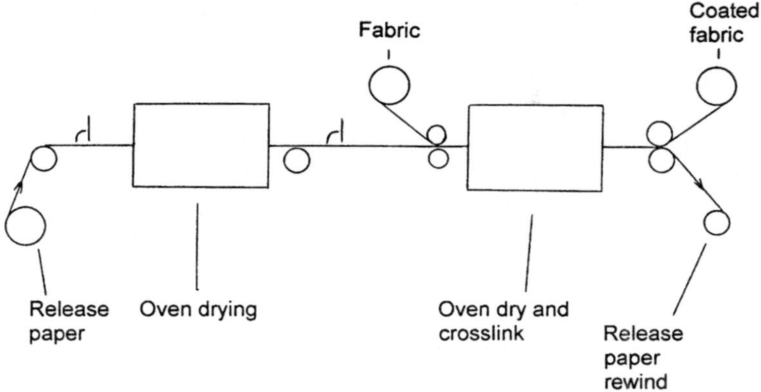

3.10 Transfer coating. This technique allows stretchy knitted fabrics or open construction fabrics to be coated because the resin is not applied directly to the fabric. The resin is first coated on to release paper to produce a film. This film is then 'transferred' to the fabric. The process is: (1) coat the resin (which will become the top layer) on to the transfer paper and evaporate off the solvent, (2) coat on the second layer, which will become the base layer (and is actually the adhesive which joins the top layer to the fabric), (3) lay the fabric on top of the coating, nip together, evaporate the solvent and crosslink the two layers together, (4) peel the coated fabric off the release paper.

straight afterwards, the fabric is laid over this base layer and joined to it by nip rollers. The paper with the coating and fabric on it then passes into a second oven, which dries and crosslinks the two layers together. The base layer sticks to the fabric, while the top layer, which was applied first to the release paper, does not stick to it, because of its release properties. After the assembly emerges from the second oven, the freshly produced coated fabric is peeled off the release paper and taken up on to a batching roller. If higher specifications are required, it may be necessary to coat on a higher add-on of polyurethane, and this can be achieved by a three-layer coating, i.e. inserting an intermediate layer in between the top and base layer. Decorative or embossed designs can be obtained using embossed paper or by further processing.

This method of coating differs from the direct method in that the coating generally becomes the face side of the material. When used for apparel, the fabric is on the inside of the garment. Because the base material is flexible knitted fabric, and resin penetration does not occur, an extremely soft and flexible coated fabric is obtained. Transfer coating is more expensive than direct coating, partly because of the added cost of relatively expensive release paper and the more expensive double headed plant. In addition, the high specifications sometimes require solvent-based polymers, especially in the top layer which becomes the outer face in a garment. In some cases the release paper can be rolled up and re-used, but the release properties deteriorate each time it is used; for top quality products, it can only be used once. For some products, the fabric is napped or raised slightly to provide more mechanical keying so that maximum coating adhesion is obtained combined with optimum handle and drape. If raising agents are required for the raising process, silicone types should be avoided because of their negative effect on coating adhesion. The amount of any raising agent used should be kept to a minimum and the effect on adhesion checked.

3.7.2 Products from transfer coating

One of the main outlets for transfer coated polyurethane fabrics is in up-market, waterproof protective clothing. Because the coating becomes the outer face of the garment, it is possible to produce coated fabric with excellent, high visibility appearances. Polyurethane transfer coatings are used in upholstery and produce an attractive product. The polymer formulations need careful consideration, however, but many of the earlier problems which arose in the 1970s appear to have been overcome. Early problems included delamination and surface cracking, some cases of which were attributed to microbe attack. These occurrences are now quite rare and the industry seems to have overcome the difficulties. Other outlets for transfer coated polyurethane include upholstery, luggage, footwear, gloves and waterproof mattress covers. Quality products are possible from transfer coating, but the process requires relatively expensive plant and high technical competence.

It is possible to load more pigment into a coating than the amount of dye which can be applied to a fabric. Moreover, certain materials, such as luminescent substances, cannot be 'dyed' on to a fabric, but can be incorporated into a coating. Luminescent materials absorb light energy and re-emit light in the dark and are useful in protective clothing. This concept of using the coating to contain useful materials has many potential applications and the fabric which, in these cases, becomes simply a support for the coating can be lightweight and inexpensive.

3.7.3 Transfer coating of PVC

PVC is also used widely in transfer coating to produce 'leathercloth' material for garments, luggage, upholstery, footwear and for automotive interior trim. Automotive seating was once one of the major outlets for PVC coated fabric, but the use of polyester has now become almost universal. A significant proportion of the upholstery transfer coated PVC is produced from expanded or 'blown' coatings which have a soft touch. The soft touch is produced by the inclusion of chemical blowing agents – fine powders, which break down under the action of heat and give off gases such as carbon dioxide or nitrogen to give a pore or foamed structure. The coating oven is normally zoned and the temperature at which the chemicals 'blow' is higher than the temperature at which the coating surface has hardened. In this way, the gases are trapped inside the coating to produce tiny bubbles giving the soft touch properties. The factors to consider are blowing agent particle size, concentration in the PVC, decomposition temperature and oven temperatures. The expanded layer may be 400–450 g/m^2, and this is covered with a non-foamed layer which may be 200–300 g/m^2 to provide good abrasion resistance. The PVC is embossable and can contain FR chemicals, biocides and pigments. PVC coated upholstery fabric is waterproof, easy cleaned and hard wearing.

3.8 Coagulated polyurethane coatings

3.8.1 Imitation leather and suede

This is a specialised process requiring special apparatus and considerable capital investment which includes solvent recovery. Development started over 40 years ago and there are several hundred patents involving many of the major textile and chemical companies, e.g. DuPont, Bayer, Toray, Kuraray and Porvair. Reproducing the properties of leather is not easy and many products have appeared over the years. The most successful ones have been those involving microfibres which were not available in the early stages of development. In addition to microfibres, selection of the base polyurethane polymer and finishing treatments is also a critical factor.[20–22] Further discussion on artificial leathers and suedes appears in Chapter 4, Section 4.9.3.

The essential process involves polyurethane which is dissolved in a solvent and is then thrown out of solution under controlled conditions to form a precipitate or coagulation.[23–25] This material is soft, has a pore-like structure and is the basis of the imitation leather. Commercial processes use polyurethane in DMF solvent which is applied to the fabric carrier. This coated material is then brought into contact with a bath containing a DMF/water mixture in which the polyurethane is not soluble and is therefore coagulated. The coagulated material is washed and dried and the potentially toxic DMF solvent has to be recovered. There are numerous variations in process design and conditions which have to be optimised, such as polymer type, base fabric design, additives, viscosity, coagulation bath conditions, washing and drying conditions, but most of the commercially available artificial leathers and suede are made using this type of process. Waterproof breathable material can be produced by the addition of fluorocarbons to the formulation to confer hydrophobic properties to the polyurethane microstructure.

An alternative and more environmentally friendly method of coagulation is thermo-coagulation using water-based polyurethane resins. This has a significant advantage over solvents, but the properties obtained so far do not appear to be of the same quality as the solvent system. However, base fabric selection and finishing skills are critical factors. An example of such a system is Stahl Chemicals Isothermal Foam Coagulation (IFC) system. This latter system has the advantage of requiring no special coating apparatus.[26]

3.8.2 Breathable waterproof microporous coatings

A related process for producing breathable waterproof coatings on ordinary direct coating machines is the evaporation, dry coagulation and phase separation technique.[23] The polyurethane resin is in a mixture of MEK/toluene dispersion in water which is coated on to the fabric and dried gradually under carefully controlled conditions. The MEK/toluene evaporates first and the polyurethane, not being soluble in the remaining water, coagulates. Finally the water evaporates off leaving behind a porous coagulated layer of polyurethane. To produce a commercial product, two layers are applied. The first layer contains adhesion promoters to secure it to the base fabric, and the second layer contains a fluorochemical to assist in water repellence and crosslinking chemicals to improve abrasion resistance. It is possible to control the degree of breathability obtained by varying the amount of water in the formulation. Following the coating process, a water repellent finish such as a flourocarbon or silicone is required to ensure good water resistance of the microporous structure.

Garments produced from microporous material can in certain circumstances allow the passage of liquid water, such as in areas of high flexing and pressure, i.e. elbows, armpits and seat of the trousers. In addition, it is believed that pores can get blocked up by soiling and that surface active agents left behind after washing can allow water through the pores. These problems can be solved by the application of a thin layer of a 'solid film' coating of a hydrophilic polyurethane resin. Hydrophilic polyurethane adhesives are also available to laminate microporous films to fabrics. As has already been mentioned, hydrophilic polyurethane coatings can be used as waterproof breathable coatings in their own right.

3.9 Back licking roller techniques

The back licking roller method can be used to apply finishes or coatings of very low viscosity to one side of a fabric, see Fig. 3.11. For some finishes,

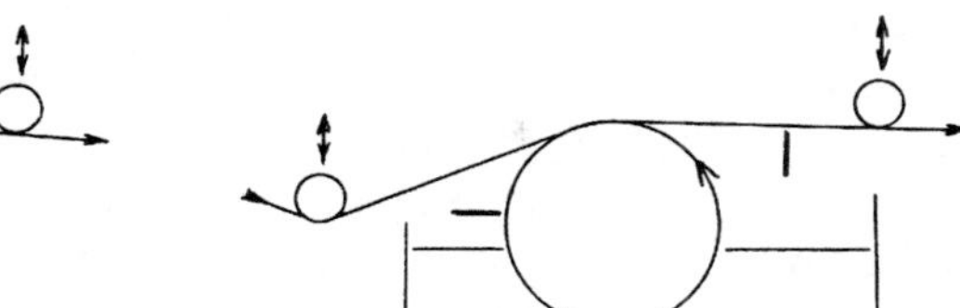

3.11 Back licking coating methods. The simplest type is for applying finishes or resins of very low viscosity to fabric. It is an alternative to impregnation (padding) when pressure on the pad mangle may damage pile or raised fabrics. Control of add-on is accomplished: by altering the speed of rotation of the application roller relative to the fabric speed (it may even be reversed to reduce add-on); by varying the 'solids' content of the resin, and by positioning of the guide rollers which can be adjusted to increase or decrease the area of contact between the fabric surface and that of the application roller.

Thicker resins may require more control by the use of doctor blades on the roller and/or on the fabric. In some arrangements a back up roller over the application roller is used to force resin into the fabric. Rollers are sometimes used in place of blades.

the technique is an alternative to impregnation, which could crush pile fabrics, or foam processing. The amount added on is controlled by the solids content of the bath, the angle, area and degree of fabric contact with the application roller, the speed and direction of rotation of the application roller and the speed of the fabric. Sometimes a doctor blade may be applied to the roller to help control and maintain uniformity of liquor or resin pick-up. Woven velvet automotive seat fabric is sometimes processed in this way to lock in the pile. Carpets and other articles are also processed using back licking techniques. Some reports claim that two different finishes may be applied to the same substrate at the same time – a foam finishing technique on the top of the fabric and a back licking technique for the other side. As in all finishing techniques, wetting agents may be required and surface condition and properties of the substrate will influence the results obtained.

3.10 Hot melt extrusion coating

This method is used for thermoplastic polymers such as polyurethane, polyolefins and PVC, which are applied by feeding granules of the material into the nip between moving heated rollers. There are two general designs in use, the Zimmer machine, which has two melt rollers, and the Bema machine, which has three.[27] They are smaller versions of calenders, but differ in that they need a fabric (or paper or film) as a substrate on which to deposit the film as it is produced. This method can apply resin (in the form of films) to fabric at a faster rate than can be achieved by direct or by transfer coating. Experienced operators believe that layers of film between 2.5 and 20 mils can be applied by this method. One 'mil' is a thousandth of an inch and is equivalent to 25 micrometres. PVC production rates of 600 pounds (272 kg) per hour for the Bema and 450 pounds (204 kg) per hour for the Zimmer are claimed at 60 inches (152.5 cm) width. Control of resin add-on is by the distance between the rollers and the pressure applied. There is no solvent or water to dry off and so high add-ons are possible. Sometimes the first base layer is applied by the direct method to seal the fabric and obtain good fabric–polymer adhesion together with a flexible handle. Subsequent layers are then applied by the hot melt extrusion method.

The hot melt extrusion of 100% molten polymer on to a fabric is also used to produce light weight coverings or tarpaulins and sack material, by coating tape woven polyethylene or polypropylene with either polyethylene or polypropylene, see below in Chapter 4. The polymer is extruded as a full width continuous sheet of material (or 'curtain') on to the fabric substrate. The slot die extruder manufactured by Nordson, which is used for application of adhesives, could also probably be used to extrude polymer sheeting. In some hot melt extrusion methods the polymer needs to be first melted in an Archimedes screw extruder and then delivered to the coating head via heated and jacketed hoses to keep the polymer molten.

3.11 Calender coating

3.11.1 Introduction

Calenders are primarily used to produce unsupported films of PVC and rubbers from compounded polymer 'dough'. The process can also be adapted to apply the freshly produced film to fabric. Calenders consist of a number of massive rollers, sometimes five or more in various configurations, which rotate to crush the 'dough' and smooth it into films of uniform thickness.[28,29] The thickness of the film is determined by the gap separation between the rollers, but there is usually a limit to the thinness of films which may be produced by this method. The more rollers, the more accurate and uniform is the film produced. Both PVC and rubbers are produced in unsupported film form on these machines in bulk, and it is generally uneconomical to process less than about a tonne of polymer. Some of the rollers are heated to keep the polymer molten, but friction of the moving rollers also generates heat, and the material is fabricated into a continuous sheet which can be brought into contact with the fabric to which it adheres,

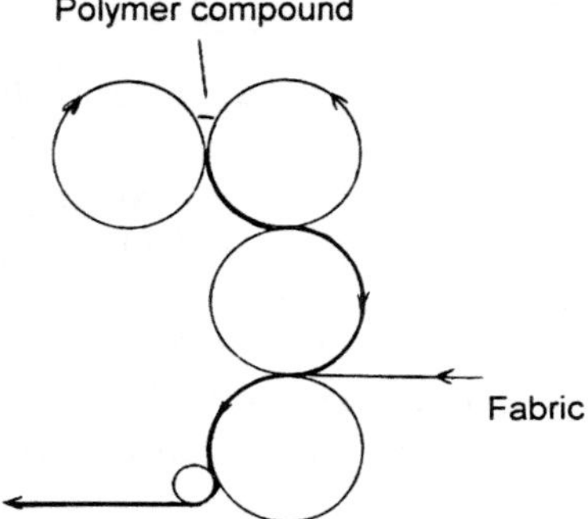

3.12 Fabric lamination directly after film production on the calender. This diagram is schematic only because there are a large number of different roller configurations, some making use of many more rollers. It is a highly technical process, requiring considerable skills and is widely used for rubbers and PVC.

see Fig. 3.12. There is no solvent or water to dry off and so high add-ons are possible. In some operations, for example when polyurethane is being applied, it is usual to apply the first base layer by the direct coating method, knife on air to obtain the best fabric–polymer adhesion and add subsequent layers by calendering. These roller calenders should not be confused with the laundry-type calenders used for smoothing or ironing fabric on which the continuous press or flat bed laminators are based, see Section 3.17.1.

3.11.2 Calender film lamination of pre-prepared films

Fabrics can be used to reinforce film produced by calenders and certain products are produced by the double sided lamination of a very open woven material by pre-manufactured thermoplastic films of polyurethane, polypropylene or PVC. The three materials which are pressed together by a hot mangle, the two films melt and join together through the open weave of the fabric.[30–32] Sometimes the fabric, more accurately referred to as a net, is not actually woven but comprises lengthways and widthways threads (or very narrow tape) locked in place by an adhesive. This material which is sometimes referred to as 'weave lock' determines the tear resistance of the material. Film lamination allows thick layers of polymer to be added without the need for multiple passes which would be required if the direct method was used. Open knitted scrims or open non-woven fabrics are also used as reinforcing material in double sided laminates of thin PVC films for end uses such as bunting and lightweight coverings.

3.12 Rotary screen coating

The rotary screen technique, see Fig. 3.13, which applies compound to a fabric by forcing it through a cylindrical screen, is used mainly for textile printing. The technique can also be used for coating polymer on to fabric with add-ons, it is claimed, from between 5 and 500 g/m^2. The add-on is controlled by the resin viscosity, the mesh of the cylindrical screen, the speed and the pressure of the squeegee bar inside the screen.[33–35] An array of dots are pushed through the perforated screen by the squeegee bar inside the screen and by centrifugal force on to the surface of the fabric. The fabric

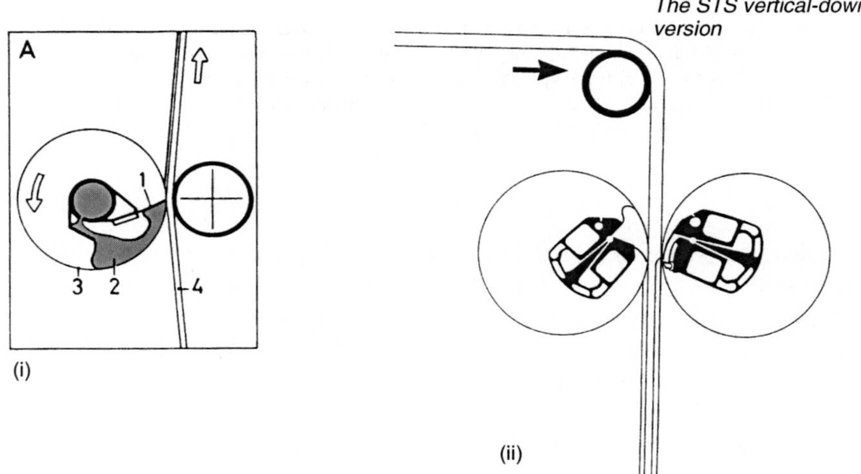

3.13 Stork Rotary Screen technology for fabric coating, foam finishing and printing. (i) shows the rotary screen application of coating resin. A squeegee (1) presses the resin (2) in the screen through the perforated screen wall (3) on to the substrate (4). Light squeegee pressure produces a small amount of resin add-on. The add-on can be increased by applying higher pressure and also by the positioning of the squeegee in relation to the centre of the back up roller. Other factors influencing add-on are the screen type, perforation shape and size and resin solids content, viscosity and rheology. (ii) shows the screen to screen application system (STS), in the vertical-down position. Here two sides of the fabric are being coated in a single pass. The vertical-down arrangement is used for coating textiles and nonwovens. The vertical-up arrangement is used to apply PVC to tarpaulins. Horizontal techniques are used for combination applications, if the lower screen works with a closed squeegee, e.g. resin coating above and finish application (impregnation) below. The screens are seamless, and because they rotate at the same speed as the fabric, the coating action is a process of placing resin on the fabric rather than one of scraping. Diagram reproduced with kind permission of Stork Brabant BV.

moves at the same speed as the rotation of the rotary screen and there is, thus, no frictional contact between the two. When deposited on to the fabric, the resin in the dots flows and merges together to form a continuous coating. However, a 'whisper' doctor blade may be required for resins where this does not happen.

This technique allows some stretchy fabrics to be coated because the resin is *placed* on to the fabric, instead of being scraped on. The rotary screen method, therefore, requires less fabric tension for uniform application, and resin penetration is minimised, allowing the production of coatings with soft handles and good drape. This method is generally restricted to water based resins because of the problems presented by the provision of solvent wash-off facilities. PVC plastisols, which can also be applied from a rotary screen, face the same solvent wash-off requirement. Hot melt powder adhesives in the form of aqueous pastes are applied using rotary screens in the form of dot coatings. Because the coating substrate is not subject to high tensions, this method is well suited to lightweight nonwoven fabrics and delicate films such as water resistant and breathable films for apparel. The adhesive add-on control and uniformity at the low weights required are excellent. A further advantage of this technique is that the materials can be dried, rolled up, stored and reactivated when required. Because the adhesive is in dot form, the resultant laminate has a soft handle and drapes well. The best known suppliers of rotary screen equipment are Stork and Zimmer. A fundamental difference between the two manufactures is that the Stork apparatus uses a squeegee inside the rotary screen to control resin delivery, whilst Zimmer equipment relies on a magnetic system.

The rotary screen method can be used to apply foamed coatings or for foam processing of fabric finishes instead of padding. It is also possible to use two rotary screens (Stork's STS – screen to screen – method, see Fig. 3.13 (ii)), each one on a different side of the fabric, and hence apply two different finishes to the fabric simultaneously. The position of the squeegee is critical to controlling add-on.

3.13 Other coating methods

Coating of lacquers and low viscosity resins can also be carried out using a number of rollers in various configurations and combinations of configurations to control the add-on accurately. These include direct methods and back-licking methods, which use both blades and rollers as the application tool. These general methods are used extensively in the film processing industry for applying very low levels of add-on very accurately, and are not used to any significant extent in the textile industry apart from the simple single roller technique. Latex is sometimes applied to the back of woven pile fabrics by this method to lock in the pile. Gravure rollers are used to apply measured amounts of both solvent and water-based adhesive for fabric lamination. The discontinuous layer produces laminates of good drapeability. Hot melt adhesives can also be applied but only for very long production runs to justify the down time required for cleaning. More recently moisture cure polyurethane adhesives have been applied by this method which allows accurate control of add-on; see Section 3.17.7.

3.14 Fabric impregnation

3.14.1 Application techniques

This technique is related to coating and, as has already been mentioned, is actually a method of producing a coated fabric, if carried out repeatedly. PTFE coatings are applied to woven glass fabrics and Kevlar woven fabrics by this technique to manufacture calender belts for lamination machines. Fabric 'padding' or 'dipping' – other terms for impregnation – is universally used in fabric finishing plants and most fabrics will have some kind of finish on them, although certain automotive fabrics are a notable exception for the reasons stated in Chapter 4. Apparel fabrics, especially outerwear materials, will generally have some shower-repellent finish on them, shirt fabrics will have easy care finishes and other indoor apparel fabrics are likely to have some type of soft finish. The process consists of passing fabric, open width through a trough containing an aqueous solution or dispersion of the chemical, squeezing out the excess liquor using a pad mangle and drying the fabric in a stenter. A wetting agent is generally added to the liquor to ensure rapid and even wetting out of the fabric; occasionally, if running at speed, anti-foam may be required. If the fabric is to be coated or laminated, the anti-foam should not be silicone based, because it may affect polymer adhesion. Commercial impregnation machines will 'dip and nip' the fabric twice in rapid succession to ensure even uptake. Some fabrics may require higher concentrations of wetting agent for uniform application of a finish and, if the material is to be coated or a water-repellent finish is being applied (see below), special non-rewettable wetting agents may be recommended. Residual wetting agent may influence coating adhesion in the case of a coating substrate and could 'work against' the water repellency of an applied water repellent such as a fluorocarbon. Examples of non-rewettable wetting agents are Fluowet UD and Sandozin NRW, both made by Clariant. The latter product thermo-degrades at about 100 °C.

Although impregnation is technically a relatively simple process, bulk production requires chemical dispensing apparatus to ensure that the pad bath is always at the same concentration and is kept 'topped-up' during long production runs. Automatic dispensing apparatus is available for this purpose. Calculations must be carried out to determine the final pad mangle 'expression', i.e. the amount of water retained in the fabric after the last squeezing process as a percentage of the dry based fabric weight. For thin, flat woven polyester fabrics this may be as low as 20%, and for a heavier weight fabric woven from textured yarns it may be more than 100%. The pressure on the nip roll determines the expression for any given fabric. Needless to say, the nip rollers must be true and parallel and the pressure gauge should be calibrated and the actual pressure used recorded. For energy economy, the pad mangle should be as tight as possible to squeeze out as much water as possible. There will then be less water to evaporate off in the stenter and this will also have the benefit of being able to run faster. A limitation, however, is raised or pile fabrics that may suffer pile crush or water marking if excessive pressure is used. A more energy efficient and environmentally friendly method of chemical finish application is foam finishing which has already been described.

3.14.2 Application of water-repellent finishes for outerwear

The importance of the water-repellent finish on outerwear garments is discussed in Section 4.1.3. The most effective and durable water repellent finishes are fluorocarbons such as Teflon (DuPont) and Nuva F (Clariant). They must be applied according to manufacturers' instructions on to clean fabric, because any contaminants or residual lubricants, especially those of a hydrophilic nature, could seriously impair their performance. The 'spray rating' apparatus (BS 3702 or AATCC 22-1989) is used to assess the quality of a water-repellent finish. A rating of 100 is the perfect result and is awarded to a test specimen which exhibits no water drop sticking or wetting out of the fabric surface. A rating of 90 indicates slight sticking or wetting, 80 indicates wetting at spray points and ratings continue through 70, 50 and down to 0 which indicates complete wetting of the face and underside of the fabric. Generally, a rating of 90 is the absolute minimum requirement, but if the finish is correctly applied, a rating of 100 is easily achieved.

Excellent results can also be achieved using silicone, wax and metal salt finishes, but these are likely to reduce coating adhesion if applied to fabric before coating. As mentioned before, silicones have release properties and must not under any circumstances be applied before coating. They have, however, been known to be applied to coated fabric after coating, but this must be done carefully and after resin crosslinking has been fully achieved. It is possible to coat resin on to fabric which has been fluorocarboned and some factories do this, especially with water-based resins, because it helps to control resin penetration and thus produces a softer coated fabric handle. However, the coating adhesion will be less than that obtained by coating over non-finished fabric and pre-tests should be carried out to establish that the peel bond is still strong enough for the end use.

3.15 Lamination

3.15.1 Introduction

Lamination, by definition, combines two materials and this very act results in modification of physical properties based on the individual characteristics of the separate components. If one component has limited stretch only in the warp and the other limited stretch only in the weft, the combined material will have limited stretch in both warp and weft. Lamination of any fabric invariably produces a laminate, which is stiffer than either of the two starting materials, although this can be minimised by choice of the most suitable lamination method and adhesive. The adhesive used to join the components may reduce stretch further and cause additional stiffening – an objective in adhesive choice is to choose the adhesive which provides the strongest bond with the least amount. A common problem with laminated fabrics is 'cracking' which is caused when one or both materials joined is not stretchy enough to allow the laminate to be curved in an arc. Use of an excessive amount of adhesive or too much foam burnt off in flame lamination may cause cracking. The suitability of fabrics for lamination should be considered when product are designed; clearly the more stretch, the better the handle and drape of the laminated product. Usually a third material is used as the adhesive, but sometimes one of the materials being joined can itself act as the adhesive as in flame lamination of polyurethane foam.

Lamination is one of the fundamental processes in the manufacture of car interior trim, and flame lamination is extensively used. Polyolefin foams, which are used in car interiors, can also be flame laminated, especially to other materials with some thermoplastic nature. When polyolefin foams are joined using hot melt films, it may be necessary to corona treat the surface of the polyolefin foam. Because the polyolefin foam is closed cell, any moisture or water condensation must be avoided because, during the lamination process, it will be difficult for the water to evaporate quickly and this could lead to bubbling, delamination or other defects. This statement also applies to films or any other material which is impermeable to moisture or is of low air porosity. Entrapped air may also cause similar problems during lamination; some machine operators believe that the problem is minimised by keeping the two materials well apart until just before the nip or point of lamination. This may be difficult in a flat bed system and suitably positioned rollers may be necessary. Recently, other factors have arisen which will influence choice of materials and chemical adhesive type – recyclability and eventual disposal of the article.

For many years, lamination was carried out by the application of both solvent- and water-based adhesives by back licking rollers or by some other coating device, and then drying off the solvent, sometimes on a Palmer unit, a large oil or steam heated cylindrical drum with an endless blanket. If crosslinking was required, this was done separately on heated rollers or in a heated chamber. This complete process, whilst reasonably fast – 20–40 m/minute was possible – was very energy intensive and solvents presented health and safety issues which would not be tolerated today.

Hot melt adhesive methods are cleaner and less energy intensive with far fewer health and safety considerations. Hot melt lamination is accomplished by two separate processes: first a means of applying the actual adhesive; and second bringing the two substrates together to form the actual bond under the action of heat and pressure. The latter process is often referred to as fusing. The method of application depends on the form of the adhesive, i.e. film, web, powder or liquid. The methods of powder application are detailed below in Sections 3.17.3 to 3.17.7 and in Fig. 3.14. The method of fusion depends on the article being produced and can be either continuous or static. Static fusion presses can also thermoform the product at the same time, and this process is used frequently in the automotive industry, see Fig. 3.15. There is a range of different fusion processes and these are illustrated in Fig. 3.16 to 3.18. Obtaining a good, durable bond is only one of a series of factors to consider, especially when raised or pile textiles are being processed, because the aesthetics, appearance and fabric handle must not be affected by the heat and pressure applied. These and other issues are discussed below and especially in Section 3.17.1.

3.15.2 Lamination of fabric for apparel

In the apparel industry, handle, flexibility and drape are of major importance, but durability to flexing and to washing are also important. The problem arises, therefore, of finding the best method and adhesive materials for a durable bond and applying them in a controlled manner to maintain the fabric flexibility and aesthetics during the lamination process. For a bond of high strength, it is generally necessary for the adhesive to penetrate the material and to cover the widest possible surface area. The challenge is to select the best adhesive and application process which has the least effect on the substrate aesthetics. Ideally, the least amount of a highly effective adhesive should be applied. Too much adhesive is a waste of resources and likely to lead to fabric stiffening and it could also result in thermal discomfort in the garment, because the adhesive itself could form an impermeable barrier to perspiration. This factor is especially important when waterproof breathable fabrics are being prepared by lamination of a membrane to a fabric. It is usual to apply the adhesive – hot melt powder or moisture curing polyurethanes – in dot or discontinuous form, as will be described below. The use of hot melt adhesives in film form would consid-

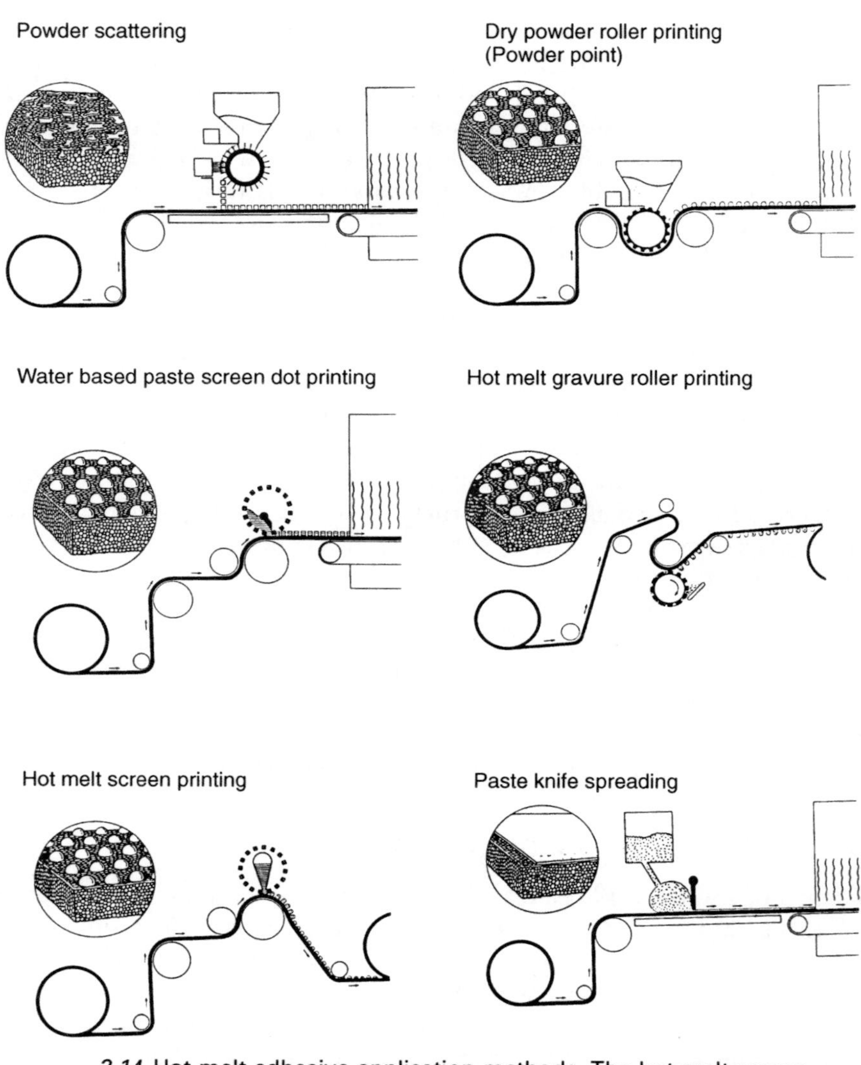

3.14 Hot melt adhesive application methods. The hot melt screen printing (not now in widespread use) and the hot melt roller (gravure roller) methods require a screw extruder to melt the powder and deliver the molten adhesive to the coating head. Diagrams reproduced with kind permission of EMS-Chemie AG (Switzerland).

erably reduce the breathablity of the membrane and cause stiffening. Even with discontinuous methods of adhesive application, it has been estimated that the adhesive can cover up to 20% of the surface area of the breathable membrane and may thus have a significant effect on breathability. Because of this breathable adhesives have been developed.

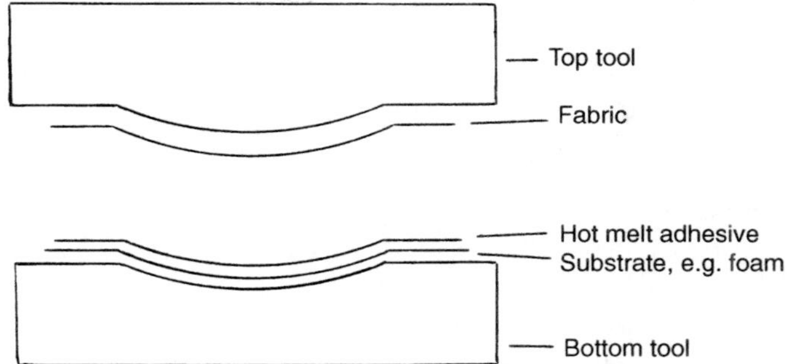

3.15 Static fusion press. The materials (hot melt adhesive and foam) situated in the bottom tool are first heated by IR heaters, which are then moved away. The fabric (held in the top tool) is placed over the hot melt adhesive by lowering the top tool, and a measured amount of pressure is applied for a pre-optimised period of time, for example 30 seconds.

In other presses, the top tool is heated. There are many types and variations. In some cases, moulding to the shape of, say, automotive components may be accomplished with the application of vacuum.

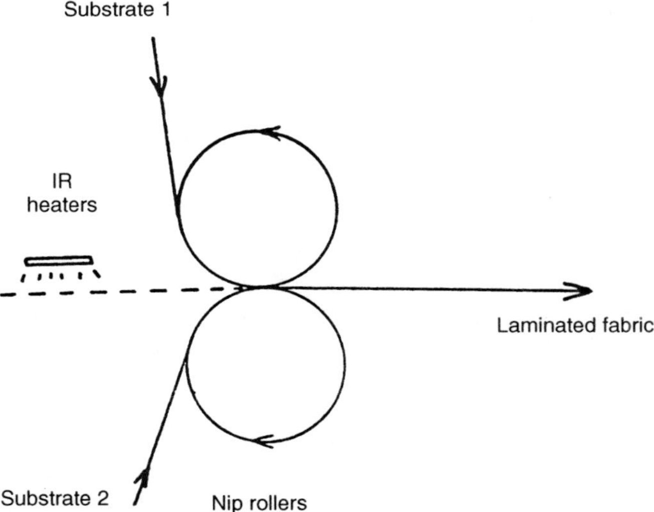

3.16 Simple lamination by nip rollers. The hot melt adhesive in the form of a film or web is heated by IR heaters and fed into the nip in between the two substrates. One substrate may be supported, and the hot melt adhesive may be powder from a powder scatterer or the substrate may be preprinted with adhesive powder. If films are being laminated to fabric, Teflon coated rollers are necessary for non-stick. The bottom roller is likely to be silicon rubber coated.

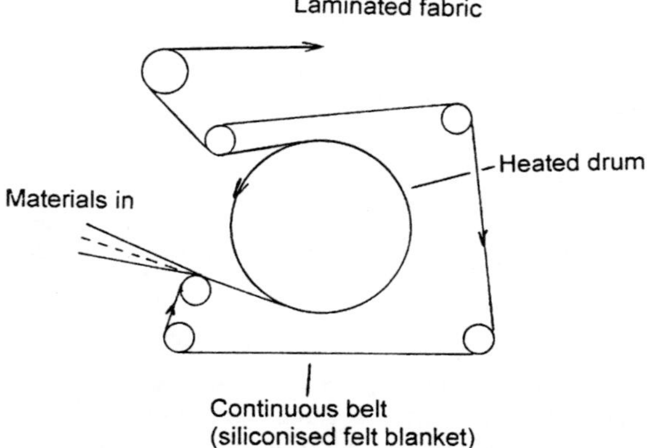

3.17 Heated cylinder calender dry laminator, e.g. Stork, Bates. The hot melt 'sandwich' is fed into the machine which functions by a continuous siliconised felt belt wrapped around a heated drum. The drum surface is sometimes PTFE coated and is non-stick. The machines are also used for transfer printing and pressure on the fabric cannot be easily controlled. Materials which may be crushed or damaged by pressure, such as pile fabrics and certain nonwovens and foams, cannot be processed on this machine.

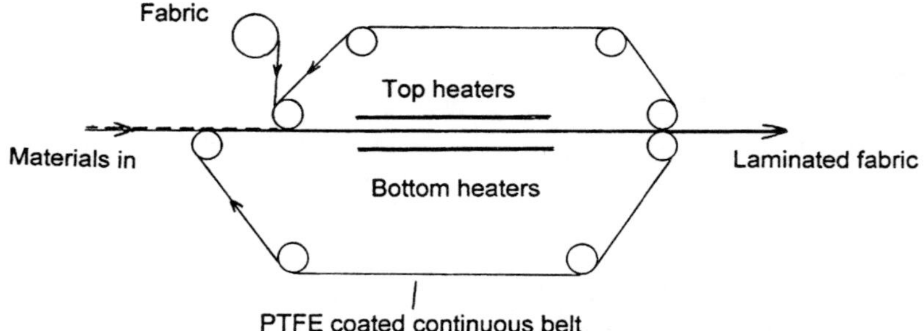

3.18 Calender (continuous flat bed) laminator. Cooling of the goods is generally necessary. In some machines this is incorporated inside the machine.

The waterproof breathable films are generally very thin and delicate materials and require considerable skill and care to handle without creasing. The film take-off feed rollers should be driven, because any pull-off tension may cause the film to 'neck-in', i.e. reduce in width, and path lengths should be as short as possible so that the film does not reduce in width under its own weight, curl at the edges or develop creases. Whether being used as an adhesive or as one of the lamination substrates, films must be supported, preferably with free rotating rollers, or even driven rollers, so that they experience as little friction drag as possible. Otherwise wrinkles or creases may be caused which can be very difficult to remove. A spreader roller, bowed bar or cord expander roller may be necessary, especially just before the nip or point of joining.

3.15.3 Lamination of car interior materials

The largest volume material in the car interior is the car seat cover which is a triple laminate made from polyester fabric joined to polyurethane foam with a thickness of anything between one and ten millimetres, with a scrim fabric on the back. Door casing and headliner fabric is also laminated to polyurethane foam to provide a soft touch, but a scrim fabric is not generally needed. A variety of methods are used in door panel assembly, with textiles being used in combination with plastic foils in polyurethane, PVC, PVC/ABS and polypropylene; textile/leather combinations appear in up-market models. Solvent spray adhesives are still widely used with 100% solids hot melt coming into use. Some manufacturers make use of pressure sensitive adhesives, which are clean and require no heating or drying and no special safety apparatus such as extraction units or spray booths. The nature and texture of the materials being joined, the performance required plus the plant available, all have an influence on which adhesive and which adhesive type to use.

The following factors must be taken into consideration:

- chemical nature of substrates to be joined, i.e. polyester, PVC, etc;
- physical condition, i.e. texture, pile and surface nature – will it be damaged, especially by hot melt processes?
- fabric construction, open or relatively closed – will adhesive penetrate?
- fabric stability – how it will affect handling – is stretching or shrinkage likely to occur?
- presence of fabric finish or residual lubricant and possible effect on adhesion;
- temperature resistance required of the resultant laminate;
- initial bond strength specified;
- bond durability, i.e. resistance to water, high relative humidity, etc;
- plasticiser migration (of PVC components);
- UV and light resistance (if applicable);
- possible effect on appearance, e.g. discoloration of face fabric.

3.16 Flame lamination

The flame lamination process is in widespread use throughout the world. It was actually invented in the 1950s (original patents were ascribed to Reeves Brothers[36]) and extensive commercial development took place in the 1970s. This lamination method was once used extensively to produce laminated fabric for garments, curtains and drapes, and it makes use of the polyurethane foam itself as the adhesive. It is a quick, economical process

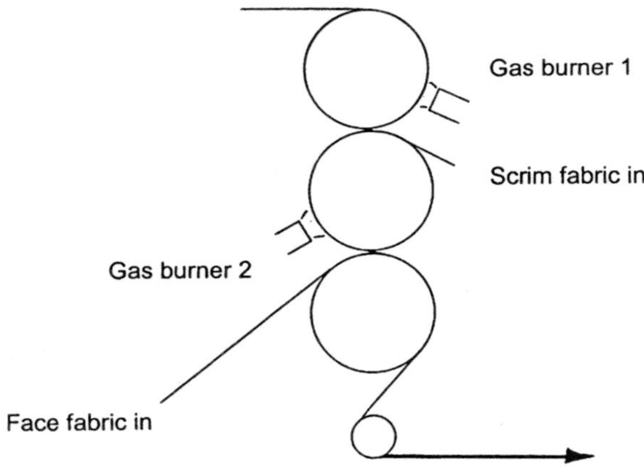

3.19 Flame lamination. The gas flame burner 1 melts the surface of the foam, which then acts as the adhesive for the scrim fabric. On the other side, burner 2 melts the other surface of the foam, which then acts as the adhesive for the face fabric. Thus, three separate materials are fed in and a single triple laminate emerges.

but requires a certain amount of technical skill, regular maintenance and abatement of the fumes produced. At the time of writing it is used almost exclusively for the production of car seat laminates. All three components, face fabric, polyurethane foam and scrim fabric, are fed into the laminator; the three materials joined together emerge at speeds between 25 and 40 metres or more per minute. A gas flame licks and melts the surface of the moving foam, which then acts as the adhesive to the fabric, which is laid over it. This happens twice in a double head machine, as shown in Fig. 3.19. To compensate for the foam burnt off, input foam slightly thicker than that specified must be used. Headliner and door casing face fabric is generally produced in the same way, but without a scrim – a bi-laminate. It is possible to flame laminate polyester face fabric to a polyester non-woven material (polyurethane foam substitute) using 'mini' foam, i.e. polyurethane foam about 1–2 mm thick. The foam adhesive is virtually all burnt off, but some readers may hold the view that this practice is not satisfactory because one of the objectives of using non-woven fabric is to replace polyurethane foam and remove the need to flame laminate!

Machine settings controlling flame temperature (gas/air ratio), burner distance, gap separation of the rollers and speed must be optimised for each quality of foam and fabric being laminated. The actual temperature of the gas flame depends on the gas used (propane or regular town gas), and also on the position of the flame itself which is applied to the foam. For consistent results for a particular quality of fabric and foam, the same settings must be used. Flame retardant grades may need to burn more off to produce a satisfactory bond. Originally, polyether polyurethane foam, see below, could not be bonded well by the flame lamination process. However, the foam manufacturers have modified it so that it can now be bonded just as well as polyester polyurethane foam, but this has incurred extra cost. Different machine settings may be needed to process polyether polyurethane foam and more 'burn off' may be necessary to achieve the bond strength required. The two types of foam, which are both polyurethanes, have slightly different properties, the main one being that polyether polyurethane foam has better hydrolysis resistance than polyester polyurethane foam. The former variety is better suited to more tropical regions of the world and is specified by some OEMs (original equipment manufacturers). The two terms sometimes cause confusion, especially when, for brevity, they are referred to as simply 'ether foam' and 'ester foam'. They are both polyurethane foams. The bonding achieved by flame lamination is believed to be a combination of both physical and chemical, and the final bond strength may need about 12 hours or so to develop fully its maximum strength.

The flame lamination process has come under environmental scrutiny in recent years,[37–40] because it produces potentially toxic fumes by the burning of polyurethane. Alternative methods have been evaluated and developed using hot melt adhesives.[41–52] However, the cost of controlling the emissions is, in many cases, covered by the economies of the process and certain large volume operators have chosen to continue to operate flame laminators, at least for the time being.[53,54] They have installed effective fume control equipment, such as carbon adsorption, which has satisfied the environmental authorities. However, their customers, influenced by pressure groups and others, may not approve of flame lamination and may at some stage apply pressure to switch to alternative methods. In addition, emission limits imposed by local authorities may also become more stringent.

Flame lamination produces a flexible laminate with high bond strength without affecting the aesthetics of the fabric in any way. A particular requirement of laminated fabric for car seats is the ability to form both concave and convex curves without 'cracking'. In-put tension needs to be controlled very carefully for the laminate to be uniformly dimensionally stable and to have the ability to lie flat on the cutting table. Panels cut from the laminate must keep their shape and not distort under the differential tensions within the material itself which are caused by the substrates, with foam in particular being joined in a stretched state. Regular maintenance and cleaning and careful vigilance during the bonding process are essential for the success of a flame lamination operation. Individual burner nozzles may become blocked with ash and cause areas of poor bond strength, and ash may build up on the burner and eventually fall on to the fabric.

The flame lamination process can also be used to heat activate hot melt films to join two non-polyurethane foam materials together, e.g. a PVC film to a polyester fabric. Only one burner with a smaller than usual flame is used, and care is taken not to over soften or even ignite the adhesive film. Polyolefin foams can also be flame laminated to certain other materials, but require a lower processing temperature. These processes are significantly 'cleaner' than flame lamination of polyurethane foam.

3.17 Hot melt lamination

3.17.1 Flat bed laminators – calenders

When the issue of flame lamination being an environmentally unfriendly process arose, thoughts turned first to calenders and the hot melt adhesives, which had been used in the garment industry for many years. The calender principle, see Fig. 3.18, is that the two materials being joined are made into a sandwich with a hot melt adhesive film, web or powder in the centre. This is then fed into the calender, which heats the materials and melts the adhesive to produce a laminate. Webs or films are available only at fixed weights and widths but, if volumes are sufficiently large, the suppliers will normally provide any width required. Much higher volumes are required if webs or films are to be specially made at a particular weight. The advantage of powder is that it can be conveniently applied at any weight, and at any width, for both short and long production runs of fabric.

Calenders are usually heated electrically, and transfer of heat by conduction is not as rapid as, say, in a textile stenter. Consequently, temperature response to controls is not as rapid as on other machines. The goods being processed take heat out of the machine, and heat is also being lost all the time to the surroundings. In addition, any moisture present in the goods, especially in natural fibres such as cotton and also in polyurethane foam, needs to be first heated and then evaporated. Because of this, bond strength should be checked frequently in a production run. The important temperature is the 'glue line temperature', i.e. the temperature in between the two substrates where the adhesive actually is, and not the temperature on the machine control panel. Depending on the thickness of the substrates being laminated and the machine speed, the set panel temperature could be 20–30 °C higher than the actual glue line temperature, and the hot melt adhesive may still not be melted. Heat sensitive paper is available to determine the actual temperature at the glue line, see Fig. 3.20 and 3.21.

Optimum heater temperature, height adjustment, pressure and speed settings must be established by thorough trials in order to determine the best conditions for producing laminates of the right quality at the maximum

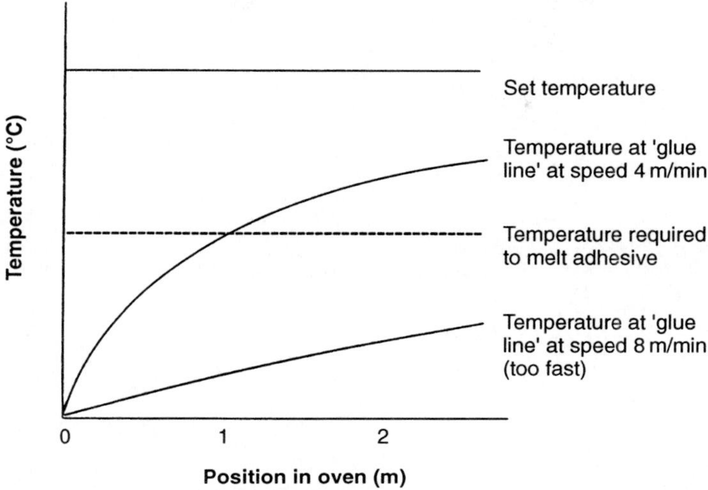

3.20 Calender/flat bed lamination – contact heat. At speed of 8 m/min bonding does not take place because the adhesive never reaches its melting temperature. The correct balance of time and speed determines the 'dwell time', i.e. the time actually in the heating zone. This must be optimised. The set temperature may be hot enough to damage the goods if the machine is stopped or is operated too slow. Heat sensitive indicator paper is necessary to determine the actual temperatures attained at the 'glue line' and at the surface of the goods.

speed for commercial production. Modern machines accurately record all processing conditions and can be computer controlled. If conditions are too mild the adhesive will not be melted and the required bond strength will not be produced; if they are too severe the fabric appearance could be damaged by glazing or by flattening pile or texture. Viscosity of the molten hot melt adhesive is important because, at the bonding temperature, it must flow to cover a certain amount of substrate area and 'key' into the substrates being joined.[55–57] If it flows too much it will strike through the substrates, causing stiffening and, in extreme cases, flow away from the surface, hence resulting in a poor bond.

Calenders are available from several manufacturers and with different designs, e.g. different layout of heater zones and heater arrangement, some with cooling inside the machine, others with the cooling unit outside. Much thought has gone into the design to enable the production of quality laminates at commercial speeds. The main drawback of the calender method is that the heat is supplied to the hot melt adhesive *via the substrates themselves*, which could be prone to damage by heat, especially fabrics with textured yarns or with a pile. This is aggravated by the fact that, when the two substrates are nipped together to form the bond, they are both hot. More-

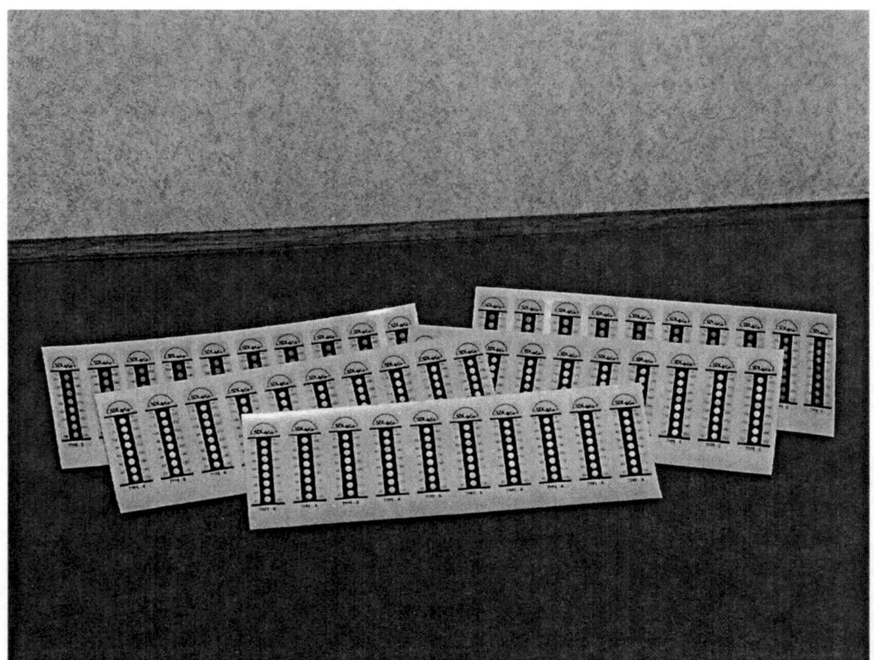

3.21 Temperature indicator paper strips. These are essential for ascertaining the actual temperature at the glue line. This could be significantly different from the temperature set on the machine controls. The faster the speed and the thicker the goods being laminated, the larger the difference in temperature. Thick fabrics, especially fleeces and foams, are good heat *insulators*. Photograph supplied by SDL International Ltd and reproduced with kind permission.

over, the materials most often being joined in the automotive textile industry, textured polyester fabric, nonwoven materials and polyurethane foams, are also good *insulators* of heat and so the process is quite slow. The 'open' lamination method shown in Fig. 3.22, being used for substrate S3, is best for especially thick materials – if the equipment is available. Infra-red heaters are used to heat the adhesive, but they should emit radiation of the optimum wavelength so that heat is absorbed irrespective of the colour of the substrate.

Fabric processed on calenders, or on any other lamination machine, must first be heat stabilised to a temperature above that to which it may be subjected during the lamination process. (See Fig. 2.1.) Fabric pile or raised surfaces can be damaged at temperatures well below the melting point of the fibre. Polyester, for example, begins to soften at around 73 °C, which is not normally noticed unless it is put under high pressure – the higher the

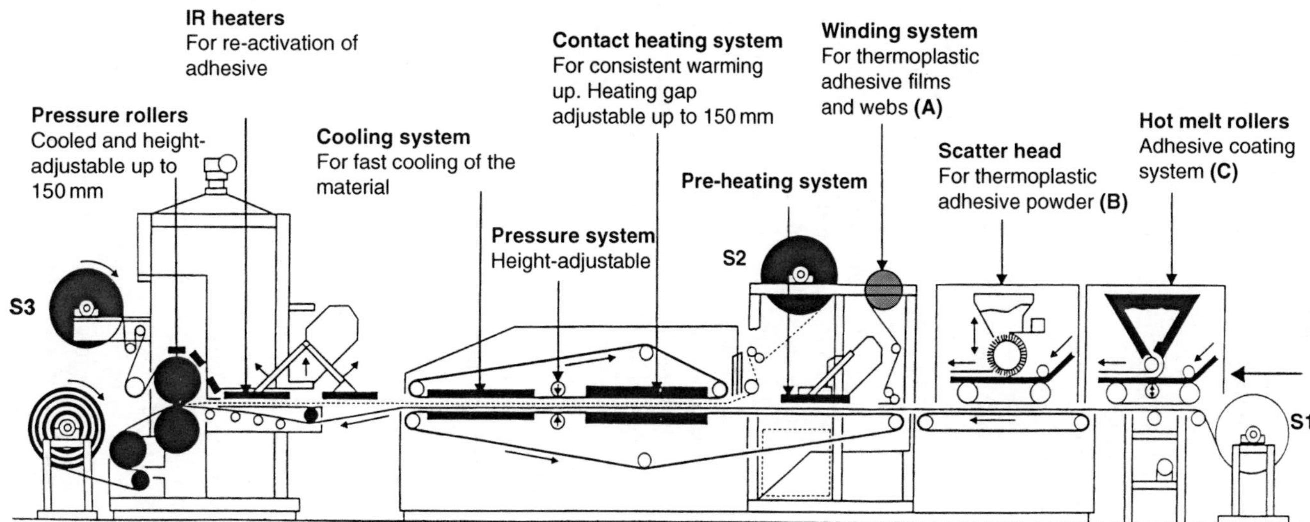

Reactivation of hot melt adhesive by IR heaters and lamination using external pressure rollers. Substrate 3 is heat insulating or heat sensitive material, not suited to contact heating.

Calender heating zones temperatures, height adjustment, pressure roller gap, pressure and speed settings computer controlled.

Three different adhesive application methods.
A. Adhesive film/web
B. Powder scattering
C. Hot melt roller.

3.22 Schematic representation of lamination of three substrates, S1, S2 and S3, in a single, continuous process, using three different hot melt systems. The modular design of the range allows it to be tailor-made for a wide range of applications and components. Reproduced with kind permission of Maschinenfabrik Herbert Meyer GmbH.

pressure and the longer the time under pressure, the more it will deform. The combined effects of a temperature in excess of, say, 95 °C and lamination pressure can cause some pile flattening. To minimise this damage, the top belt on some calenders can be set to a precise distance from the bottom belt. This facility of being able to set a gap between the belts is also useful in reducing the loss of thickness by crushing, which can occur especially when pressure is combined with heat. But, having said this, both polyurethane foam and thick nonwoven fabric materials (as has already been noted), are good insulators of heat, and some compression is necessary for rapid heat transfer. Polyurethane foam is likely to recover, but nonwoven fabric may not – depending on the temperature and pressure applied – and this must be checked beforehand. Any number of layers of material can be joined simultaneously provided that a multiple feed system is available, but the limiting factor is likely to be speed, because heat has to penetrate through all the layers to reach the glue line and activate the hot melt adhesive. Generally, the choice is low temperature machine settings, which will preserve material properties but at low production speeds or higher temperature settings, giving higher production speeds but with the risk of thermal damage to the fabric or other substrate. High temperatures may also produce unsatisfactory results because of thermal shock, shrinkage of the goods and strike through of adhesive.[55–57] Long heating zones, such as those on the newer Reliant calenders, allow lower temperatures to be used for more gradual and gentler heating, thus overcoming these problems and also allowing reasonable production speeds. Lightweight and surface sensitive materials could be marked at regular intervals by the belt join. Although belt joins vary in quality, continuous belts are to be preferred in spite of being substantially more expensive than joined belts.

Calenders, also referred to as flat bed laminators, are used extensively for headliners and other textile automotive components, because several layers of materials, each with an adhesive layer in between, can be joined with one pass. The use of calenders is not an alternative to high volume flame lamination for seat covers or other fabric/foam joining operations because of the relatively slow speed. Calenders can be used for laminating non-roll goods, such as leather hides, to foam and for small scale production lamination. True flat bed machines – a very small number have curved heating paths – can be used for processing rigid materials. They are especially useful for development and preparation of samples, when anything from A4 size pieces to thousands of metres can be conveniently produced. Belt joins may produce a mark on the goods, which can sometimes be overcome by balancing the conditions of temperature and pressure. Belts with very flat joins are available, but the most satisfactory remedy is use of continuous belts which have no join. However, these are considerably more expensive than joined belts.

3.17.2 IR heaters

Infra-red heaters are used widely in hot melt lamination machines to heat or pre-heat materials and adhesives before nipping together. Manufacturers of IR heaters, such as Krelus or the local electricity, supplier can advise on which type is best for the application. Heaters can emit radiation in the short, medium or long wave sector of the IR electromagnetic spectrum and may be influenced by the colour of the substrate. Lighter materials may reflect back energy instead of absorbing it.[58] Commercial heaters used for textiles emit radiation in the IR medium wave sector, and are generally designed to minimise variations. They are claimed to be 'colour-blind' and therefore heat up brightly coloured and darker areas equally. The main limitation to IR heaters is response to controls, especially if fabric speed varies or production is stopped for some reason. This will cause the fabric and the hot melt adhesive to be over- or under-heated, depending on the circumstances. Although switching the power off instantly lowers the heater temperature to below the melting point of most materials, some factories mount the heaters on rails which are automatically moved away from the fabric should the production line stop. Some control may be exercised by using heaters in banks with individual controls on each 'strip' of heater.

Commercial medium wave IR heaters may take 10 seconds to attain maximum temperature, but some manufacturers claim 66% output within five seconds. Medium wavelength IR heaters can attain approximately 800 °C maximum. Higher temperatures may be possible with short wave IR and the response time is quicker, but the heating element lifetime is usually considerably lower and the radiation is more sensitive to the colour of the material being heated. Needless to say, care is needed by operatives when working close to IR heaters and they must not be used for drying flammable solvents. Infra-red heaters are also used in stenter operations to partially preheat fabrics and water-based coatings to allow production speeds to be increased.

3.17.3 Powder scattering

Powder adhesive lamination is the most versatile, and probably the most economical, method of hot melt lamination, because powder can be applied at any optimised weight and width, and also because powder is not as costly as the corresponding web or film. Careful thought and pre-trials are needed to determine which of the substrates the powder should be scattered on. The usual procedure when laminating automotive fabrics to foam is to scatter the powder on to the face fabric first, because the unsupported foam on some machines is not capable of being self supporting. When the scrim is laminated, the powder must be put on to the foam because, in most cases, the scrim construction is too open for powder to be scattered on it. In this situation, however, powder may sink into the foam pores and be wasted and so the choice of particle size needs careful consideration. Smaller particles are not only wasted; they can also cause the foam to lose resiliency and reduce its porosity. Sometimes hot melt powders are supplied treated with an optical brightening agent so that the small particles are more visible under UV lamps that are fitted in the vicinity of the processing machine. The machine consists of a hopper containing the powder with a gravure roller at the bottom, the effective length of which can be controlled by the use of blanking off plates. The roller rotates and picks up powder, which is scraped off by a wire brush outside the hopper. The powder then falls onto the moving substrate below. A commercial machine is shown in Fig. 3.23.

The amount of powder applied is controlled by the speed of rotation of the gravure roller and the speed of the moving substrate. The substrate, with powder on it, then passes under IR heaters which melt the adhesive. The speed must not be excessive or the powder adhesive will not be melted sufficiently. The second substrate is then placed over the molten adhesive and the two materials are joined by bringing them together at a pair of nip rollers, or alternatively the substrates pass into a calender. When IR heaters and nip rollers are used, the same factors, relating to hot melt adhesives mentioned in connection with calenders (see Section 3.17.1), apply. The molten adhesive must have the correct viscosity at the temperature of bonding for satisfactory results, see Fig. 3.24. In addition, the correct balance

3.23 Powder scatter coating head. Photograph supplied by Web Processing of Whaley Bridge, UK and reproduced with kind permission.

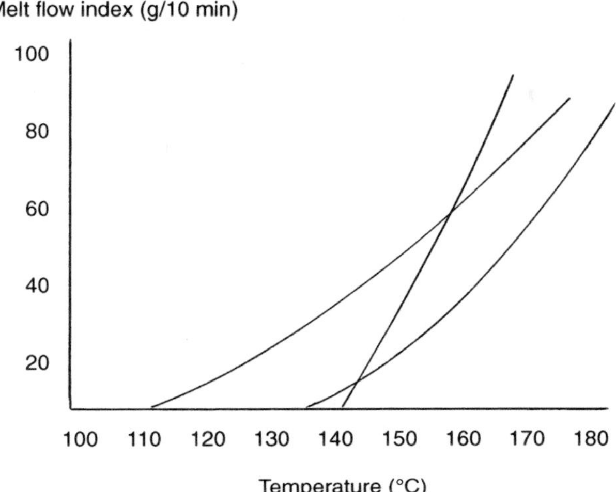

3.24 Typical melt flow index versus temperature curves for hot melt adhesives. Different adhesives can have very different flow rates which can influence the quality of the bond obtained. If flow is too high, at the temperature of joining, the adhesive may flow away from the interface, causing excessive penetration and a weak bond. The adhesive manufacturers will advise on suitable products.

of time (speed), temperature and pressure must be established for the actual substrates being joined and the powder adhesive being applied. As with calenders, too much pressure and time could cause the adhesive to strike through the substrates resulting in problems of appearance and stiffening. If the temperature is too low, the speed too fast and the pressure too low a poor bond could result, see Fig. 3.25. Careful cooling may also be necessary in order to avoid curl in the completed laminate – commercial apparatus is available for this. Any powder which does not fall on to the substrate can be collected in a tray underneath the machine and reintroduced into the hopper, thus minimising waste.

3.17.4 Dry powder printing (powder point)

Powder can be applied directly to the fabric by a dry printing technique using a gravure roller. This method is sometimes termed 'powder point' and is also known as the Intaglio Process. Hot melt powder of size 0–200 micrometres in a hopper fills the recesses of a gravure roller. This gravure roller is sometimes heated or warmed at temperatures just below the powder melting point to hold the powder together. A squeegee doctor blade removes the excess material so that powder is present only in the

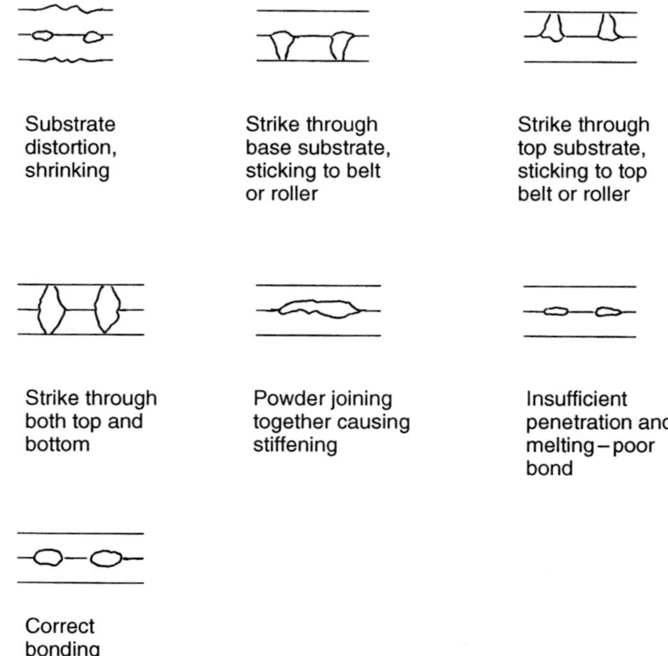

3.25 Fusing problems encountered with powder hot melt adhesives. Reproduced with kind permission of EMS-Chemie AG (Switzerland).

recesses of the roller. The fabric or substrate to be joined is heated to a temperature over and above the melting point of the powder by contact with a heated roller. As it touches the gravure roller, the powder in the recesses is melted and 'lifted out' on to the moving fabric in the form of an array of loosely sintered dots. The fabric then passes into a heated chamber or under IR heaters, which melts the dots of powder which then become firmly attached to the fabric. Some operators lightly calender the dots to smooth them. The material printed with adhesive dots is now ready for further processing, or it can be rolled up and stored for use at another time. This method is used extensively for preparation of garment interlinings.

3.17.5 Rotary screen dot printing

Powder can also be compounded into a paste for dot printing through a rotary screen, e.g. Stork apparatus. The benefits of this method have already been mentioned – lightweight delicate material can be processed, because the resin is placed on to the substrate as it moves forward at the same speed as the rotation of the rotary screen. A significant advantage is that there is minimum heating of the substrate and less risk of stiffening or discoloration. The application of heat after dot printing is only to dry off the water in the paste and not to sinter or melt the hot melt adhesive powder. With this process the fabric can also, if required, be rolled up after drying for reactivation at a later date by another unit or customer. This technique is widely used in the nonwovens industry. Preparation of the paste is a skilled compounding process, because it is necessary to produce a paste which has the correct viscosity, stability and flow properties as well as giving the adhesion required. The finest particle size powder is normally used, 0–80 micrometres, and mesh size must be pre-decided to obtain a laminate with the required handle and bond strength.

3.17.6 Doctor knife application of adhesive paste

Fine adhesive powder compounded into a viscous paste can also be applied by the direct coating method using a doctor knife. If the paste adhesive is selected carefully, it may be possible to replace the acrylic coating on automotive fabric with this hot melt adhesive and then use the same adhesive coating to laminate the fabric to foam or other substrate. Thus, the single hot melt adhesive coating does the work of both the acrylic coating (e.g. binding the tufts of velvet fabrics or improving the dimensional stability of woven or needlepunched fabrics) and the adhesive in the lamination process. In this way two processes are combined into one, saving time and expense. EMS-Chemie have hot melt adhesives and pastes suitable for this application.[59]

3.17.7 Melt print gravure – roller

In this process hot melt powder or granules are melted into a trough or transported to the trough via heated pipes. The 'trough' in many machines is the space formed by a doctor blade held against the roller, see Fig. 3.14. The doctor blade also scrapes off excess adhesive as the roller rotates. The gravure roller, which has point dots or a gravure pattern appropriate to the add-on required, picks up the adhesive and transfers it by a print process to one of the substrates. This material is then joined to the other substrate by nipping together.

Hot melt, moisture cure polyurethane adhesives, in the form of a gel, can be applied by this method. To prevent premature adhesive crosslinking, an inert gas blanket of nitrogen may be necessary. However, the latest hot melt, moisture curing polyurethanes are claimed to have an open time of two hours or more and may be processed without the need for an inert gas blanket. The moisture crosslinking polyurethane adhesive is piped to the blade/trough from a drum unloader. Adhesive add-on is controlled by adjustment of the roll blade angle, pressure and viscosity (accurately from about $3\,g/m^2$), and the adhesive can be applied in a variety of different dot sizes or patterns. Higher add-on is possible with

3.26 Moisture curing, hot melt polyurethane adhesives application head. Photograph supplied by Web Processing of Whaley Bridge, UK and reproduced with kind permission.

different rollers, and the commission laminator may need more than one gravure roller to cover a wide range. Clean down for moisture cure polyurethane adhesives is claimed to be much quicker and simpler than for hot melt adhesives, which may char if overheated and solidify when cold. Manufacturers of melt print gravure rollers believe that this method is the leading contender to replace flame lamination in large scale lamination of automotive fabric for car seats. A single roller probably covers the add-on range for seating fabric. A commercial machine is the subject of Fig. 3.26 and 3.27.

3.17.8 Hot melt – screen application

Hot melt molten adhesives have also been applied using rotary mesh screens. Mesh size and pattern are critical in producing a laminate with the

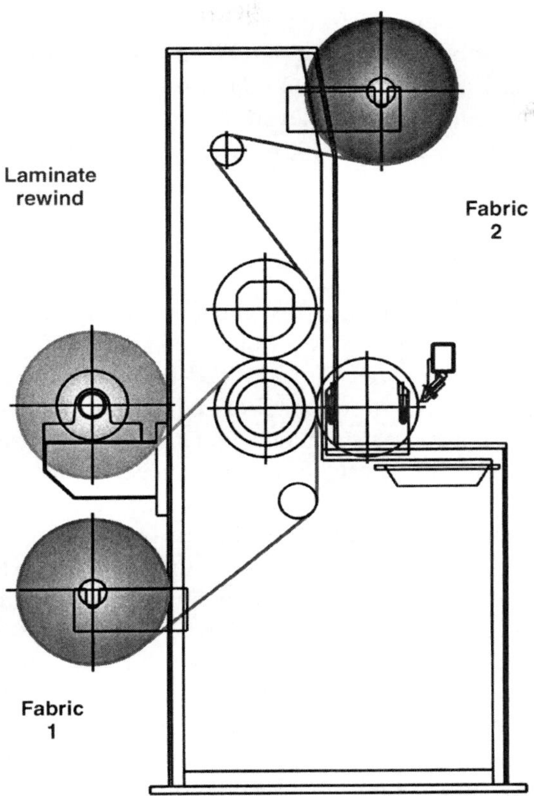

3.27 Schematic diagram of a moisture curing, hot melt polyurethane adhesives application head (supplied by Web Processing). The moisture curing, hot melt adhesive is first delivered to the coating head via heated jacketed pipes from a drum unloader, which is not shown in this diagram. The hot liquid adhesive is fed into a full width enclosure formed by a doctor blade and the gravure roller. Adhesive flows into the recesses of the gravure roller, and any excess is removed by the doctor blade as the roller rotates. The adhesive is 'printed' on to Fabric 1. Fabric 2, the second material (foam in an automotive fabric lamination process), is fed into the nip rollers, and the laminated material is taken up on the centre roll. In an actual commercial lamination system, the fabrics would be delivered to the coating head on fabric accumulators or even a stenter pin arrangement. For operators wishing to change from flame lamination to a moisture curing, hot melt system, it is only necessary to replace the actual flame lamination head with the gravure roller applicator. The existing feed and take-off apparatus can largely be retained – see also Fig. 3.30.

required bond strength and handle. Molten adhesive viscosity and temperature control are other factors to consider. One drawback to this process is the amount of down time necessary to clean the machinery when the adhesive or screen is being changed. It is believed that few – if any – operators are presently using rotary screens for hot melt adhesion application to textiles.

3.17.9 Slot die extruder

This type of machinery includes pumps capable of delivering liquid molten adhesives, jacketed or heated hoses and a coating head capable of delivering adhesive uniformly across the width of the goods. A drum unloader with a heated platen is required for adhesive in jelly form and a screw extruder for adhesive in powder or granule form. The machine is capable of extruding a continuous film of material but, to produce a flexible laminate, the adhesive is extruded in a discontinuous array of small dots or small streaks with add-ons as low as 2 g/m^2. This is achieved by reducing the delivery of polymer to the extruder head to a low rate – 'starving' the slot die – so that the polymer emerges in a discontinuous manner. Adjustment of add-on from very low levels is controlled simply by reducing the rate of delivery to the slot die. Again, the right balance of time, temperature and pressure are necessary in order to obtain just the correct bond strength without laminate stiffening or adhesive strike through, and again the adhesive must have the correct viscosity. The molten adhesive is applied to one of the substrates just in front of a pair of nip rollers and just before the second material is introduced. An important advantage of this method is that the substrates being joined are not themselves exposed to heat during the lamination process, and so there is minimal risk of damage to fabric aesthetics of texture and pile. The principles of this method are shown in Fig. 3.28 and 3.29.

Because the hot melt adhesive is totally enclosed right up to the moment before use, moisture curing polyurethane adhesives can be used with minimal risk of premature crosslinking. These adhesives are activated by moisture from the atmosphere and in the substrates themselves. The chemical crosslinking allows high bond strengths at low levels of add-on and, because they are also 100% active material with no solvents present, they are environmentally friendly.[60] Machine down time in this process is believed to be minimal because the adhesive is totally enclosed within the system and all parts are heated. A blanket of an inert gas such as nitrogen is sometimes necessary to prevent premature crosslinking of the adhesive during down periods, although the latest adhesives are believed not to require this. The Porous Coat™ system is a specialised slot die patented by Nordson for applying less then continuous films of material with add-on in the region of 2–30 g/m^2.[41]

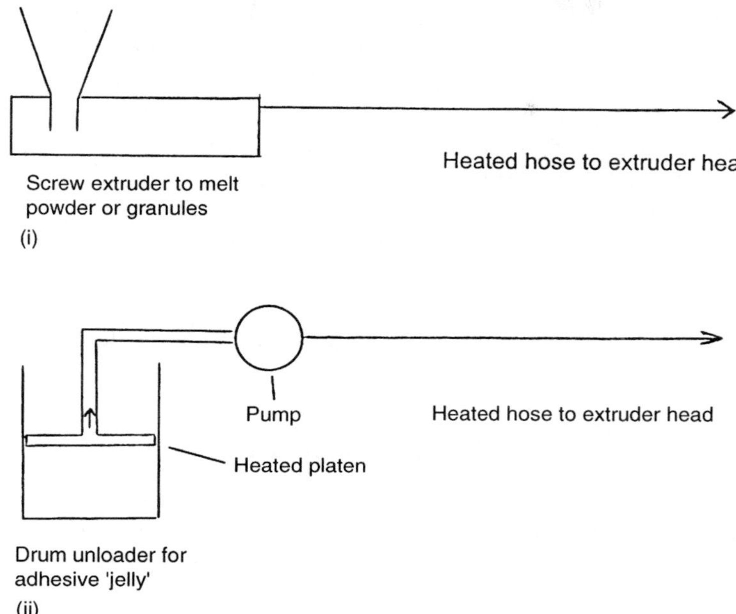

3.28 Slot die extruder applicator: (i) arrangements for adhesive powders or granules; and (ii) for moisture cure polyurethane adhesives.

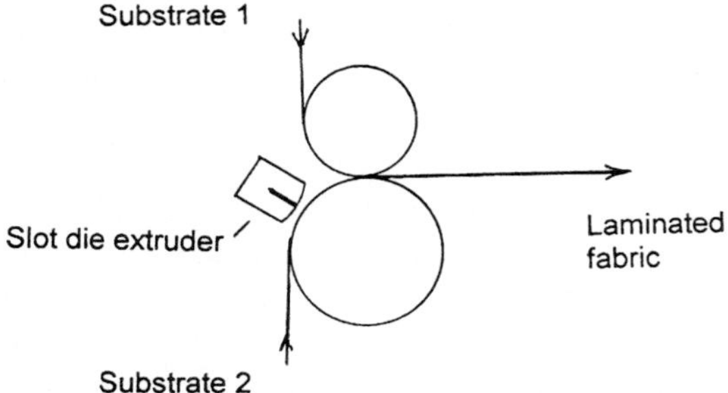

3.29 Schematic diagram of a slot die extruder head. An advantage of this method is that the substrates themselves are not heated directly.

3.17.10 Spray application

The problems usually associated with spray applications are uniformity and precision of application, penetration of the fabric substrate, occasional blocking of a spray nozzle, control of the liquid being sprayed – usually a solvent – and continuous drying of the liquid. In theory, all types of liquid adhesives can be sprayed – hot melt, solvent-based, water-based and high solids versions. In practice, however, hot melts need expensive apparatus to ensure that they do not solidify prematurely or char, solvents present problems of flammability and water-based adhesives may not dry at commercial speeds. In recent years, reactive polyurethane adhesives have been developed which allow high bond strengths with low levels of add-on. Moisture cure polyurethane adhesives do not need a high temperature to initiate crosslinking, are available as a jelly and are virtually 100% solids content.

Machtex of Holland have specialised in spray lamination equipment over many years and their machines can be used to process materials such as raised velvets or velours which would be damaged by high temperatures and pressure if laminated on a calender. Their CMW-3 machine is a multipurpose model suitable for several adhesive types.

3.18 Discussion of the various methods

A common feature of all lamination methods is that the substrates must be delivered to the lamination head, where they are actually joined together in a flat, tension free condition, if a quality, dimensionally stable laminate is to be produced. This is not easy with polyurethane foam and lightweight knitted scrim fabric, both of which stretch easily, especially at the lamination speeds necessary for a commercially viable process. Lamination is a critical process and must be carried out with precise control for consistent stretch and set properties, thickness and bond strength, and the ability to lie perfectly flat for accurate panel cutting. The machine feed and take-up units are just as important as the lamination unit itself. They need careful design and set up if lamination is to continue at commercial speeds without stopping the process when fresh rolls of face fabric, foam and scrim are joined on. These material handling logistics are common to all methods of lamination. Fabric surface geometry, handling considerations, volumes being processed, and available plant and personnel skills all contribute to the hot melt application method. A schematic multi-purpose lamination range is shown in Fig. 3.22 and the elements of a full lamination range are shown in Fig. 3.30. In all hot melt processes one of the main problems is rapid cooling causing loss of stickiness ('tack') of the hot melt adhesive in air before the second material can be introduced. This is particularly relevant for powder adhesive in the form of small individual particles which lose heat to the surroundings very quickly. The time the adhesive remains molten and tacky is referred to as 'open time'. Sometimes small secondary heaters, an IR bar or even a small ceramic heated bar, are used to provide extra heat just before the joining nip to prolong the open time.

A summary of the various lamination methods appears in Table 3.1. When the different methods are costed, *all* factors must be taken into consideration, i.e. cost of adhesives, energy consumption, speed of process, time

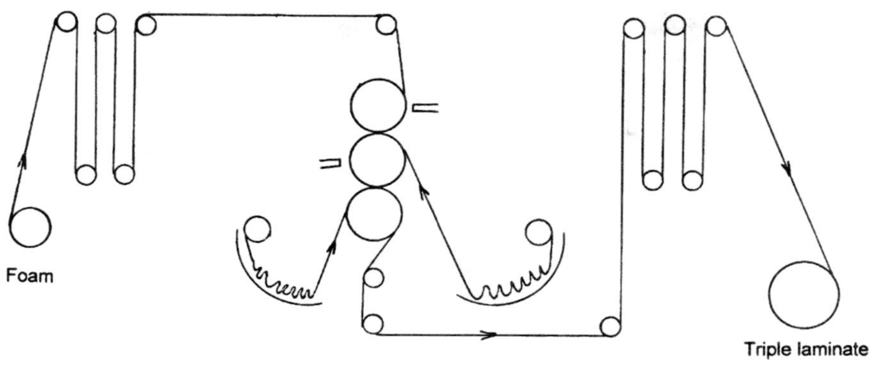

3.30 A schematic complete lamination range – example flame lamination for automotive tri-laminate fabric. A lamination range is made up of three elements (four if fume abatement is included), feed system, lamination head and take-up unit. This range shows a double headed lamination unit, but single heads are used and two passes are required for a tri-laminate. The actual flame lamination head can be replaced with a gravure roller, slot die extruder or some other method and the laminator operated with the existing feed and take-up systems.

Feed systems can include accumulators (only two units are actually shown in each accumulator but generally ten or more are used in practice) to allow batches of material to be joined on while the machine is in motion. Also included may be brake rollers, driven guide rollers and other tension control devices, scrays to hold fabrics in a relaxed state, stenter pins for knitted fabrics, anti-static bars to prevent fabric sticking to rollers and causing creasing, finger uncurlers for knitted fabric, bowed bars and cord expander rollers for films. The objective is to present the substrates to the lamination head in a crease-free and tensionless condition. Substrates joined with tension can produce dimensionally unstable laminates which have a tendency to curl and 'crack' when folded, and an inability to lie flat for panel cutting.

Take-up systems can also include accumulator units, tension and fabric guide rollers and a driven 'A' frame. It is important that crosslinking is complete, or at an advanced stage, before the joined materials are disturbed in any way. They must be held firmly together while the bond develops. Cooling systems are sometimes necessary with calender systems and some powder adhesive laminations.

spent on cleaning and maintenance, time spent when adjustments are made to produce different qualities of product, manning levels, 'burn off' of foam or reduction in thickness. Fume abatement – both initial installation and running costs – and costs resulting from the production of second quality material must not be overlooked. Other ancillary items which may also be required should be noted and included, e.g. storage facilities, especially for

Table 3.1 Summary of lamination machine types

Machine type	Adhesives used	Advantages	Disadvantages
Calenders, Flat bed laminators	Hot melt adhesives Powders Webs Films	Versatile – any shape or length of material may be laminated from leather hides, A4 size samples to short or long lengths of material. Process can be intermittent or continuous. Relatively inexpensive plant – but cooling plant likely to be necessary.	Risk of heat damage – heat is supplied to the adhesive through the substrates. Webs/films vary from moderate to high cost. Also webs/films available only in finite weights and widths. Care needed to prevent stiffening. Speed relatively slow.
Spraying	Water-based Solvent-based Hot melt	Good handle if penetration is controlled.	Water needs energy to dry off. Solvents need health and safety control. Precise control of spray area difficult. Risk of blocked jets resulting in delamination areas and down time for cleaning. Can be expensive.
Powder scattering	Hot melt powders of varying particle size (0–800 μm or more).	Generally good fabric handle. Powders generally inexpensive. No waste as add-on can be any amount and any width. Relatively inexpensive plant – but cooling may be required.	Powders may penetrate into substrate, e.g. foam pores, causing stiffening and waste.
Rotary screen	Water-based pastes Fine powders (0–80 μm) compounded into an aqueous paste. (Solvents/ hot melt adhesives are possible but rarely used.)	Good fabric handle. Delicate materials (e.g. lightweight fabrics and nonwovens) may be processed. Substrates may be pre-printed with adhesive for reactivation elsewhere. Moderately expensive plant.	Several screens may be needed to cover a range of adhesive add-ons and applications. (Solvents – wash off requires special area/ facilities/hot melt cleaning may be laborious – rarely used)

Dry powder roll printing (Powder point-Intaglio)	Hot melt powders 0–200 µm	Powder slightly less expensive than paste requirements. Good handle and breathability – used in garment industry for interlinings. Moderately expensive plant.	Fabric substrate is heated – risk of discoloration, stiffening. Cleaning time consuming. More than one roller necessary for wide production range.
Gravure roller printing	Hot melt adhesive granules or powder – melt pump required. Moisture cure polyurethane gels – drum un-loader required.	Good control of low add-ons. Minimum heating of fabric substrate. Good fabric handle. Reasonable production rates. Moderately expensive plant.	Cleaning may be laborious. Down times and stoppages may necessitate cleaning due to hot melt adhesive solidifying. Moisture cure polyurethane cleaning easier. More than one print roller may be necessary for wide production range.
Slot die extruder	Hot melt adhesive powders or granules – melt pump required. Moisture cure polyurethane gels – drum un-loader required.	Add-on easily varied to any amount from lowest to continuous film sheet. System totally enclosed – less risk of adhesive solidifying or premature crosslinking. Minimum heating of fabric substrate. Good fabric handle. Substrate aesthetics (pile/raised) unaffected – no actual heating of fabric substrates.	Expensive.
Flame lamination	Polyurethane foam Polyolefin foams. Hot melt webs/films	Excellent handle. Economical method at high production volumes. Convenient method making use of flame laminator if already available.	Requires careful maintenance and regular cleaning. Fumes from polyurethane foam need abatement. Polyolefin foams may require corona pre-treatment.

Notes: 1. Hot melt films and webs may be applied (with care and reduced flame size) by flame lamination.

2. Calenders and certain powder application ranges require additional plant for cooling of fabric and substrates after lamination. This is usually not necessary with flame lamination and with the slot die extrusion and gravure roller methods where minimum heating of the substrates occurs – but it is important to keep the substrates in close contact until a strong bond is formed.

flammable solvent adhesives, drum unloader apparatus, hot melt extruder and pumps in the case of hot melt adhesives. The higher energy charges required for drying off water-based adhesives should not be underestimated, and higher insurance premiums which may be required for flammable solvent-based adhesives should not be ignored.

For high production levels involving polyurethane foam, such as in the automotive industry, flame lamination appears still to be the most economical method for most producers, despite the increasing cost of emissions control. However, an important factor when comparing costs of a particular method with flame lamination is the cost of the foam which is burnt off. Speciality foams, such as high density types, are significantly more expensive than regular quality foams and the cost of burn off for these foams may significantly exceed the cost of some hot melt adhesives. Of the alternative methods, gravure roller and slot die extruding appear to be the most viable for high volume production, for example of automotive seats. The capital cost of gravure roller equipment is significantly cheaper than that of a slot die extruder, but the adhesive add-on is probably easier to control with the latter mentioned method.

3.19 Other related coating and joining processes

3.19.1 Flocking

Fabrics flocking is a relatively economical way of imparting an attractive finish and pleasant touch to fabrics.[61–67] Basically, the process attaches very short fibre lengths (from as short as 0.2 mm to over 8.0 mm), each fibre length standing on end, to a base fabric to produce a pile fabric which can also resemble a suede or velvet material depending on pile height. Virtually any fibre, natural or synthetic, can be made into flock and the feel of a flocked fabric is strongly influenced by its length and denier and also by the method of cutting. Quality flock is precision cut to a specified pile height, while lower quality flock used for toys and packaging is random cut. Precision cutting is achieved by cutting fibre tow with a guillotine or rotary cutter. Sometimes the fibre is cut wet to prevent build up of heat and the wet lubrication is helpful. Finishes are applied to the flock so that it performs satisfactorily during the flocking process, and it is dried in a manner for optimum moisture content.

The production method involves coating adhesive on to the base fabric and applying the flock by a continuous process, using either a mechanical beater bar or an electrostatic technique. If the flocking process is electrostatic, the electrical conducting properties of the adhesive are important. Water-based and solvent-based adhesives used include crosslinking acrylic, polyurethane, plastisol and latex PVC, SBR and butadiene acrylonitrile. They can be applied by any coating method, and by silk or rotary screen

3.31 A 6-hopper flocking machine. Photograph supplied by Web Processing of Whaley Bridge, UK and reproduced with kind permission.

printing methods for decorative effects. The adhesive chosen must have the required durability, dry cleanability and machine washability for apparel and furnishing fabrics. Early flocked fabrics suffered flock loss in cleaning treatments and discolored and emitted undesirable odours. If the adhesive layer is excessively thick, however, it can act as a barrier to perspiration in apparel and in upholstery fabrics and this may lead to thermal discomfort. After the flocking process, excess flock is removed by suction, the adhesive is dried and cured and then a final cleaning and removal of excess flock takes place. For three-dimensional articles such as gloves, a windblown technique is applied in which an air stream blows the flock on to pre-adhesive coated goods. Flocked fabrics can be further processed by brushing, embossing or printing. A commercial flocking machine is shown in Fig. 3.31 and a photomicrograph of a cross-section of flocked material in Fig. 3.32.

Recent development in car seat fabric involves polyester flock 0.5–1 mm long, 1.5–3.5 dtex, flocked on to polyester fabric. Novalis Fabrics (associated with Fiat and Rhone Poulenc) believe that flocked car seat fabric improves thermal comfort because millions of small flock fibres support the passenger, thus making it easier to transfer away sweat. Flocked fabric has many industrial uses, such as window seals in cars and application where flocked surfaces help to reduce noise. Domestic applications include curtains and upholstery fabrics, apparel and soft toys. Flocked inner surfaces of large tents and marquees are said to reduce dripping condensation.

Flocked yarns, not to be confused with chenille yarns, can also be produced by first coating a base yarn with an adhesive. The flock yarn and the

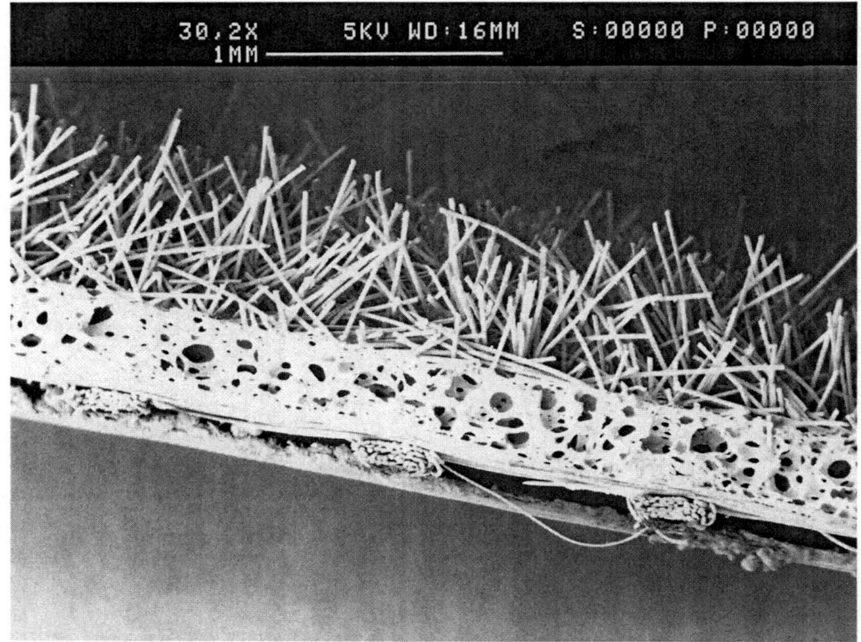

3.32 Flock on crushed foam coating. Photograph supplied by Noveon and reproduced with kind permission.

base yarn are then charged with opposite electrical charges and the flock yarn attaches itself vertically to the base yarn. For optimum abrasion resistance, such as for automotive applications, the flock fibre is nylon because it is more flexible and resilient than polyester.

3.19.2 Yarn coating

Some mention must be made of yarn coating which is used for the production of threads to be used in high speed sewing operations. Other products produced by yarn coating include dental floss, shoelace threads, braids, tyre cords and other industrial yarns. The Bates yarn coating and bonding machine, shown in Fig. 3.33, essentially comprises a yarn creel, a trough to contain the resin, a squeeze roller and an IR drying zone.[68] There are stainless steel bars inside the trough to control the resin pick-up by the yarn, and an oscillating full width comb which separates the yarns and prevents grooving of the nip rollers. The yarns are dried in a profiled, three zoned heater, i.e. the first zone is at a relatively low temperature, the middle at a higher one and the third and last at the highest. There is facility for the

3.33 The Bates Yarn Coating and Bonding Machine (YCB). Photograph supplied by Bates Textile Machinery and reproduced with kind permission.

yarns to be stretched by up to 30% by increasing the output speed relative to the input speed.

3.19.3 Welding

Welding of textiles and plastics is essentially a hot melt process and only thermoplastic materials can be welded directly.[69,70] The processes commonly used in textile and plastic processing are ultrasonic welding, where localised heat is produced by vibrational mechanical energy and HF (high frequency) and RF (radio frequency welding). In the latter two cases – which are similar processes – the localised heat is produced by molecules of the substrates vibrating rapidly in an applied alternating electrostatic field. The thermoplastic materials must first melt and then flow and form a bond. Materials which weld well by one method do not necessarily weld well by the other methods. PVC is by far the best material for HF welding but produces only acceptable results with ultrasonic welding. Hot melt adhesives can also be activated by welding techniques and used for materials which cannot be satisfactorily joined together directly. To be suitable for RF welding the material must be thermoplastic with a high 'dielectric loss factor' and must have the correct melt flow properties.

Welding is the most effective way of joining water-resistant materials together, because there are no sew holes. The seams of quality, water-resistant garments are covered with tape which is welded on using thermoplastic material to seal the sewing holes against penetration by water.

The Welding Institute has carried out a feasibility study on laser welding of garment panels to replace sewing in clothing manufacture. The process, called Clear-Weld™, is a joint development with the Gentex Corporation. The possibilities for use in water-resistant clothing have been recognised.[71]

3.19.4 Hook and loop fastenings (Velcro™ type products)

These relatively new and novel ways of joining components together are finding many uses in apparel and shoes, and have a wide variety of industrial applications. A significant advantage is ease of disassembly, making them attractive for such applications as display and exhibition board coverings and for fastening covers on to seating in aircraft. Hook and loop type fastenings, sometimes called 'touch and close', examples of which are the Velcro™ branded products, are especially suited to the car where ease of disassembly has become important. These fasteners are much stronger than many would believe and can be used for permanent joins – permanent until the end of the car's life. These materials are generally made from raised, knitted nylon 66, although polyester is sometimes used. They have many applications in the car, and a method of seat making has been developed which has the advantage of producing sharp, well defined, deep contours without any of the lifting or bridging problems sometimes associated with stretch fabrics. The hook part of the fastener is attached to the seat foam cushion and the loop part sewn to the cover. When brought together, a very strong join is produced. It is believed that hook and loop fastenings are being considered as a means of holding instruments to the dashboard in goods vehicles, which is likely to make disassembly much simpler. They could possibly have other applications in automotive construction, thus simplifying disassembly and recycling.

3.19.5 Other coating processes

There are material coating processes used in the film and paper industries which require expensive equipment and which must be carried out on a large scale to be commercially viable. Amongst these are powder coating, vacuum deposition, electrostatic deposition and sputtering techniques.[72] It is possible, for example, to produce highly reflective surfaces by these methods, but a smooth continuous surface is required and fabrics may not be suitable. However, papers or films which can be processed by these techniques may be laminated to fabrics to provide a smooth continuous surface. This can be produced on to the surface of a fabric by the transfer method, or by the direct method if sufficient polymer is applied. Fabrics have the performance advantage over both films and paper of being much more resistant to tear propagation. Alternative curing methods, such as UV curing, have also been researched and explored, but these have not so far been commercialised in the textiles area.[73]

3.20 References

1. Woodruff FA, 'Environmentally friendly coating and laminating; new processes and techniques', *JCF*, Vol 21 Apr 1992, 240–59.
2. Woodruff FA, 'Environmentally friendly coating and laminating; developments in machinery and processes', *Progress in Textile Coating and Laminating*, BTTG Conference, Chester 2–3 Jul 1990, BTTG, Manchester.
3. Lomax GR, 'Coated Fabrics: Part 1 – Lightweight breathable fabrics', *Textiles*, Vol 14 No 1 Spring 1985, 2–8 and 'Part 2 – Industrial uses', *Textiles*, Vol 14 No 2 Summer 1985, 47–56.
4. Lomax GR, 'Coating of Fabrics', *Textiles'*, Vol 2 No 2 1992, 18–23.
5. van Parys M, *Coating*, Euratex (EEC Comett), Guimaraes 1994, 91–101.
6. Whiteman R, 'On line gauging, control and benefits for the coating and laminating process', *JCF*, Vol 23 Oct 1993, 87–104.
7. Clifford CF (Gaston County), 'Foam finishing technology: the controlled application of chemicals to a moving substrate', *Textile Chemist and Colourist*, Vol 10 No 12 Dec 1978, 37–40. Also in *American Dyestuff Reporter*, Jan 1979, 32–6, 41.
8. Namboodri CG and Duke MW, 'Foam finishing of cotton-containing textiles', *Textile Research Journal*, Mar 1979, 156–62.
9. ICS_Texicon Technical Information Brochures.
10. Anon, 'A foam finishing success story (Datacolor International Autofoam finishing), *Textile Month*, Jul 1993, 38–9.
11. Leigh I, 'Foam textile backcoatings – putting the technical performance into textiles', *TTi*, Feb 1998, 18–21.
12. Davids J (Rohm and Haas), 'Crushed foam coating', *Textile Manufacturer*, Dec 1974, 47–9.
13. Rohm and Hass, *Speciality Chemicals for the Textile Industry*, Brochure, May 1995.
14. Woodruff FA, 'Some developments in coating machinery and processes,' *Developments in Coating and Laminating*,' Shirley Institute Conference, Manchester 27 Mar 1981, Shirley Institute Publication S41, BTTG, Manchester.
15. Werner M (Ciba-Geigy), *Foam Coating*, Technical Information Brochure, 1985.
16. Kubin I (Ciba) 'Function and fashion coating for apparel', *Melliand English*, Jun 2001, E124–E127.
17. Durst P (Texas Adhesives), 'PU transfer coating of fabrics for leather-like fashion products', *JCF*, Vol 14 Apr 1985, 227–41.
18. Keeley V, 'Transfer coating with polyurethanes', *Progress in Textile Coating and Laminating*, BTTG Conference, Chester 2–3 July 1990, BTTG, Manchester.
19. Lerner A (Millmaster Onyx), 'The effect of physio-chemical characteristics of urethane polymer solutions on the transfer coating process', *JCF*, Vol 5 Apr 1976, 248–67.
20. Linden B and Schmel F, 'Soft leather substitute materials and their impact on the international leather and leather products trade', *JCF*, Vol 14 Jul 1984, 9–35.
21. Borri C, 'Comfort for car interiors – Alcantara – features and advantages of a unique product', *IMMFC*, Dornbirn 22–24 Sept 1993.
22. Tanaka N and Tanaka J (Kuraray), 'New man-made leather for automobiles', *IMMFC*, Dornbirn 15–17 Sept 1999.
23. van Parys M, *Coating*, Euratex (EEC Comett), Guimaraes 1994, 146–50.
24. Durst P, 'The PU coagulation process and its success in PUCFs', *JCF*, Vol 13 Jan 1984, 175–83.

25. Zorn B, 'Porous polyurethane films and coating', *JCF*, Vol 13 Jan 1984, 166–74.
26. Hemmrich J, Fikkert J and van der Berg M (Stahl), 'Porous structured forms resulting from aggregate modification in polyurethane dispersions by means of isothermal foam coagulation' *JCF*, Vol 22 Apr 1993, 268–78.
27. Mann A, 'Hot melt coating techniques of the Bema and the Zimmer machine', *JCF*, Vol 3 Jul 1973, 31–7.
28. Crowther BG and Edmonson HM, 'Calendering', in *Rubber Technology and Manufacture* (Editor Blow CM), Newnes-Butterworth (Plastics and Rubber Institute) London 1971, 277–82.
29. Stern HJ, *Rubber, Natural and Synthetic*, Second Edition, Maclaren, London 1967, 325–31.
30. Meulen BH Ter (BF Goodrich), 'Industrial TPU applications in Europe', *JCF*, Vol 12 Jan 1983, 148–59.
31. Boron FT, 'Thermoplastic polyurethanes – a versatile material for industrial coated fabric applications', *AATCC Symposium, Coated Fabrics Update*, Boston MA, USA, 31 Mar–1 Apr 1976, AATCC, Research Triangle Park NC, USA.
32. Pittenger F, 'Preparation of polyurethane coated fabrics by film laminates', *AATCC Symposium, Coated Fabrics Update*, Boston MA, USA, 31 Mar–1 Apr 1976, AATCC, Research Triangle Park NC, USA.
33. Zimmer JPM and Mayer K, 'Versatile machine developments in coating technology', *JCF*, Vol 26 Jan 1997, 188–211.
34. Cuff JJ, 'An up-date of rotary screen technology for coated fabrics; case studies in the USA market', *7th Annual International Symposium on Coating and Laminating*, Charlotte, NC USA 17–18 Nov 1997, Technomic, Lancaster PA, USA.
35. Bell JE, 'The rotary screen coater and protective surface coating decoration-creation', *JCF*, Vol 13 Jul 1983, 35–49.
36. US Patents 2 957 793 (1956) and 3 057 768 (1957), Dickey JW, assigned to Reeves Brothers Inc., New York.
37. Webb J, 'Polyurethane plant emission control', *Urethane Technology*, Oct/Nov 1995, 20–22.
38. Webb J, 'Flame laminators – watch out you are on the list', *Urethane Technology*, Oct/Nov 1995, 23.
39. Garner C, 'The low down on flame laminating, *Inside Automotives*, May/Jun 1995, 23–5.
40. Garner R, 'Flame or dry? the debate is on', *Automotive & Transportation Interiors*, May 1994, 52–4.
41. Gillessen G (Nordson), 'Flame lamination and alternative technology', *6th International Symposium on Coating and Laminating*, Dusseldorf 4–5 Nov 1996, Technomic, Lancaster PA, USA.
42. Miles DC (Dermil), 'Dry powder bonding adhesives in automotive trim laminates', *JCF*, Vol 20 Apr 1991, 229–39.
43. Hopkins J (Nordson), 'A comparative analysis of laminating automotive textiles to foam', *JCF*, Vol 24 Jan 1995, 250–67.
44. Woodruff FA, 'The impact of legislative and technological changes on coated derived processes and machinery', *Industrial, Technical and High Performance Textiles*, World Textile Congress, Huddersfield University 15–16 Jul 1998.
45. Halbmair J (Bostik), 'Overview of hot melt adhesives application equipment for coating and laminating full-width fabric', *JCF*, Vol 21 Apr 1992, 301–10.

46. Rossitto C, 'Laminating with hot melt adhesives', *JCF*, Vol 16 Jan 1987, 190–98.
47. Schnerring K, 'Hot melt horizons,' *Textile Month*, Mar 2001, 17–20, 48.
48. Deubelbeiss J, 'Hot melt, the reliable route for technical textiles', *Textile Month*, Mar 2001, 22–4.
49. Field I, 'Why is the industry turning to dry heat lamination technology?', *TTi*, Sept 2000, 21–4.
50. Painter CJ, 'Waterproof, breathable fabric laminates; a perspective from film to market place,' *JCF*, Vol 26 Oct 1996, 107–30.
51. Luchsinger MH, 'General overview of fusible powder equipment', *International Textile Bulletin*, 1/1989, 5–16.
52. de Jong E, 'Polyamide hot melt adhesives for bonding textile fabrics', *International Textile Bulletin*, 1/1978, 73–8.
53. McBride R and Sellers J, 'Flame lamination meets environmental challenge', *Automotive & Transportation Interiors*, Apr 1994, 60.
54. Lebovitz R, 'Lamination process meets the global manufacturing challenge', *Automotive & Transportation Interiors*, Dec 1997, 40–41.
55. Griltex EMS, Hot Melt Adhesives Manual, 500/3.95/SuT.
56. Bandwise Reliant, Technical Information Brochure, *Coolstream LSTF*, May 1997.
57. Virkna A, 'The adhesive properties of interlinings', *Textiles Magazine*, Issue 3 2000, 8–12.
58. Krelus AG, Switzerland Technical Information Brochure.
59. EMS Chemie Sales Service Letter TF & A No 18, *Back Coating of Decorative Textiles*, Aug 1998.
60. Green D, 'A new beginning in the fabric laminating industry – reactive PUR adhesives', *JCF*, Vol 28 Oct 1998, 116–25.
61. Bolgen SW, 'Flocking technology', *JCF*, Vol 21 Oct 1991, 123–31.
62. Smith TL, 'New technologies expand applications for flocked materials', *Automotive & Transportation Interiors*, Oct 1995, 38–9.
63. Maag U, 'Flock-textile coating, a solution for many technical problems', *IMMFC*, Dornbirn 15–17 Sept 1999.
64. Higginbotham RS, 'Flocking', *Textiles*, Vol 7 No 3 Oct 1978, 58–64.
65. Bolgen SW (American Flock Association), 'Flocking technology', *JCF*, Vol 21 Oct 1991, 123–31.
66. Bianchi JP (Novalis Fibres), 'Flock technology for car interiors', *IMMFC*, Dornbirn 17–19 Oct 1997.
67. Bianchi JP, 'Flock technology for car interiors for automotive textiles', *IMMFC*, Dornbirn 17–19 Sept 1997.
68. YCB, Yarn coating and bonding machine. Technical Information leaflet. Bates Textile Machine Co (Leicester) Ltd.
69. *Dielectric Heating for Industrial Processes Handbook*, Union Internationale d'Electrothermie, Paris 1992, 9–11 and 32–3.
70. Anon, 'How physical properties affect ultrasonic welding', *BPR*, Jun 1988, 41–5.
71. Anon, 'Welding process for clothes manufacture', *TTi*, Jan/Feb 2001, 7.
72. Bellemare D, 'Powder deposition techologies give advantages, options', *Textile World*, Aug 2000, 88–91.
73. Bluestein C, 'Alternative curing methods', *JCF*, Vol 25 Oct 1995, 128–36.

3.21 Further reading

1. *AATCC Conferences on Coating and Laminating Papers*, AATCC, Research Triangle Park NC 27709, USA.
2. *Annual International Conference on Coating and Laminating* papers, organised by Technomic, Lancaster PA, USA from 1990 to date.
3. Anon, 'A new era of surface effect finishing', *International Dyer*, Apr 1996, 29–33, 40.
4. Anon, 'New wet sueding development', *International Dyer*, Apr 1996, 38.
5. Barden B, 'Coated fabrics', *Kirk Othmer, Encyclopaedia of Chemical Technology*, Fourth Edition, Vol 6, John Wiley, New York 1993, 595–605.
6. BTTG Conference, *Progress in Textile Coating and Laminating*, Chester 2–3 July 1990, BTTG, Manchester.
7. Carty P and Byrne MS, *The Chemical and Mechanical Finishing of Textile Materials'*, Newcastle upon Tyne Polytechnic Products Ltd, Newcastle 1987.
8. Cohen ED and Gatoff EB, 'Coating processes', *Kirk Othmer, Encyclopaedia of Chemical Technology*, Fourth Edition, Vol 6, John Wiley, New York 1993, 606–35.
9. Fung W and Hardcastle M, *Textiles in Automotive Engineering*, Woodhead, Cambridge 2001.
10. Hoffman WR, 'Industrial coating and laminating; functionality and limitations', *JCF*, Vol 23 Oct 1993, 124–30.
11. Jordan M and Tyson M, 'High performance coating technologies in the 21st century', *Industrial, Technical and High Performance Textiles*, World Textile Congress, Huddersfield University 15–16 July.
12. Lomax GR, 'Coated Fabrics': Part 1 – Lightweight breathable fabrics', *Textiles*, Vol 14 No 1 Spring 1985, 2–8 and 'Part 2 – Industrial uses', *Textiles*, Vol 14 No 2 Summer 1985, 47–56.
13. Rosato DV, *Plastics Processing Data Book*, Second Edition, Chapman & Hall, London 1994, see 'Coating', 392–416.
14. Shirley Institute Conference Papers, *Developments in Coating and Laminating*, 27 Mar 1981, Shirley Institute Publication S41, BTTG, Manchester 1981.
15. Tatlor DJ, *Lathams Technology of Clothing*, Third Edition, Blackwell, Oxford 2000.
16. Teumac FN (Reeves Bros), 'Coated fabrics', *Encyclopaedia of Polymer Science and Engineering*, Second Edition, Vol 6 (Editors Mark HF, Bikales N and Kroschwitz JI), John Wiley, New York 1986, 632–45.
17. Tortora PG and Collier BJ, *Understanding Textiles*, Prentice-Hall, New York 1997.
18. Van Parys M, *Coating*, Eurotex (EEC CometT Program), Minho, 1994.
19. Williams J, 'Advanced bonding systems – a review', *JCF*, Vol 26 Apr 1997, 286–92.
20. Woodruff FA, 'Environmentally friendly coating and laminating: new processes and techniques', *JCF*, Vol 21 Apr 1992, 240–59.
21. Woodruff FA, 'Coating, laminating, bonding, flocking and prepregging', unpublished book chapter, Web Processing, Whaley Bridge, UK 1994.
22. Wypych J, *Polymer Modified Textiles*, John Wiley, New York 1988.

4
Products from coated and laminated fabrics

4.1 Protective clothing – sports and industrial

4.1.1 Introduction – general requirements of garments

A principal requirement of all garments, whether they are worn for protection for sports activities or for work, is *comfort* in all its interpretations, physical and psychological.[1] Comfort is a well-researched area, with many articles, research papers and textbooks (see further reading) having been written on the subject or on aspects of it. Some of these are discussed in Chapter 5 and also later in this chapter in connection with various coated products. To summarise, the garment should be light in weight, flexible without restricting body movements, the correct size for the person wearing it, in a design and colour appropriate for the activity and acceptable to the wearer, and breathable with minimum restriction on the passage of perspiration. Finally it should possess all the properties needed to give the type and degree of protection required. 'Mass customisation' requirements of exact garment fit, garment accessories such as hood, pockets, zips, etc. in the fabric design and colour requested by the individual customers, combined with quick delivery, are likely to arise in the foreseeable future. This will apply not only to the fashion and retail market, but also, eventually, to the industrial protective garment sector.

Protective clothing is used to protect the human body from the elements of sun, wind, rain, chemicals, heat and fire, biological agents, microbes, molten metals, impact, sharp cutting edges, spikes, etc. Motorway workers wear garments for high visibility, the military wear uniforms with disruptive print designs, for low visibility. Breathable waterproof materials have been developed for thermal comfort, but some researchers believe that imaginative garment design can assist with this as well as with comfort in the wider sense. In addition to all the above-mentioned factors, the garment should be easily made up, affordable, durable when worn, washable and should not deteriorate in storage. As will be seen, meeting any single one of the criteria is relatively simple; it is combining all or most of these requirements into a marketable garment that presents the challenges. These comments apply to all protective clothing, including that for medical and military use which is discussed in Sections 4.7 and 4.8.

British Standard BS 3545 discourages the term 'waterproof' for protection from penetration by liquid water and recommends instead, 'water resistant'. The reason stated is that it can be confused with the hydraulic bursting strength of the fabric. However, waterproof is used everywhere in the trade and in American standards and is therefore used in this book.

Some mention should be made of skin cancer which seems to be on the increase. Research into protection against skin cancers associated with UV radiation is being carried out, notably at the University of New South Wales in Australia. White-skinned populations in tropical regions of the world are especially at risk, but skin cancers are also quite high in Scandinavia. Depletion of the ozone layer is likely to increase the risk. Australians have the highest incidence of sun-induced skin cancers in the world, and it is estimated that two thirds of the population of that country will develop some form of skin cancer during their lives. Understandably Australia is the first country in the world to develop a standard for UV protection by apparel fabrics.

4.1.2 Apparel fabrics, leisure and sports protective clothing

Fabrics for clothing are coated or laminated mainly to enhance their ability to protect the human body from the elements, especially wind and rain. The degree of protection is sometimes classified into three levels of performance, showerproof, rain repellent and waterproof. The last mentioned gives the highest level of protection. 'Showerproof' fabrics should be expected to offer protection only against light drizzle, whilst 'rain repellent' should be capable of withstanding 'fairly heavy' rain.[2–4] The terms rainproof, stormproof and weatherproof are also used. All of these terms must be used with a certain amount of caution, especially by the seller of the material, because they could imply total protection from wet weather. The outdoor garment trade, however, recognises that material with a hydrostatic head of at least 100 cm of water is necessary for the term 'waterproof' to have any real meaning. Hydrostatic head, the test for water resistance, is explained fully in Chapter 5.

Protection from the elements is required in different degrees, from a light shower of drizzle to prolonged heavy rain, by everyone in everyday life. People who have to work in poor weather conditions, such as electricity repair persons, soldiers or sports persons, require the highest level of protection. Quality fabric sold by reputable companies is generally tested to a defined standard of performance. Of the three levels, shower resistant and waterproof are the best known; 'rain resistant' seems to be rarely used. The three levels describe different degrees of protection from the weather.[2–4]

Keeping dry is important not only for comfort, but health, safety and actual survival may depend upon it. Clothing may lose much of its thermal insulation properties when wet because the filling loses its thickness, and also because heat is conducted away faster by water than by air. A water-resistant garment depends not only on the quality of the material but also on the way the garment is made up. It is especially important that the seams are sealed effectively, preferably by tapes which are themselves waterproof. The tapes are produced from hot melt material. Tape sealing machines function by heat and pressure settings which must be optimised for best results. Some garment makers 'dope' the seams, i.e. paint them over with a compound, which is usually solvent based. This is labour intensive because the garments have to be laid out to dry before they can be packed.

Waterproof protective clothing is produced from nylon or polyester fabric coated with either acrylic, polyurethane, PVC, or a rubber-based resin such as Neoprene. The least expensive material is acrylic, usually coated on 64 g/m^2 ('2 oz') fabric with a total weight of perhaps 85–90 g/m^2. Middle order material is 128 g/m^2 base fabric, generally coated with polyurethane or PVC, whilst heavy duty material is 220 g/m^2 ('7 oz') base fabric coated with polyurethane, PVC or a rubber. These polymers are direct coated on to woven fabric using either water-based or solvent-based resins, usually in two or three layers. The coating is applied to the side of the fabric that becomes the inside of the garment. These 'regular' coatings have good abrasion resistance and it is not necessary to protect the coated surface with fabric lining. This is not the case with 'breathable' materials as will be seen below. Polyurethane is also applied by transfer coating to knitted fabrics to produce much softer and flexible coated fabric but, in this case, the coating is applied to the side which becomes the outside of the garment.

All coated fabric is generally treated with a water-repellent finish on the other side of the fabric, i.e. the outer face side. The water repellent is usually a fluorocarbon finish, but sometimes it is a silicone. Note that, because of its release properties, the silicone must be applied after the completion of coating. The water-repellent finish is necessary so that water runs off the surface with minimum soaking into the fabric. When the garment is new, rain water 'pearls off' without any wetting out of the surface. This finish, however, loses its effectiveness after a period of time, either through wearing and abrading off or because of soiling contamination. Manufacturers sometimes recommend its refurbishment by using finishes sold in aerosol cans or by immersion in a proprietary product. This point is discussed further in Section 4.1.3.

It is important that waterproof material remains waterproof after wear and tear for a reasonable length of time and after washing or cleaning.

Quality control tests for water resistance are, therefore, carried out on the coated fabric, 'as received', after flexing on a Schildknect or Crumple-flex machine, after abrasion and sometimes after a washing or dry cleaning test. The weight is checked to ensure that the correct amount of coating has been applied and a peel bond test is used to check for good adhesion. Poor adhesion results in delamination of the coating; the flexing or washing tests carried out during quality control should reveal this. If the coated fabric is stiffer than usual, this may indicate excessive penetration by the resin which could mean lower tear strength and possibly lower hydrostatic head.

4.1.3 Breathable waterproof materials

Regular polyurethane, PVC or rubber-coated fabric is waterproof but does not 'breathe', i.e. it does not allow perspiration to pass through to enable body cooling. This perspiration condenses on the inside of the garment making the wearer quite wet from his or her own perspiration. During a shower, this liquid can be very easily be mistaken for rain leaking into the garment, and many people are surprised at the quantity of water exuded from their own bodies. The hotter the weather, or the more strenuous the activity, the more the body sweats and produces perspiration. This was an insoluble problem for many years, until the early 1970s when WL Gore introduced their Gore-Tex material, which has a very high standard of water resistance combined with a high level of breathability. Gore-Tex is a PTFE film with nine billion tiny micropores in every square inch ($6.45\,cm^2$); these are 20000 times smaller than a drop of liquid water, but 700 times larger than a water molecule. Thus they are too small to allow liquid water to pass through, but large enough to allow the passage of molecules of water vapour, see Fig. 4.1 and 4.2. Gore-Tex, however, was – and still is – an expensive product, and so, over the next decade or so, there was a period of intense activity by chemical material manufacturers to develop an alternative material with a similar high standard of performance but at a lower price.

The next breathable waterproof material offered for sale was probably Entrant, a microporous polyurethane coating produced by Toray Industries. Soon after, a large number, probably dozens, of other materials appeared for sale, all claiming to be both breathable and waterproof. When tested, however, many could not be classed as 'waterproof', having a hydrostatic head of only 60 cm of water or even lower – it was not surprising that they had a high level of breathablity! The next commercially successful breathable waterproof products on the market were two polyurethane coatings, one developed and sold by Union Chimique Belge (UCB) (Belgium) and one invented by the Shirley Institute (later British Textile Technology Group – BTTG) and marketed by Baxenden Chemicals. These last two

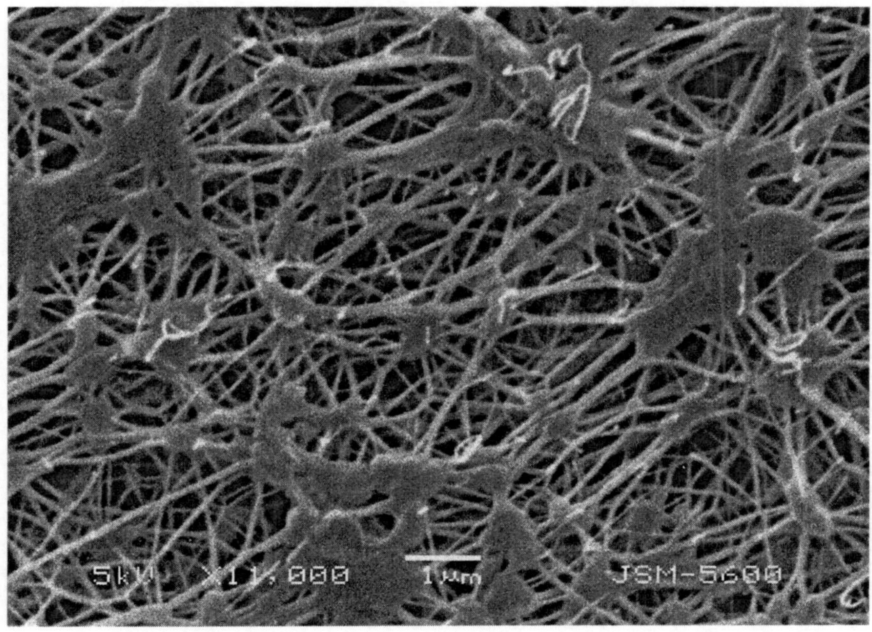

4.1 PTFE microporous film magnified 11 000 times showing that the pores are too small to allow drops of water to pass though but large enough to allow the passage of molecules of water vapour. Photograph supplied by BTTG and reproduced with kind permission.

products can be applied in the same way as regular solvent-based polyurethane, i.e. by direct coating. The UCB product is microporous but the BTTG product has no micropores. It is a solid hydrophilic polyurethane coating through which water molecules move using hydrophilic groups in the molecular chain, rather like 'stepping stones', to reach the other side, see Fig. 4.3 to 4.5. The driving force for this action is the high relative humidity near the human body on the inside of the garment and the lower humidity on the outside. Baxenden now have both water-based (Eco Dry) and solvent-based (Witcoflex Superdry) resins for the production of breathable waterproof garments by direct coating. Witoflex Superdry can also be used in transfer coatings.

In the late 1970s, AKZO launched their Sympatex polyester film material, a 'solid film' type product which has proved to be a commercial success. A short while later, Porvair launched their Porelle microporous polyurethane film. Since this time there has been a large number of new coatings and films with waterproof and breathable properties. When waterproof breathable products first appeared in the late 1970s and early 1980s, there was much discussion on the test method for breatheability.

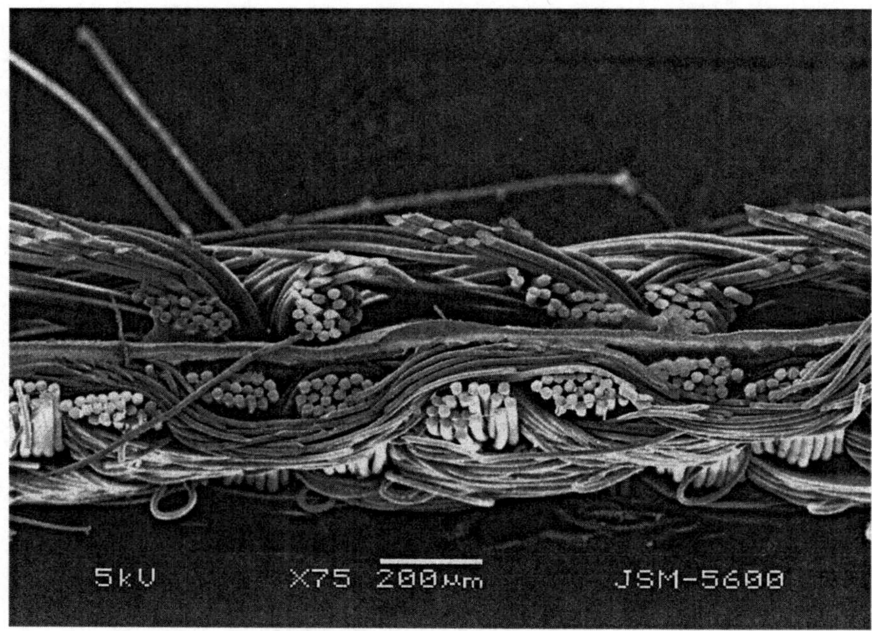

4.2 Cross section of a tri-laminate incorporating a PTFE microporous film magnified 75 times. Photograph supplied by BTTG and reproduced with kind permission.

Different methods produced different results and could even change the ranking when a series of different products were tested. Naturally, manufacturers favoured the test method which showed their particular product to be the best or to rank highly. The new materials became the subject of numerous articles in the popular outdoor sports press and 'the battle of the breathables' was a feature of the Camping and Outdoor Leisure Association exhibition held at Harrogate in October 1985.[5–10] There was a certain amount of confusion amongst both customers and manufacturers regarding test result figures and performance claims. Eventually, the British Standards Institute (BSI) convened a panel of experts to produce an impartial test method for breathability and to set durability standards; the result was BS 7209; 1990. The Hohenstein Institute developed their sweating hot plate method which has been shown to agree with actual wear tests. More products have been introduced and existing ones improved but, even at the time of writing, confusion still exists and test method development and assessment still continue to occupy the time of research laboratories.[11–20] The whole concept of comfort continues to be researched in detail and, furthermore, dynamic test methods under simulated conditions of actual use and whole clothing assemblies are being investigated. Methods of mea-

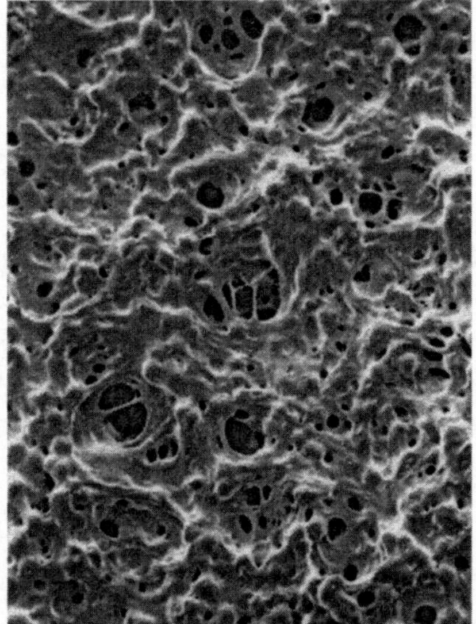

4.3 Surface of a typical poromeric film or coating showing the micro pores. Photograph supplied by Baxenden Chemicals and reproduced with kind permission.

(a) 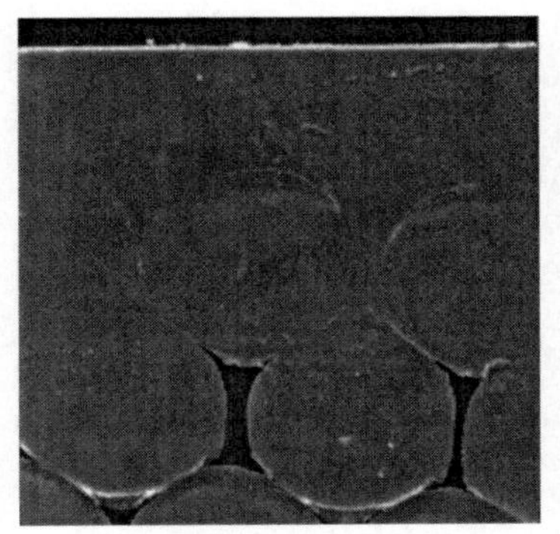

(b) 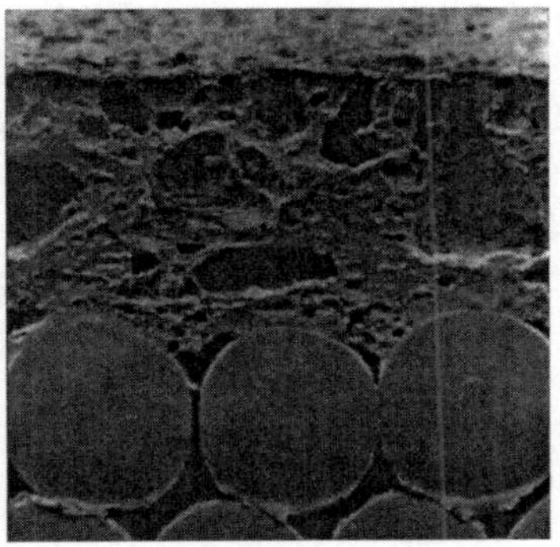

4.4 Cross section of (a) Witcoflex SuperDry 'solid film' coating and (b) a typical microporous coating. Both coatings are waterproof and breathable, but breathe by different mechanisms. Photographs supplied by Baxenden Chemicals and reproduced with kind permission.

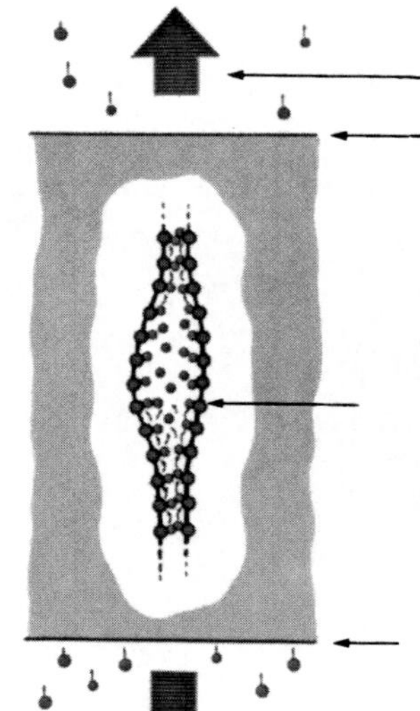

4.5 The breathing mechanism of 'solid film' coatings and films. Artwork reproduced with kind permission of Baxenden Chemicals.

suring the breatheability of fabrics and factors influencing breatheability and thermal comfort are discussed more fully in Chapter 5.

Manufacturers still debate the relative merits of the two types of 'breathing mechanisms', microporous film and solid film, which is also referred to sometimes as 'monolithic' film. It is claimed that the micropores can become contaminated with body oils or surfactants and thus allow water through. In addition, micropores can become enlarged when garments are stretched at elbows and knees and again let water through. In addition to performance, the durability and physiology of protective clothing are being assessed, together with the ecological impact of both its manufacture and eventual disposal.[21,22]

Many of the up-market garments now produced are made from films such as Gore-Tex and Sympatex (AKZO). They are laminates made from waterproof breathable film joined to the garment face fabric for protection and support. In some cases, a second scrim fabric is joined to the other side of the film to produce a three-component laminate. This is necessary to protect the thin film from damage. However, in recent years this practice seems to have been replaced more and more by a method in which the third component is not joined to the fabric/breathable waterproof membrane (two-component laminate), but is used as a loose lining material. Generally, it

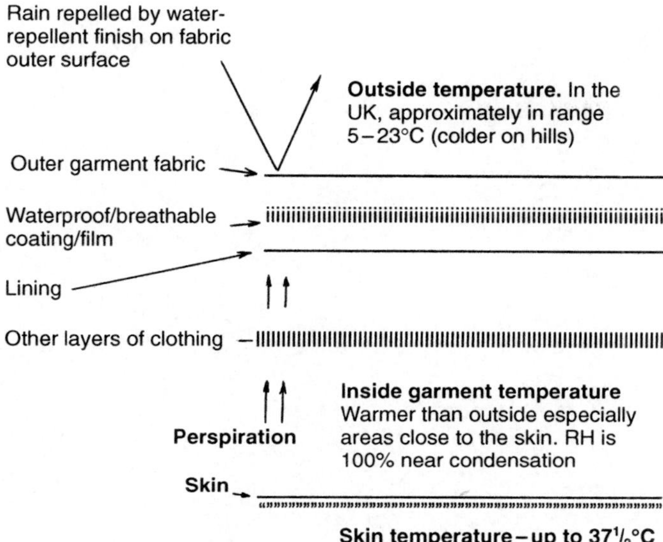

4.6 The clothing assembly for protection from the elements. The first line of defence is the outside surface of the fabric. This offers protection from milder wind and rain. The water-repellent finish (when new) repels water and prevents it being absorbed by the fabric. The coating or film protects against more severe wind and rain. The human body perspires to promote cooling, and for thermal comfort perspiration must evaporate and be removed from the vicinity of the skin to alleviate dampness discomfort. Moisture management considerations become important, i.e. wicking and transport of liquid water to remove it from the skin to the inner surface of the outer garment and then passage in the form of water vapour through the breathable coating or film to the outside atmosphere. This process may be complicated by other layers of clothing, and these and the lining of the outer garment play a role in moisture management in removing water away from the skin surface; fibre type and whether it is hydrophilic or hydrophobic is also relevant.

consists of an open mesh knitted fabric which is soft and flexible and has an open construction which does not hinder the passage of perspiration. In fact, the lining material is believed to help the overall performance of the microporous membranes by soaking up some of the condensation on the membrane and removing it from the surface (see Fig. 4.6 and 4.7). Moisture management is important for comfort and is the subject of research.[23,24] Microporous membranes are thought to function less effectively when the surface is covered with a layer of condensation. Solid film hydrophilic films, on the other hand, are thought to work less efficiently if condensation is drawn away from the surface. The lining material should be chosen to optimise the overall performance of the garment. Bi-laminated fabric

4.7 Leisure protective clothing, the 'Lady Linn 370'anorak by Vander of Newton Abbot, UK. The jacket is in breathable Cyclone (Carrington) material and features taped seams and a mesh lining. Photograph supplied by Vander and reproduced with kind permission.

has a significantly softer and more flexible handle than the traditional trilaminated material. In some cases, the waterproof breathable membrane is laminated only to the loose lining material. Furthermore, more garment makers are using 64 g/m^2 (2 oz) woven fabric as the base material, sometimes in a rip-stop construction for added tear strength. In this way, a lightweight garment with excellent drape and softness coupled with high performance is produced. It is a very significant improvement on the stiff and poor draping trilaminated material which was generally on offer in the mid-1980s. The lamination process needs to be very carefully considered because, in addition to causing stiffness, it can significantly reduce breathability. This has been discussed in Chapter 3 and, for optimum performance, breathable adhesives are being developed. Waterproof and breathable membranes are now being used in footwear, gloves and even in socks.

4.1.3.1 Closely woven fabrics

There is a third category of breathable and waterproof material produced from water-repellent finished but uncoated closely woven fabric. Although not coated or laminated, these are mentioned for completeness. The best known is the Ventile^R material which was developed for the British Military during the Second World War. It is produced from Egyptian cotton yarns woven into a fabric which allows breathability through gaps in the weaving when dry. On wetting, the cotton yarns swell to close up the gaps in the fabric and this prevents the penetration of water for up to 20 minutes when the wearer is submerged. Because it is not coated, the breathability is high, but the fabric is not waterproof in prolonged heavy rain. Waxed cotton jackets have similar properties. Fabric tightly woven from polyester microfibres works on the same principle, although the polyester does not actually swell when wet. The use of microfibres ensures that the gaps in the fabric are very small even when dry. In recent years the Eta proof fabric, woven from long staple cotton by a Swiss manufacturer, has appeared on the market.[25] These garments from closely woven fabrics are comfortable to wear and are soft, flexible and quiet with less of the rustling sound produced by many coated or laminated fabrics. 'Noisy' garments are a problem for the MOD and for activities such as bird watching and even fishing which require quietness.

4.1.3.2 Face fabric water repellency

The performance of these uncoated materials is very dependent on the quality of the water-repellent finish applied to the outside of the fabric. The method of assessment is the spray rating tester, see Fig. 4.8. If rain does not have the opportunity actually to soak in, but the drops of water 'pearl off',

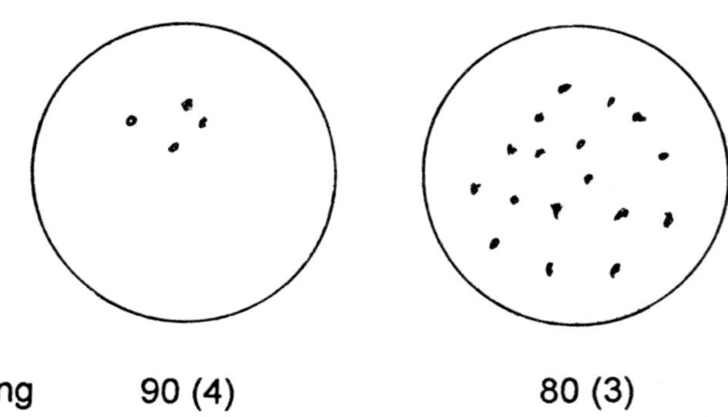
Spray rating
90 (4)
80 (3)

the coating does not have much work to do. However, in a dynamic situation with fabric flexing and folding, the water does eventually penetrate the surface of the fabric and the coating is needed to prevent it progressing further. The breathability of a coating or a film is believed to be reduced if the supporting fabric is 'wetted out' and there is a layer of water on the outside of the garment caused by reduced efficiency of the water-repellent finish. Thus, an effective water repellent on the outside of the garment is an important feature for water resistant/breathable fabric coatings and laminates. However, even the best water-repellent finishes eventually lose at least some of their effectiveness through abrading, washing off or contamination. For this reason, it is important not to wash waterproof garments in detergents which could harm the water-repellent finish. Also, trace amounts of detergent left behind on the fabric would help water 'wet out' the surface which is the first step in sinking into the fabric.

An industry has grown up on products to improve, refurbish and restore the water-repellency finish of waterproof garments. Cans of silicone and fluorocarbon finishes are available to spray on to garments that have lost some of the water repellency they possessed when new. An alternative and probably more satisfactory method is to apply a wax finish by full immersion in a wax solution such as Nikwax products.

4.1.4 Water repellency – shower resistance

As previously stated, most outerwear is treated with a water-repellent finish on the outside of the garment. This is applied to cause water to 'pearl off'

4.8 The spray rating tester determines the resistance of fabric to surface wetting by water. A specified amount of distilled water is sprayed on to the test fabric mounted at an angle of 45° and the appearance of the fabric is assessed by comparing it to standards or the Association of American Textile Chemists and Colourists (AATCC) Spray Test Rating Chart (photographs). No drops of water sticking to the fabric surface and no wetting of the surface is rated 100. Slight sticking or wetting is rated 90, and wetting at the spray points is rated 80, see above. More significant wetting is rated 70, complete wetting of the upper surface is rated 50, while complete wetting of both upper and lower surfaces is rated 0. In practice, it is quite easy to achieve a rating of 100 if the water repellent is applied properly. A 90 rating is usually acceptable, but 80 is rarely acceptable for 'as received' samples although it may be acceptable after the sample has been washed or dry cleaned. BS EN 24920 rating is *generally* similar to AATCC 22 except that the ratings are on a 5 to 1 scale. Thus 100 = 5, 90 = 4 and so on.

Photograph of apparatus supplied by James H Heal & Co Ltd and reproduced with kind permission. The ratings 90 and 80 are sketched from the AATCC 22 Chart.

rather than wet out and soak into the fabric. There are several different classes of chemicals that can be used for this purpose, including certain waxes, silicones and fluorocarbons. When correctly applied, virtually all products give good results initially, and when water is poured over the fabric, it runs off without sticking to or wetting the fabric surface (see Chapter 5). After a period of wearing, however, the finishes become less effective and rain begins to wet out and sink into the fabric. At the time of writing, there are no completely durable water-repellent finishes for textile fabrics; fluorocarbons are generally regarded as the most effective. Waxes mark off and can feel slightly sticky, but they can be replenished quite easily by the garment owner. Silicones are effective but they can attract oil-based soiling. Care is required when silicones are used on coated fabric, because their release properties can affect the adhesion of the coating. A fabric with a silicone already on it cannot usually be coated or laminated, but it is possible to apply the silicone to one side of a coated fabric, for example by a back licking process or a rotary screen.

Another test used is the Bundesmann apparatus (Fig. 4.9) which attempts to simulate actual rain situations by subjecting the fabric surface to a shower of drops of water, while at the same time lightly abrading the other side. The abrading action has the effect of 'encouraging' water to seep through the fabric and simulates movements of the body when a garment is being worn. Any water which does penetrate through the fabric is collected in a receptacle and measured to determine the efficiency of the material as a barrier to the 'shower' of rain. The drops of water used in the Bundesmann apparatus (0.32 cm radius) are in fact much larger than those normally encountered, even in a 'cloudburst' (0.15 cm radius), and the kinetic energy (force) of the drops on impact is nearly six times as great as those in the 'cloudburst', 20 000 ergs compared to 3460 ergs. For comparison, drops of 'heavy rain' are quoted as having a radius of 0.075 cm and a kinetic energy of 220 ergs.[26] Despite this, some researchers consider the Bundesmann too mild; others believe it to be unrealistic in certain circumstances, because

4.9 The Bundesmann water repellency test apparatus. A shower of simulated rain falls 150 cm from the nozzles at the top of the apparatus on to the surface of the test fabric, which covers the test cups. The drops of 'rain' are about 0.07 ml and there are 300 identical drop formers. The drop rate, i.e. the 'rain' intensity, can be varied. Arms in the form of a cross rub the undersides of the test fabrics held in the cups. This makes the test more severe, by simulating the situation of clothing on a moving body. The amount of water passing through the fabrics is assessed by measuring the amount of water collected in the cups. Photograph supplied by James H Heal and Co Ltd and reproduced with kind permission.

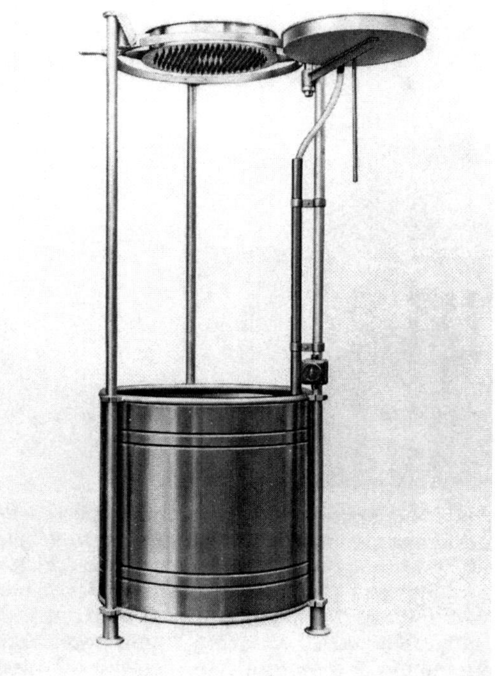

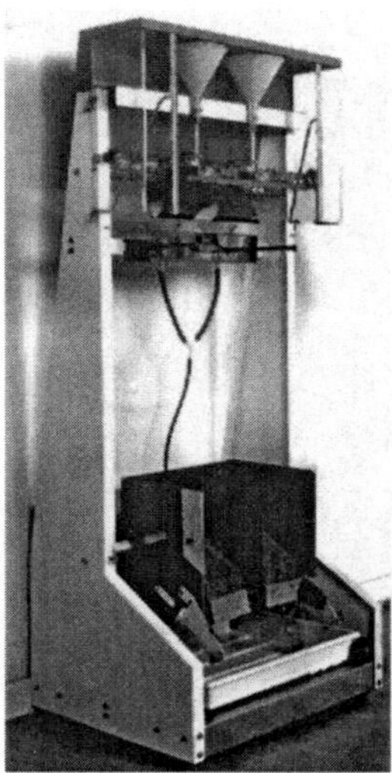

4.10 The WIRA shower tester measures resistance of fabrics to water absorption and penetration and conforms to BS 5066: 1974 (1993). Pre-weighed fabric samples are exposed to a shower of 500 cm^2 of distilled water and the time taken for the first 10 cm^2 of penetration is observed (if it occurs). The amount of water absorbed by the fabric samples is obtained by weighing. Photograph supplied by WIRA Instrumentation and reproduced with kind permission.

rain does not always fall from directly above the fabric but can hit the surface of the garment at all angles. Rain simulators such as the Wool Industries Research Association (WIRA) apparatus (Fig. 4.10) have been designed to reproduce conditions of actual rain more closely. The WIRA Shower Tester tests two fabric samples simultaneously and conforms to BS 5066;1974 (1993). Even more sophisticated rain simulators have been designed and built for more intensive research.

The most frequently used test for waterproofness is the hydrostatic head which measures the pressure of water required to penetrate through the fabric, see Fig. 4.11. For many years, 100 cm hydrostatic head of water has been taken to be an acceptable level of performance. Some manufacturers now require 150 cm or higher. In fact, many quality waterproof coatings, when new, have hydrostatic heads well in excess of 700 cm and Gore-Tex and Sympatex have levels significantly higher than this. In practice, the higher the initial hydrostatic head, the better the durability and service life of the garment. Many manufacturers require a minimum hydrostatic head of 150 cm after the quality control tests of flexing and abrasion. If the initial waterproofness was only, say, 200 cm, a reduction of 30% would result in

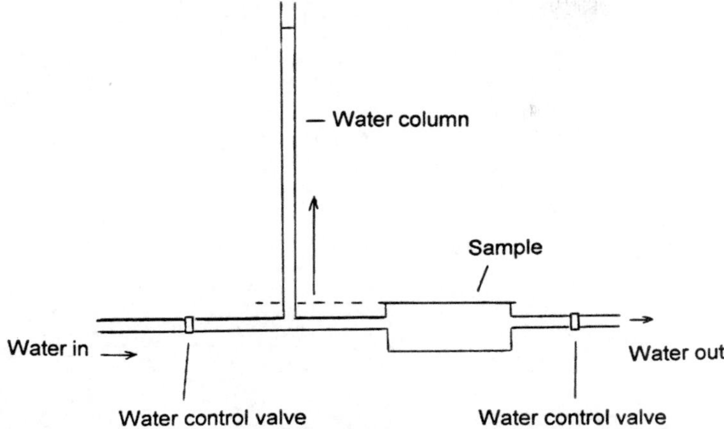

4.11 Schematic diagram of the hydrostatic head tester for water penetration resistance (waterproofness). The principle is to attempt to force water through a fabric sample held in a clamp. The pressure is measured by the height of the column of water.

rejection, whereas if the initial hydrostatic head was over 700 cm, the fabric would still be acceptable even with a reduction of 75%.

There are several other tests conducted on entire made-up garments to simulate actual water protection under dynamic condition. BTTG and several other laboratories have a 'rain simulator' which directs showers of 'rain' from all angles on to a garment to simulate real conditions. Even this is regarded by some researchers as unrealistic, because the garment is stationary, but a person would be moving and flexing the material.

4.1.5 Industrial and specialist protective clothing

This area includes all clothing worn for protection by persons doing a particular job and excludes sports and leisurewear which were discussed above.[27,28] Some examples of protective clothing are shown in Fig. 4.12 to 4.15. Further comments on medical and military uses are made in Sections 4.7 and 4.8. There is of course some overlap as far as protection from the weather is concerned, because the same considerations generally apply in terms of technical performance. The sports person usually has the choice whether or not to proceed when the weather is poor, but people who have to do a job of work, sometimes in the event of an emergency, do not have this luxury. Aesthetic and fashion considerations are very important in sports and leisurewear, but not generally as important in workwear. However, industrial workers need to be comfortable both physically and mentally and fashion demands are becoming more relevant, especially in a

4.12 Specialist industrial protective clothing. Fuel man Jean-Paul Bresson is clothed in material satisfying all safety requirements at the Ariane rocket facility at Kourou, French Guiana. Photograph supplied by Bobet Coated Fabric Company and reproduced with kind permission.

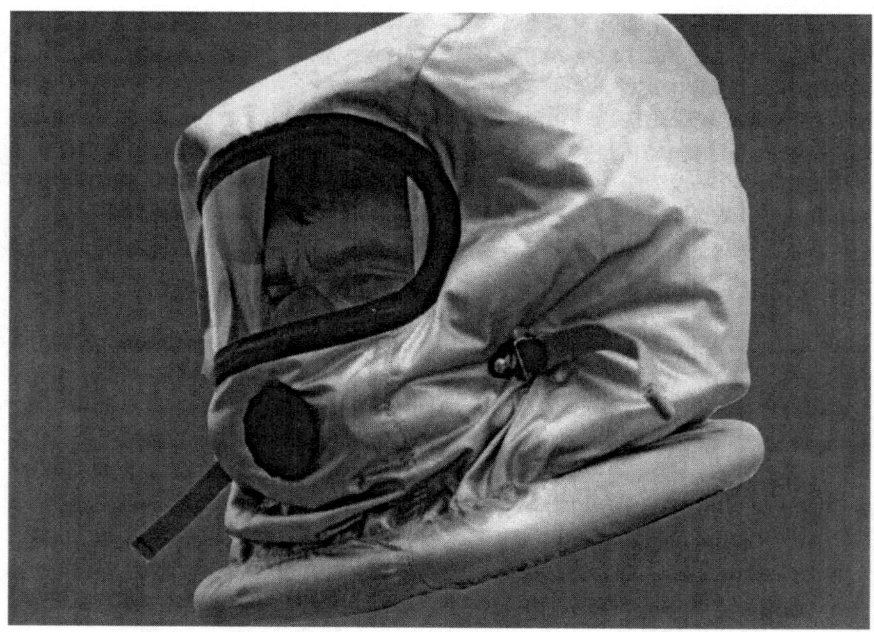

4.13 Specialist industrial protective clothing produced from aluminised FR Hypalon coated on to Nomex. (Both coating and fabric substrate are DuPont trade names.) Photograph supplied by Bobet Coated Fabric Company and reproduced with kind permission.

4.14 Passenger Immersion Coverall Suit. This has been designed in conjunction with the MOD (UK) for passengers in helicopters and tactical and transport aircraft. It is produced from waterproof, lightweight synthetic fibres and has been specially designed for rapid donning. Photograph supplied by Beafort Air-Sea Equipment Ltd, Birkenhead and reproduced with kind permission.

4.15 Industrial protective clothing in breathable waterproof Exeat™ fabric (coating applications) by Vander of Newton Abbot, UK. The clothing is equipped with reflective strips for high visibility and safety. Photograph supplied by Vander and reproduced with kind permission.

world where the overall quality of life is continuously improving. Moreover, comfort contributes to safety and is therefore of first importance. A person subjected to stress through discomfort may not be able to concentrate fully on the work in hand. In addition, if the protective garment is uncomfortable, the worker may choose not wear it even though he, or she, is then exposed to the particular hazard.

Concern for health and safety at work has led to the development of specialist clothing to protect against chemicals, fire, heat, radiation, biological agents and against mechanical hazards such as sharp edges. Personal protection systems (PPS) and equipment (PPE) are now an essential feature of occupations where a potential hazard exists and they are being continuously improved. Standards and test methods began to be formulated in the mid-1970s, especially in the USA where the ASTM (American Society for Testing and Materials) formed Committee F-23 on Protective Clothing. This initially dealt with protection against chemicals, but later became responsible for all aspects of protective clothing.[29,30] In Europe, the EC Directive on PPE (89/686/EEC), applicable from July 1995, lays down the basic safety requirements for manufacturers of equipment to ensure the health and safety of users. The Directive applies to industrial applications, such as protective clothing and life jackets, but not to equipment for military, police or private individual use. There is also a Directive aimed at the users of PPE.

Choice of which polymer to use for a coated fabric as a barrier to a particular hazard may be relatively simple – the difficult part is fabricating it into a flexible garment which will be comfortable to wear, durable and affordable for use by anyone who requires it. The ideal garment should not only be lightweight and comfortable, but should not restrict body movements in any way.[1,31] Much research work has been done on the comfort of clothing, and there are numerous factors to consider. All these apply to industrial protective clothing, and there are additional technical, production and commercial considerations. These factors can be summarised:

(1) comfort considerations, both physical and psychological, which include thermal comfort, touch and next-to-the-skin sensations, overall aesthetic appearance, design and colour, current fashion, prejudice, suitability for the occasion and correct size and fitting;
(2) technical factors of performance, cleanability, storage and disposability;
(3) production and commercial factors which include availability and cost of materials, ease of fabrication and production, e.g. cutting and sewing of garment panels, effective seaming, manufacturing, storage and distribution cost considerations.

There are many articles in the literature on specialist protective clothing.[32–40] See also Further reading (Section 4.11).

The choice of material and coating resin for particular applications can be derived from Tables 2.1–2.3 in Chapter 2. PVC is used in many applications because of its low cost, but more specialist protective clothing is produced from Neoprene, nitrile rubber, butyl rubber and Viton. Base fabrics are varied and include polyester and nylon and, for comfort in hot climates, sometimes cotton and polyester/cotton blends.

Tyvek, a spunbonded nonwoven high density polyethylene produced by DuPont, is used in many types of protective clothing, and it can also be coated or laminated to improve its barrier protection properties further. Some qualities of Tyvek have an initial hydrostatic head of about 75 cm, and they have excellent resistance to many chemicals, including acids and alkalis. Tyvek is also available coated with polyethylene which makes it more impermeable to liquids, but also to perspiration, so that it is uncomfortable if worn for long periods. A material which can provide protection against a wide variety of chemicals is produced by the lamination of Tyvek to a film called Saranex^R (Dow), which is itself a triple laminate comprising low density polyethylene (LDPE), Saran (PVDC/PVC) and an ethyleve vinyl acetate (EVA)/LDPE copolymer.[40]

Specialist protective clothing ('turnout' clothing) for fire-fighters in the USA consists of three layers of material laminated together. The outer layer is an aramid fabric or an aramid blend; the middle layer is a barrier against moisture, steam or corrosive liquids; and the inner layer consists of thermal insulating material, such as aramid, wool felt or quilting. The middle layer is Neoprene coated aramid fabric or a Gore-Tex film laminated to nylon. While it is believed to be generally effective, the typical weight of a coat is 70–80 pounds (32–36 kg) which must reduce mobility and be tiring to wear. Concerns have been raised by the International Association of Fire Fighters on possible deterioration of the barrier film during wear, storage or cleaning because, if this happens, the wearer is vulnerable not only to harmful liquids but also to bloodborne pathogens.[41] The last mentioned factor is relevant because fire-fighters' duties include rescue operations of persons in situations such as vehicle accidents and aircraft disasters.

For protective clothing for heat radiation protection combined with fire protection, such as in the molten metals industry, heat reflectance aluminised coatings on FR cotton, FR rayon, Zirpro wool, pre-oxidised acrylic fibre (e.g. Panox – LUCF) or aramid are used. Some degree of reflectance can be produced by incorporation of aluminium powder into coatings which may be transfer coated on to fabric. Decorative and fashionable aluminised coatings are sometimes produced in this way, but the actual amount of light or heat reflected is quite low. Much better results can be obtained by lamination of a pre-prepared aluminised polyester film, such as Mylar[R] (DuPont) on the fabric, and up to 90% reflection of radiant heat is believed to be possible. Thermal insulation is dependent on the thickness of the fabric and felts provide the best bulk.

4.1.6 High visibility garments

There are three main categories of high visibility material. Two of these can be readily produced by textile coating through incorporation of the appropriate pigment into either PVC or polyurethane resin which is then transfer coated on to a base fabric. 'Day glow' or fluorescent pigments which produce a high visibility effect in daylight are available in a number of colours, but orange and especially yellow are the most used. Photoluminescence materials such as zinc sulphide actually give out light in the dark for a limited amount of time (up to six hours is claimed), after being 'charged up' by daylight.[34,42] Alternatively, pre-prepared film containing these pigments may be laminated on to fabric. The third category are reflective microprisms which are incorporated into a film in such a way that they reflect back the light which falls on to them in a the same manner as road signs. Commercial examples are the Scotchlight[R] products made by the 3M company. Reflective coatings are normally used as strips on high visibility clothing worn by fire-fighters, the police (Fig. 4.15), motorway construction workers, etc. Joggers and cyclists also make use of this material. Recently Reflec plc has developed water-based printing inks which can be applied to textiles to produce retro-reflective effects with good abrasion resistance.[40] High visibility garments and reflective strips are a valuable contribution to safety, especially road safety, and there are both American and European standards.

4.1.7 The spacesuit

Some mention must be made of the use of coated fabric in this most exotic of end products which protects astronauts from a wide variety of hostile and harmful agents, including extremes of temperature, space dust (micrometeoroids), pressure fluctuations and radiation.[43] The spacesuit, which encloses a habitable mobile atmosphere, is pressurised at 4.3 psi (0.3 kg/cm) and on Earth weighs about 112 kg. It is also equipped with cooling water, food, waste collection – including removal of carbon dioxide – electrical power and communication systems.

The spacesuit comprises many different layers of fabric and films, including Nomex, Kevlar, aluminised polyester film, polyester and nylon. The coated fabrics are nylon coated with Neoprene for protection against micrometeoroids, and a pressure bladder is produced from polyurethane coated nylon which retains oxygen against the vacuum of space. The aluminised polyester films required for thermal insulation are reinforced by lamination to polyester scrim fabric. The space programme also uses static free coated fabric in the space shuttle cargo bay. More recently, all metal space suits are pressurised to 8.3 psi (0.58 kg/cm).

4.2 Industrial and functional products

4.2.1 Garment interlinings

This is a very high volume business because interlinings form an integral part of many garments, for example in the collar, waistband, cuffs and lapels.[44,45] Hot melt fusing is used in place of sewing for joining materials, although some sewing is still required for long term durability and aesthetic purposes. Interlinings are necessary for stiffening and shape retention, and materials can be fused together much more quickly and with greater accuracy than by sewing. Sewing requires skill to avoid puckering and to produce quality products at speed.

The hot melt adhesive used is selected according to the garment fibre type, drapeability requirements and durability requirements of dry cleanability and severity of washing. The most common chemical types used are polyethylene, polyamide and polyester. Polyvinyl acetate is sometimes used for low temperature fusing of leathers and pile fabrics, but it has limited durability to cleaning. Plasticised PVC is used for large areas, such as coat fronts. Methods of hot melt powder application are scatter coating, dry dot printing (powder point) and paste coating through a rotary screen.

Processing is usually on a flat bed, static fusing press or a continuous press, and conditions must be optimised to preserve fabric aesthetics and properties, e.g. no shrinkage, distortion, pile crushing, glazing, discoloration, etc. Note that for heat to reach the adhesive 'glue line', it must pass through the material being processed. The variables of time, temperature and pressure must be optimised for each particular job being done, and this is best accomplished using temperature sensitive indicator paper to measure the temperature at the glue line. Hot melt fusing was discussed in detail in Chapter 3.

4.2.2 Tarpaulins

HGVs use large numbers of tarpaulins (Fig. 4.16) of various types, but the one most used both in Europe and the USA is PVC plastisol coated high tenacity woven polyester or nylon.[46,47] The weights vary quite considerably, from lightweight coverings – weighing a little over 100 g/m^2 – to those over 1000 g/m^2. They are produced from base fabrics weighing from less than 100 g/m^2 to over 300 g/m^2 with the PVC being built up in several layers, usually on both sides of the base fabric. In Europe some of the highest quality tarpaulin manufacturers are members of the Complan 'club' – a

4.16 Tarpaulins are a main outlet for PVC coated fabric. PVC accepts artwork readily and, with the advantages afforded by new printing technology, the sides of vehicles may well be exploited as prominent display areas for advertising – especially in the USA.

quality assurance body established by the manufacturers of Trevira polyester (Hoechst).[46] The equivalent body in the USA is Isoplan – Trevira. Top quality tarpaulins are lacquered with an acrylic or polyurethane resin to produce a high gloss. The lacquer improves UV degradation and abrasion resistance and also, by preventing plastisol migration, reduces soiling and makes cleaning easier. An important advantage of PVC is that it can be heat and RF welded to produce good watertight seams. In common with all materials used in transportation applications, producers strive to obtain the optimum performance, with the lowest possible weight to reduce fuel costs. In addition, the lower the weight of the vehicle and essential equipment, the more weight is available for the payload. This consideration is relevant to all materials used in transportation applications and is especially important in aircraft.

Quality tests on tarpaulins include PVC peel bond adhesion, flexing resistance, cold cracking, reduced flammability, waterproofness, tear and tensile strength, and dimensional stability over a range of temperature and relative humidity. In addition, they should be resistant to oils, engine fuels and common chemicals. Two special polyesters, Trevira HT Type 711 (Hoechst) and Diolen 174 SLC (AKZO), which are resistant to micro-organisms, have been specially developed for tarpaulins. These are intended to overcome the problem of migration of micro-organisms into the tarpaulins in water which can enter the tarpaulin by a wicking mechanism via the exposed edges of the coated fabric or small cracks in the PVC coating. Tarpaulins must have especially good tear and flexing resistance and preferably be repairable in the field. Research to improve tarpaulin durability, to study mechanisms of degradation and to develop accelerated tests which reproduce conditions of wear in actual use continues.[48,49] Some non-PVC tarpaulins such as polyethylene have appeared, because some pressure groups, notably Greenpeace, believe that PVC is harmful to the environment, especially during manufacture and disposal. Used in conjunction with tarpaulins are narrow fabric fastenings, which also need to be able to withstand UV degradation and weathering. Although polyester has largely replaced cotton as the fabric substrate in tarpaulins, an article recently appeared extolling the advantages of cotton – better durability through resin adhesion and better appearance.[50]

Tarpaulins can reduce the air drag on goods vehicles by producing smoother surfaces, and fuel consumption improvements of up to 7% have been reported. More colourful eye-catching tarpaulins with advertisements have appeared in the USA recently; some analysts believe the sides of trucks on the move are ideal areas for exploitation by the advertising industry. Tarpaulins have been printed making use of the most up to date developments in digital printing.[51,52] Finally, some development is believed to be in progress to develop knitted tarpaulins which would conform to the shape of the load – although this may actually increase air drag in some cases.

4.2.3 Other applications of PVC coated fabric

PVC is a versatile material and coatings can be specially formulated for a wide variety of products, including luggage, handbags, toys, 'bouncy castles', camping articles, aquatic sports items, flags and banners, tent and picnic ground sheets and gazebos. There are also caravan, marine and pleasure craft applications and industrial articles, such as welding screens and even hose pipes. One major advantage of PVC is the ability to weld pieces together to produce joins which are impermeable to gases and liquids. Also, it can be produced in a wide variety of colours and can be lacquered to produce a high gloss and to improve cleanability. Banner and bunting material is sometimes made from two sheets of lightweight PVC sandwiching a very open scrim fabric, to improve physical strength and tearing resistance. PVC can be made to be very receptive to printing inks and paints and this is an advantage over substitute materials, especially for advertising and promotional purposes, see Fig. 4.17.

PVC coated fabric is still used for seat upholstery in some public transport vehicles, ferries and other marine uses and in public buildings. This

4.17 PVC coated fabric accepts artwork readily and is used as information banners.

'leathercloth' is generally produced using 'blowing agents' creating a pore structure which results in a soft touch. It is available in a wide variety of embossed designs and colours and has the advantages of being waterproof and easy to clean. This material and varieties of it were once extensively used on car seats before fabric coverings were developed. Similar PVC material is used for bags and luggage.

4.2.4 Non-PVC tarpaulins and coverings

Lightweight coverings can be produced by extrusion coating polyethylene fabric woven from narrow tape (or polyethylene film slit into narrow tape). The coating material is generally low density polyethylene and the woven fabric tape is usually high density polyethylene. This combination of materials generally gives the optimum results. If necessary, coating can be applied to both sides to produce a heavier product of higher quality. If higher abrasion properties or higher temperature resistance are needed, polypropylene is used in place of polyethylene. If the processing conditions are correct, a good bond is produced because the two materials, base fabric and coating, are both polyolefin. However, if another fibre is used, polyester for example, there could be delamination problems because of the chemical difference between the coating and the fabric substrate.

Extrusion coated polyethylene coverings have the advantages over PVC of lower cost, easier recyclability, flexibility in cold climates (as low as −40 °C), and significantly lighter weight.[53] They contain no plasticiser to leach out or to migrate, resulting in less surface tack and, in some cases, easier cleaning. The material can be made FR and UV resistant by addition of the appropriate chemicals and it is resistant to microbes. The disadvantages are a relatively low melting temperature, a relatively stiff texture and that printing is not easy. Compared to PVC, the material is not as durable, has lower abrasion resistance and, being thinner and lighter in weight, it does not appear to the customer to be as robust as PVC, even though the measured physical strength may be the same. In addition, fabrication is more difficult because polyolefin cannot be welded by RF but requires heat sealing. Polyethylene coverings and tarpaulins in marine uses require a high standard of UV protection.

Nevertheless, polyethylene is versatile and has many uses in temporary structures, as general coverings, wrappings and paper reinforcement, and in sacks and bales. It is also economical enough to be disposable when necessary, for example when it is used as wrapping for steel or timber or when it is used for transportation for goods and then as a covering during storage before use. Another advantage is that it can be produced in very wide widths – up to 4 m (156 inches). Developments continue in order to improve abrasion, tear strength and printability by surface modification such as corona discharge, and to produce variants which will be breathable or have anti-static properties. Polyethylene can also be extruder coated on to non-woven fabrics to produce barrier materials for disposable protective clothing against chemicals or microbes. Coverings produced from all polyolefin materials are, of course, easier to recycle than coverings produced from a fibre coated with a resin of dissimilar chemical type.

Lightweight tarpaulins are also available made from polyurethane coated fabric, which the manufacturers claim is easier to handle than PVC because of its lightness, 13 oz/yd^2 (about 430 g/m^2), i.e. about 28% lighter than the equivalent article in PVC. The polyurethane tarpaulin is hot air and RF weldable, flexible at low temperatures, has good abrasion and flexing resistance and is more easily stored than the more bulky PVC.

4.2.4.1 Intertape Polymer Group

In the USA, the Intertape Polymer Group™ is the first manufacturer to use a co-extrusion coating process which allows exceptional colour uniformity through dissimilar layers of polyolefin polymer. This gives excellent flexibility and lends itself to a wide variety of products and effects, including the production of blackout curtains, see Fig. 4.18. The company's Nova-Shield™ PVC replacement product range uses their patented 'double-stacked' scrim which is coated on both sides with a blend of LDPE containing UV and FR stabilisers and pigments. The 'double-stacked' scrim is woven from polyolefin tape yarn which also contains UV and FR

4.18 Co-extrusion technology. The Intertape Polymer Group is the industry's first to utilise a co-extrusion coater in its manufacturing process. This allows IPG to produce unmatched colour uniformity through dissimilar layers, offering the ultimate flexibility and widest variety of end-use fabrics, including black-out layers on white fabric. Photograph supplied by the Intertape Polymer Group (USA) and reproduced with kind permission.

4.19 Polyolefin coated lightweight fabric is widely used to contain dust in building and demolition operations.

stabilisers and the resultant coated fabric has tear strengths at least equal to much heavier PVC coated material.[54]

Figures 4.19 to 4.22 show a variety of different coated fabric applications, from inexpensive coverings for demolition and building work to high tech coverings for hazardous material pipelines. Interconnecting bellows on public transport vehicles and trains may be made from rubber compounds, such as Neoprene.

4.2.5 Flexible intermediate bulk containers (FIBCs) and bulk bags

These containers, which can hold up to 1500 kg, bridge the gap between sacks, containing 25–100 kg and road tankers which require about 20 000 kg for a full load. They are designed to hold all types of dry goods and powders for transportation. Goods include chemical additives, dyes, fertilisers and flour. Containers of this size cannot be lifted manually and became feasible only when fork lift trucks and other such handling apparatus became standard equipment in factories. A critical factor is that the lifting handles must be strongly made and capable of supporting the weight when fully loaded, see Fig. 4.23. Most are now made from woven polypropylene yarn, or tape yarn which is then coated with polypropylene, and weigh only about

4.20 Even road sweepers make use of coated fabric.

2.3–4.6 kg (5–10 pounds) when empty. Some FIBCs are reused but, because they are difficult to clean and are relatively cheap to produce, others are used only once. However, an industry has been established cleaning and refurbishing them.[55] The Western European market for polypropylene FIBCs and big bags grew from 52 000 tonnes in 1992 to 108 000 tonnes in 1997 and further appreciable growth is expected. In comparison, the market for polypropylene sacks declined slightly over the same period, from 25 000 tonnes to 23 000 tonnes.

Some early varieties are made from high tenacity polyester yarn, typically 1100 dtex, plain woven into fabric of approximately 270 g/m^2 in weight and coated with PVC, polyurethane or a speciality rubber such as chloroprene or butyl rubber. These FIBCs are expensive to make and are designed to be reused. The actual choice of which coating to use depends on the intended contents and fluorinated rubber is actually used in some cases. During development work by Hoechst to develop FIBCs which were reusable and capable of outside storage, the usual laboratory tests of tear strength, tear propagation, coating adhesion and coating suitability were used, plus a number of tests specially designed to simulate conditions of actual use. These include hot storage (80–120 °C for one year), outdoor storage (one year) and a 'rolling' test in which a polyethylene bottle is two thirds filled with a small sample of the coated fabric and rolled 1 million times over 15 days. The samples were inspected after testing to determine

4.21 Specialist fire protection and thermal insulation pipe coverings in silicone coated glass fabric. Photograph supplied by Bobet Coated Fabric Company and reproduced with kind permission.

4.22 Flexible bellows used to join two carriages of a Manchester Metro tram. Similar bellows are used all over the world in trains, trams and 'bendy-buses.'

any deterioration in properties. The outdoor test assumed that the container would spend about one-fifth of its life outdoors, and so if it was satisfactory after one year's continuous exposure, it was deemed to be satisfactory for five.[56]

When they are being filled with or emptied of dry powders, FIBCs need protection against static explosions; this is done by external earthing with a metal wire network in the fabric. Recent developments have allowed the metal wire in some cases to be replaced with a static dissipative yarn such as Negastat (DuPont), a polyester yarn which has permanent antistatic properties. However, it has been reported that the bag with the polyester network, while not capable of giving off a high energy spark itself, could induce voltages on neighbouring conductors, which in turn may be capable of giving off a high energy spark. Developments continue in order to produce more effective and convenient ways of overcoming this problem.

4.2.6 Other textile containers

Sacks woven from polypropylene tape yarn have, to a large extent, taken over the sisal and jute fibre sack market. This fabric is generally consolidated by curtain or extruder coating with melt polypropylene, and is lighter,

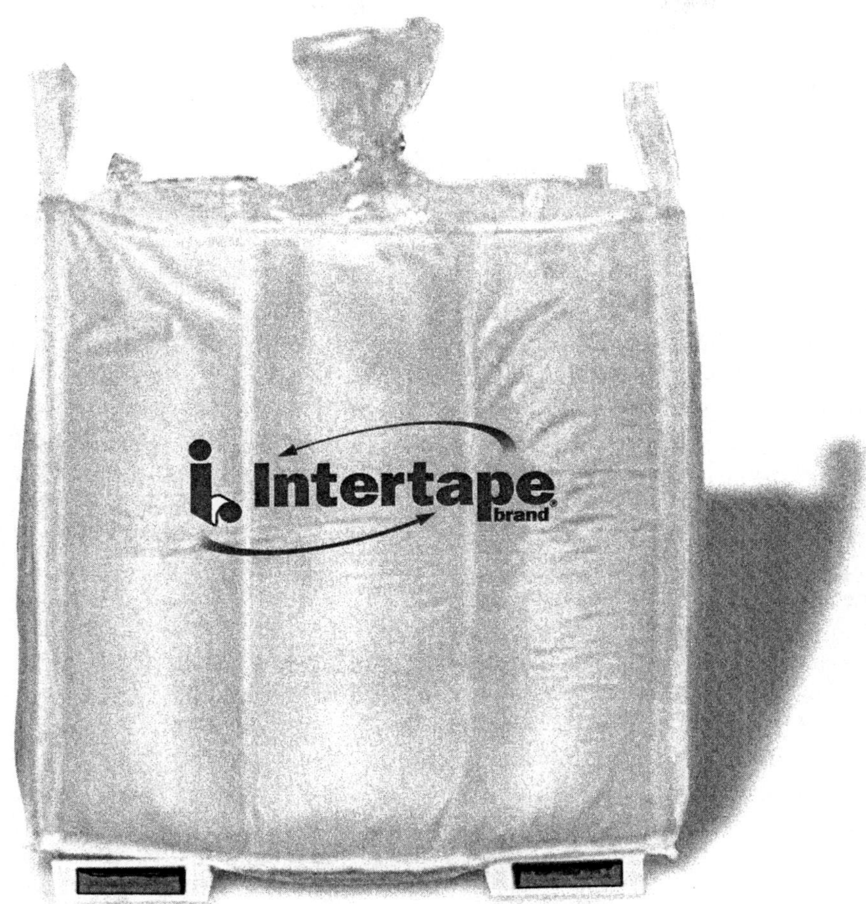

4.23 Flexible intermediate bulk carrier (FIBC) produced from all polyolefin material which is both FR and UV resistant. This is a Pallet-Free™ product (Intertape Polymer Group) and features reusable forklift bottom channels, eliminating the need for a pallet, thus allowing cost savings of up to 20%. Photograph supplied by the Intertape Polymer Group (USA) and reproduced with kind permission.

and more resistant to chemicals and micro-organisms than natural fibres. This fabric – more resembling a plastic – is also used in bags and inexpensive coverings. The international sack market is estimated to be about 23 billion units, 20 billion of which are made from polypropylene. As has been noted, the market for polypropylene sacks appears to have stabilised (after a period of high growth), presumably because of the growth of the big bag which has the capacity of ten smaller sacks, but uses proportionately less material. A return to natural renewable materials may also be a factor, but in some countries sacks made from these traditional materials now require a layer of polyolefin laminated on the inside as a barrier.

4.2.7 Bulk liquid containers (Flexitanks)

These articles first appeared around 1950 but are believed to have been used more often in recent years. The largest can hold 26500 l (7000 gallons) and are used for transporting liquids, which may be lubricants, potable water or even wine. They are made from polyester fabric coated with PVC, PVC/polyurethane blends or other polymer blends. Coated fabric weights vary from 40 oz/yard^2 to 20 oz/yd^2 (1356 g/m^2 to 678 g/m^2). The smaller ones are inexpensive enough to be disposable while the larger ones are reusable. An advantage of these 'flexitanks' is that, when empty, they can be folded up and occupy very little space. They enable liquids to be transported on one journey and other types of goods on the return journey. A 23000 l capacity container when empty weighs less than 100 pounds (45.4 kg) and can be folded into a parcel measuring 48 inches by 18 inches wide and 12 inches high (1.2 × 0.45 × 0.3 m). Much smaller 208 l (55 gallon) containers have appeared and are made from polypropylene. Some manufactures believe that flexitanks could become a growth industry, but have noted that standards need to be established to ensure that only quality products are made.[57,58]

Aircraft and military vehicles such as tanks use flexible fuel tanks made from coated fabric. These articles allow irregular and awkward space on the aircraft or vehicle to be used to carry fuel. Large containers made from coated fabric, called 'dracones', which can be towed behind ships have been made from heavy duty nylon – about 600–700 g/m^2 base fabric, coated both sides with a total of 5000 g/m^2 of polymer. The outer surface is generally Neoprene, but the inner surface is compounded according to the contents, e.g. polyurethane for potable water or nitrile rubber for oils.

4.2.8 Conveyor belts

This extremely useful device is a surprisingly recent invention. The earliest reference, in a 1795 publication, is to a 'broad endless strap of thin pliant leather or canvas revolving over two pulleys'. The earliest documented textile/rubber belt in the UK was associated with engineers working for the Mersey Docks and Harbour Board. By 1868 they had installed vulcanised rubber belts to convey 50 tons of grain per hour at 500 ft/minute. Conveyer belts are now in widespread use in virtually every industry, from conveying coal underground to service at supermarket checkouts.[59,60] The earliest belts used cotton as the load bearing fibre but, while its actual adhesion to rubber was good, the cotton was vulnerable to attack by moisture and bacteria, and this eventually led to belt failure. Moisture could enter the cotton via the edges, which were capped with either rubber or PVC in attempts to prevent its entry into the belt carcass. Later, nylon/cotton fabrics were used, but modern belts are now almost entirely made from either continuous filament nylon or polyester. This eliminates the need to cap edges, because synthetic fibres are not vulnerable to wet rotting, and allows belts to be made in wide widths which can be cut to that required by the customer. Synthetic fibres have high strength to weight ratios and good fatigue resistance which allows thinner belts to be used and, therefore, smaller pulley diameters. Polyester is generally used as the warp fibre in plain or Oxford woven fabric, because its extensibility at around 1% is significantly lower than that of nylon which is around 2–2.5%. Nylon is used as the weft yarn and contributes to good impact and tear resistance. Long-haul conveyer belts requiring very high tensile strength are sometimes made from steel cords and more specialist yarns. Kevlar was first used in conveyer belting in 1978 and DuPont claim that more than 250 000 m of belts have been made since. Kevlar offers low elongation and high strength at light weights, thus saving on energy.

The coating applied is generally either a rubber specially formulated for the purpose or PVC which has the advantage of good fire resistance. The compound ingredients depend on the conditions of the working environment and the goods to be conveyed. PVC belts came to be extensively used in coal mining following fires involving loss of life with rubber coated belts. Rubber is applied on the traditional 'spreading' machine from a mixture of ketones and layers built up by laminating layers together. Sometimes an extra layer of an open weave fabric is inserted between the top cover and the first ply to provide extra shock loading resistance.

The design of a conveyer belt depends on the layout and specific details of the system, the material being conveyed and the rate at which it needs to be delivered. The basic requirements include adequate tensile strength and modulus to transmit the power and carry the load, flexibility to run around the pulleys and low elongation and dimensional stability to run straight and true, even when there are wide temperature variations. The top surface should be suitable for the load being carried, which may be coal, packaging, chemicals or even food, and the lower face should provide sufficient friction on the pulleys for movement. For a long service life, tear strength and tear resistance should be high, and for a minimum risk of delamination, there should be good adhesion between the layers.

4.2.9 PTFE coated belts

PTFE coating is used in applications where high temperature and non-stick properties are required, such as in the food industry, in calenders for the garment industry and for fabric lamination. For high specification articles the fabric substrate is glass fibre or Kevlar in order to meet the requirements of heat resistance, i.e. to operate continuously at temperatures of 260 °C, and give very high tensile strength coupled with dimensional stability. Glass is a heavier fibre than Kevlar, but it has a thermal conductivity about 20 times higher and this is an advantage where good heat transfer is important. Kevlar belts, however, generally have a longer service life than glass-based ones. PTFE is an expensive material and the coating process is lengthy; needless to say the base fabrics must be of the first quality with no weaving faults or broken fibres which may project through the coated surface. If the construction is not suitable, the surface smoothness of the finished article may be affected and the coating process may take longer than necessary.[61]

Fabrics are coated with PTFE by dipping repeatedly in a water-based dispersion of about 45–50% PTFE, together with a wetting agent such as the non-ionic Triton X-100 (Rohm and Haas) at a concentration of about 6–9%. Each layer must be dried at about 85–95 °C, baked to remove the wetting agent and then sintered to coalesce the PTFE. In order to prevent 'mud cracking' or unevenness or other defects caused by uneven drying, only small amounts of PTFE can be applied at each dip. Each layer is dried and sintered before application of the next. Temperature control is critical to ensure good results and also to protect the base fibre, because the sintering process entails temperatures around 400 °C. If there are any losses in tensile strength of the base fabric caused to heat each time a layer is sintered, this can add up to a significant loss after a number of passes.[62]

Thinner fabrics mean more flexibility and the ability to pass around rollers of small diameter. Depending on specific requirements, belts can be made microwave compatible and anti-static. Joining of belts has to be accomplished using special techniques and endless belts are the ultimate answer. However, truly endless belts are much more expensive for the customer to buy and there are limitations on belt length. PTFE coated fabrics are used for the slots on some swing-wing aircraft.

4.2.10 Hot air balloons and airships

Hot air balloons and airships were the first devices that enabled man to fly – even Leonardo da Vinci had a design for an airship. There was much activity in France during the nineteenth century and by Count Zeppelin in Germany at the beginning of the twentieth century. More of these craft have been seen in recent years as the advertising value has been recognised. 'Balloons' of all shapes and sizes are now seen, from motor cars to packets of cigarettes to Donald Duck. The basic difference between balloons and airships is that the former simply follow the direction of the wind, while the latter are powered and have some means, usually a rudder, to control their direction. In addition, balloons are simply envelopes of hot air or gas with little or no framework, while many airships have some framework. Those that do not are called 'blimps' – a term originating from the US Navy during the Second World War. The US Navy had airships and 'blimps' during the First World War using helium which was commercially available in the USA. This gas was not available in Germany and so hydrogen was used instead. This was an all important factor which resulted in some of the well-known disasters which have come to be identified with airships. The fire which destroyed the hydrogen filled Hindenburg is believed to have been started by a spark from static electricity. The envelopes for airships were made from various types of rubberised fabric until the mid-1950s when synthetic fibres and plastics began to be used. The companies of BF Goodrich and Goodyear both had contracts to supply the US military. The framework for airships is made from aluminium, although more advanced materials such as composites and carbon fibres are now also used.

The 'lift' of balloons and airships is produced by 'lighter than air' gas. Heated air inside the balloon or airship is less dense than the cooler air outside and has a lifting power of approximately 300 g for every cubic metre of hot air (i.e. 1000 cubic feet of hot air lifts 20 pounds of weight). Helium, a very light gas, has more 'lifting power' and one cubic metre will lift one kilogram of weight, i.e. 1000 cubic feet will lift about 60 pounds of weight.[63] (Hydrogen, which is not used now because of safety considerations, has roughly 10% more lifting power than helium.) Helium is a small molecule and is much more difficult to contain than hot air – it is also quite an expensive gas and any coating must therefore minimise leakage. Helium is an inert non-flammable gas of which there are believed to be plentiful natural supplies in the USA, Algeria, the Arctic and other regions. There are about 20 manufacturers of airships in the world at the time of writing, but there has been a resurgence of interest in recent years due to road congestion and environmental considerations.

Base fabrics can be either high tenacity nylon or polyester. The actual type used may be influenced by the availability of the particular yarn required, e.g. decitex. Nylon is favoured by some manufacturers in certain applications, because it has more extensibility than polyester and can cope with shock better. Polyester generally has higher UV resistance properties than nylon but, when the fabric is coated with polyurethane, the coating generally fails first by becoming porous before the base fabric has degraded to unsafe levels.

Cameron Balloon's Hyperlast™ fabric is produced from a closely woven DuPont nylon 66 base fabric (2.3 oz, 70 g/m^2), coated both sides with a silicone elastomer. The silicone coating is more durable than polyurethane, and produces exceptional tear strength. The material is treated to improve

4.24 A Cameron airship. Cameron Balloons invented the hot-air airship in 1972. Their Hyperlast™ fabric provides unbeatable heat tolerance coupled with incredible strength. Photograph was taken from the Cameron Balloons website and is reproduced with kind permission.

UV and light degradation further and to increase resistance to the growth of mold and mildew. These treatments are essential to preserve the bright attractive colours of the balloon which are subject to prolonged and rigorous weather conditions and damaging UV radiation, see Fig. 4.24. The volume of balloons can be anything between 30 and 320000 cubic feet (0.85 and 9062 m^3) or even more. Silicone elastomer coatings, however, are difficult to paint or otherwise apply artwork to. In addition, they are slippery to handle, are stretchy on the bias (diagonally to the weave) and tend to attract dust.[64,65]

For helium containing balloons, a heavier add-on of polyurethane coating is required, because helium is a small molecule and can leak away gradually through the polyurethane. It is, however, becoming more common to laminate polyurethane films and other specialist films to the fabric – these contain the helium gas better than polyurethane coating.[66–68] Film laminated fabrics, however, tend to be stiff and can, in certain circumstances, be difficult to handle.

Some hot air balloons are produced from 300 g/m^2 polyurethane coated high tenacity polyamide fabric, sometimes with a further coating in the form of a layer of polyurethane resin as an extra barrier against UV light and air leakage. The 'round the world' balloon which may stay airborne for over three weeks requires an extra layer of polyester film laminated to the outer surface as a still further barrier against leakage. The balloon fabrics must be able to withstand conditions at 30 000 feet where the temperature can be as low as minus 60 °C, with a higher level of UV radiation than encountered at ground level. In addition, the balloon speed could reach speeds of up to 200 knots in the jet streams.

Quality tests for balloon fabrics include tear and tensile strengths, puncture and tear propagation and air and gas permeability. Permeability is assessed by measuring the time taken for a known quantity of gas to permeate through a known area of fabric under given atmospheric conditions.

4.2.10.1 CargoLifter

CargoLifter AG in Germany plans to start commercial production in 2004 of airships which can transport large heavy objects weighing up to 160 tonnes, see Fig. 4.25. These airships are semi-rigid, sometimes called keeled airships and are similar to blimps in that the envelope is not stretched over a rigid structure but maintains its shape through internal pressure of the gas. Service will be from 'door to door' with almost no infrastructure and without the need to dismantle objects into smaller components, to be transported by road and then reconstructed at the new site. The CL 160 model is 853 feet long (256m) and will fly at a height of 1500 feet and an average speed of 90 knots transporting loads like a 'flying crane'. The envelope material is made from several layers of Vectran fibre (by Celanese) and is

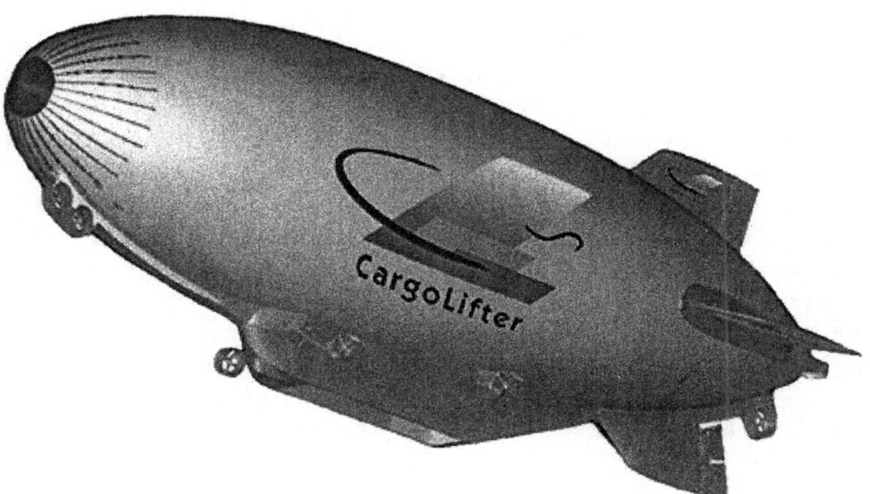

4.25 The 260 m long CargoLifter CL 160 airship which is expected to be in production in the 2004/2005 business year. The CL 75 AirCrane transportation balloon (diameter 61 m) should be ready for commercial use in 2002/2003 and will be capable of transporting loads weighing up to 75 tonnes. Source: CargoLifter website and reproduced with kind permission.

coated with polyurethane. Mylar film is laminated to the outside to reduce helium leakage and Tedlar film is also used for protection from UV rays. CargoLifter is quoted on the German stock exchange and, at the time of writing, is already in the top 100 companies' rankings. The company plan to have global coverage with a depot on each continent.[69–71]

Another German company is believed to be developing airships for the tourist industry. A company in the UK is developing the vessel as a communications platform for mobile phones and television. It will operate at height of 60000 feet using solar cells.

4.2.11 Printer's blankets

These items have always required the most specialist rubber compounding skills and knowledge.[72] The rubber coating has to be able to withstand all the various solvents used in printing without swelling which would cause distortions in print definition and quality. There are four general methods of printing: relief, gravure, lithographic and screen. Printer's blanket requirements will differ for all four. Until the 1960s the relief method was dominant, but this position has been taken by lithographic methods which do not require mechanical setting of relief print set. Modern lithographic off-set printing techniques require blankets made to an even greater degree of precision. The blankets cover rollers and in off-set techniques the rubber blanket itself transfers the print image from the plate to the paper. It must, therefore, be made to a high degree of accuracy to render a clear image, especially the surface which lightly 'kisses' the paper being printed.

The other main requirements, apart from solvent resistance, are dimensional stability and rubber resilience. Printing is a highly dynamic operation with speeds up to and exceeding 20 miles per hour (32 km per hour). Uniformity of blanket thickness and flatness is critical for print quality and definition. If the blanket stretches, it may then be uneven in both flatness and thickness and, in addition, it is likely not to run true but to move about during the printing process. Precision is especially important when colour is being used, because the same image may be printed over several times with different colours. Fabric woven from special, long staple length cotton was used as the base material, because it was at one time the most dimensionally stable material available. A typical quick quality control test was to measure the stretch of a 2.5 cm wide sample of a single ply under a 50 pound (22.7 kg) weight – it had not to exceed 2%. Rubber stuck well to this cotton fabric without any pre-treatment, but modern blankets made from low shrinkage, high tenacity polyester require a bonding agent.

There are many different grades and types of printer's blanket for each of the different printing machines and products to be printed. A typical printer's blanket for newsprint is made up from three layers of woven fabric and many layers of specially formulated rubber compound. The three-layer blanket is produced by laminating together three separate layers of the single coated fabric. The rubber covering the top fabric is the most critical and must be flat to a high degree of precision. This is accomplished by the application, knife over roller, of up to 40 layers of coating with the gap between product and blade being set accurately before each layer. The middle layers of rubber are generally Neoprene, but the top layer, which needs to be solvent resistant, is generally nitrile rubber. Each coating recipe can contain over 20 different ingredients, and there may be two or three different types of rubber within the mix. The solvents used are usually a mixture of ketones such as propanone, butanone and MEK, the last having a very low boiling point. Mixtures of petroleum solvents can give a wide band of boiling points up to 180 °C. The mixture of solvents promotes gradual and even evaporation which helps produce a flat, uniform coating which is free from porosity. The spreading machine used is basically not very different from the machine pioneered by Hancock in 1837, although the instrumentation and control have improved significantly since the middle of the nineteenth century. The blankets are produced as batch lengths of say 50–100 m and the multiple layers may take 10 hours or more to complete on the spreading machine. This process is then followed by vulcanisation and inspection. Printer's blankets are thus one of the most sophisticated and costly to produce of all coated textile products.

The temperature of certain printer's blankets increases during use in relief printing, and this can add to the wear on mechanical properties and induce shrinkage of the textile. Aramid fibres are used in some of the latest types to combine light weight, flexibility, temperature resistance and dimensional stability. Glass fibre has also been used because of its dimensional stability and strength but, if faults develop, it can become brittle and fail quickly in use.

4.2.12 Exhibition board coverings

Notice boards and exhibition boards can be produced from raised knitted fabric and items can be attached via loop fastenings (an example is Velcro™) which mechanically adhere to the loops in the knitted fabric. These loop fastenings are much stronger than may be thought and relatively heavy items as well as posters and picture frames may be attached. Other applications are in signs, noticeboards, window displays, point of sale displays and educational and conference aids. The advantages are that the items can be positioned anywhere on the board, are easily removed and repositioned and the same boards and equipment can be used time and time again. The fabric itself is usually attached to the board via hot melt adhesives that need first to be coated on. The fabrics normally used for this purpose are generally dimensionally stable and have sufficient cover to be direct coated. A suitable PVA type adhesive is selected for the required durability, thickened and direct coated on to the back of the fabric. The coating is dried on the stenter and rolled up for despatch to the customer who can reactivate the adhesive by the application of heat to fix it to the exhibition board. The adhesive selected should not be tacky when dried and should be reasonably flexible so that it does not 'crack off' the fabric when it is rolled up. In addition, it must not cause discoloration or have unpleasant odours, and it should last the lifetime of the demonstration board without peeling off or delaminating. Flame retardant chemicals may be added to the resin formulation to meet any required standard.

4.2.13 Labels

Every textile article is likely to have a label sewn in or on it to convey information. Some of this is required by law, such as the fibre content and, in some cases, information on the flammability. Care and cleaning instructions and usually the country of origin of the article are also included. In many case the information is presented in more than one language. All this means that a lot of text has to be printed on a small piece of fabric, and some labels are larger than they were, say, 15 or 20 years ago. Label fabrics are sometimes coated to improve opacity and to allow the label to be printed on both sides, hence reducing the amount required. Coating also reduces edge fraying and may facilitate and improve label quality, both in the ability to accept printing ink and also in label fabric handling – the label must be sewn in straight and in the correct position in the garment. The sewn in garment label must last for the lifetime of the garment, which may be 50 or more machine washings at 60 °C. However, the main drawback to coating of labels, especially labels for undergarments, is the stiffening it causes which is likely to irritate the skin of the wearer. Labels can be produced by direct coating of soft, carefully selected, water-based acrylic or polyurethane resins or solvent-based polyurethane resins. The resin compound normally needs a filler base such as titanium dioxide to confer opacity and whiteness.

4.2.14 Greetings cards

Up-market greeting cards sometimes make use of optically brightened satin woven fabric, which is frequently coated with a resin to improve the flatness and smart appearance and to reduce edge fraying after cutting. For greetings cards there are generally no durability requirements such as washing, but the resin must not change the usually high degree of whiteness of the fabric. The resin must be odourless and nothing must leach out of the coating to stain or mark the paper or cardboard. The usual method of manufacture is knife on air direct coating and resins are generally acrylic or polyvinyl acetate based.

4.2.15 Iron on coverings (modelling) fabric

This material can be produced from closely woven plain weave nylon or polyester coated with a resin such as polyvinyl acetate, which has hot melt adhesive properties and melts at around 140 °C under the action of heat supplied by a hot iron. If the fabric is not heat set and the resin is coated on under conditions of high warp and weft tension, the material can be ironed on to curved and round surfaces, shrinking to shape as it is applied and sticking to the surface via the hot melt adhesive. The material can be used for model aircraft, model boats or for any other similar purpose. Two layers of coating are usually required; the first layer to act as a barrier and the second layer to act as the adhesive. It is possible to produce 'solvent resistant' fabric by a selective choice of resins for the barrier layer – two barrier layers may be necessary for high specifications – followed by application of the hot melt adhesive as the top layer. The durability is determined by the hot melt adhesive and polymers with higher performance than polyvinyl acetate could be used.

4.2.16 Filter products

It has already been mentioned that fabric porosity can be modified by coating, and this property can be exploited in filter applications. Crushed foam coatings can modify the porosity from zero upwards, depending on the add-on of resin and the severity of crushing, see Fig. 4.26 and 4.27. Foam which has collapsed into the fabric and is not on the surface will present the least resistance and will be more suited to high air flow requirements. Acrylic and polyurethane resins are currently used, but development work is being carried out to produce coating formulations which have heat resistant properties and thus extend the range of applications in industrial filtration.

4.2.17 Geomembranes

The product area of geotextiles saw quite spectacular growth when they were first introduced – 5 million to 115 million m^2 expansion in North America alone during the period 1976 to 1983. Even in the twenty-first century well above average growth is expected. Geotextiles basically fulfil the engineering roles of land or soil drainage, filtration, separation or reinforcement.[73,74] When the material is required to form an impermeable barrier to liquids, for example in reservoirs, canals or irrigation channels or

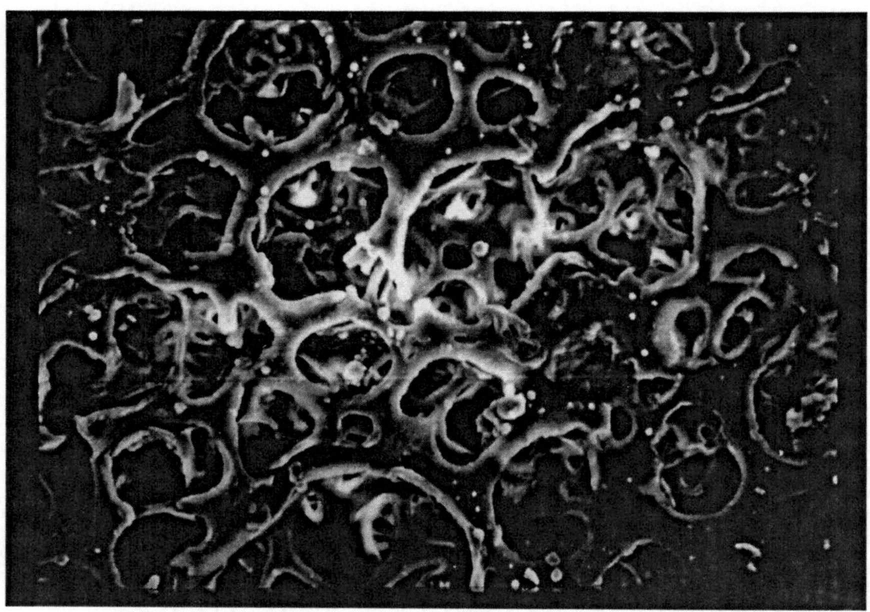

4.26 Photomicrograph of a crushed foam coating for industrial filtration. This sample has an air permeability of 60 l per m^2 per minute and is produced from Performax 'sit up' coating. Photograph supplied by Noveon (formerly BF Goodrich) and reproduced with kind permission.

to prevent leakage of toxic waste liquids, a geomembrane is required. Geomembranes have been formed by spraying the geotextile on site with acrylic or bituminous materials. A major problem is controlling the thickness of the sprayed coating. Unsupported films of Hypalon, PVC, chlorinated polyethylene (CPE) and high density polyethylene (HDPE) are used as geomembranes and, where necessary, have been reinforced with layers of woven and nonwoven fabrics. The fabrics also assist with abrasion resistance and energy absorption. Pre-coated geomembranes are used, but they are expensive and must be easily joined on site to form watertight seams. The polymers used must be resistant to weathering and resistant to the chemicals with which they will come into contact and, for minimum fabrication in the field, the coated fabric must be as wide as possible. Some water vapour permeability of the film or coating is sometimes required, for example in earthwork stabilisation, to prevent build up of excessive hydraulic pressure. Base fabrics are usually nylon, polyester or polypropylene. A more recent application of geomembranes is containment of

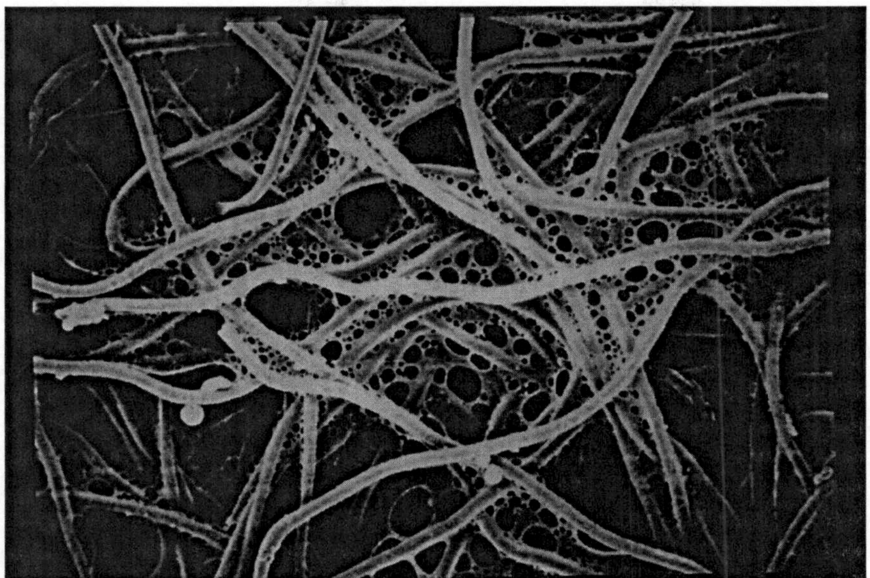

4.27 Photomicrograph of a crushed foam coating for industrial filtration. This sample has an air permeability of 125 l per m^2 per minute and is produced from Performax 'sit in' coating. Photograph supplied by Noveon (formerly BF Goodrich) and reproduced with kind permission.

leachate from landfill sites and there should be opportunities for development in this area.

4.2.18 Tyres and hoses

These articles are mentioned only very briefly here because they are not mainstream coated fabrics. However, the technical requirements involved are similar, e.g. rubber or polymer adhesion to fibre must be of the highest standard and adhesion promoters are generally required. Durability requirements of hoses depend on what is being carried in them, which may be liquid chemicals, oils or fuels. Hoses used in automotive or any transport application need to be as light and as space saving as possible. Only yarns with high strength to weight ratios, e.g. aramids, may be used. Tyres become hot in use and only yarns with good thermal stability may be used – viscose is probably the best fibre for this. Both articles must be tested to rigorous standards, because failure in use can have serious consequences. Further information is available in most rubber textbooks, see Further reading.

4.3 Automotive applications

4.3.1 Introduction

The car is the largest single market for technical textiles with about 20kg in each vehicle. Much of this fabric is coated or laminated or in the form of composite materials, i.e. seat covers, door panels, headliner modules, hoses and belts, tyres, etc. Only components produced by actual coating or laminating of fabric are described in any detail, but a brief mention is made of certain other items because their manufacture involves similar technology. The car interior has assumed much greater importance, especially since the 1990s, with the emphasis on design aesthetics, comfort and durability of the textile. The automotive industry is undergoing a period of rationalisation and reorganisation with global sourcing of materials and components. It is seeing the emergence of large companies, called Tier 1 suppliers, making components for several different automotive companies (termed OEMs). Critical factors in materials for use in automobiles are material cost, lightness in weight (for energy saving) and consistency in properties to enable modern large-scale production techniques to be carried out as efficiently and at as low a cost as possible. The intense competitive nature of the industry demands continual cost downs, just in time deliveries (JIT) and faster time to market for new developments. This is leading to new manufacturing techniques that require higher standards of consistency and new textile properties, such as impermeability to liquid polymer for use in direct moulding methods. At the time of writing environmental concerns about reduced emissions, recylability and disposal are influencing both methods of manufacture and choice of materials. Many OEMs are already using recycled fibres and are examining the use of natural fibres because they are a renewable resource. Specialist reports and volumes are available on automotive textiles that deal with these subjects in more detail, see Further reading at the end of this chapter and also Chapter 8.

4.3.2 Seat covering materials

Car seats and other parts of the car interior are covered with fabric, real leather, artificial leathers and also polymer foils. The main durability requirement of car seat covering material, i.e. resistance to abrasion, light and UV radiation, must be met.[75–80] The material must still be in almost first class condition after two years' use, so that the car will have a good resale value, and must last the life of the car, which may be as much as 15 years or more. Conditions inside a car vary from below 0° to well over 120°C, depending upon where in the world it is sold. Even in England, the car interior in a normal summer can exceed 80°C. In addition to the extremes of

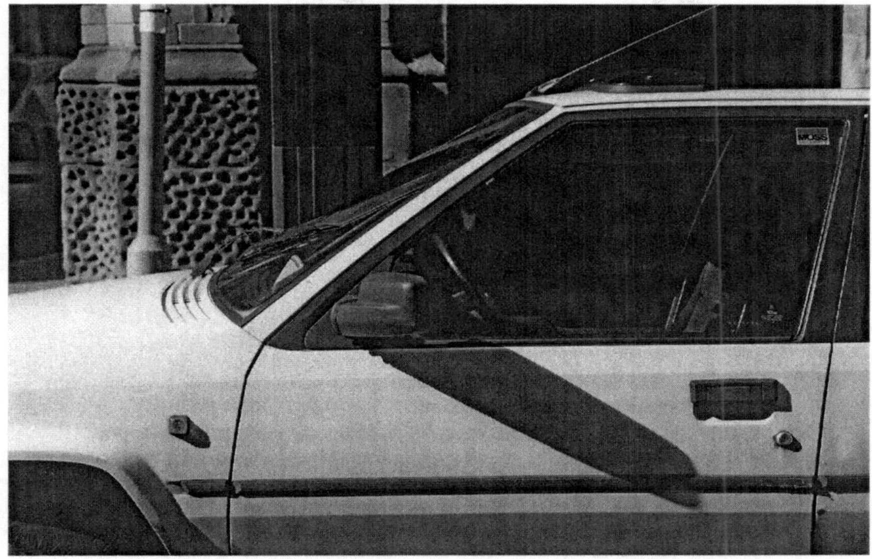

4.28 Modern cars are more aerodynamic with sloping windscreens. This results in a higher surface area composed of glass which leads to hotter car interiors because of the 'greenhouse' effect. The car interior fabrics must stand up to a daily cycle of temperature changes from 130 °C or more in hot tropical sunshine to perhaps below freezing at night. In the confined space of the car interior, the relative humidity will also vary widely. Polyester is now used for over 90% of fabric car seats because it alone has the required standard of UV degradation and abrasion resistance combined with economical cost.

temperature, the relative humidity of the closed confined area inside the car could vary from 0 to 100% continually in daily cycles. All materials used as seat coverings must be able to withstand these conditions, in addition to being sat upon for many years. When synthetic fibres were first introduced into car interiors, the highly demanding conditions had not been recognised nor had realistic test methods been developed, see Chapter 5. In addition, at about the same time (early 1970s), the oil crises demanded that cars be more aerodynamic. This led to the use of more slanting glass in cars which made car interiors hotter in sunlight due to the greenhouse effect, see Fig. 4.28.

4.3.3 Seat fabrics

Over 90% of all fabric coverings are in polyester because only this fibre combines the requirements of high performance with reasonable cost. Polyester is more resistant to the particular wavelengths of UV light which

Automotive seat fabric is almost invariably a *tri-laminate* – polyester is used on about 90% of all car seats. Wool, wool blends, leather and man-made leather are also used.

polyester fabric
polyurethane foam
nylon/polyester scrim backing

4.29 Car seat fabric is almost invariably in the form of a *trilaminate*: face fabric/polyurethane foam/scrim backing. More than 90% of all car seat fabric in the world is polyester, because only polyester has the required standard of high abrasion resistance combined with UV and light resistance. The laminate provides a soft touch, prevents creasing or bagging over many years of use and also gives deep attractive sew lines. The scrim backing helps control stretchability of knitted fabrics, helps seam strength and acts as a 'slide aid' during sewing and making up. Leather and man-made leather are also foam backed with a 'scrim slide aid.'

penetrate the glass windows of a car. Wool or wool/polyester blends are sometimes used in up market models. Both woven warp and weft knitted fabrics are used, but all fabrics are almost invariably laminated to polyurethane foam together with a scrim fabric on the other side of the foam. This trilaminate car seat fabric is produced to impart a soft touch to the fabric surface and to prevent creases or 'bagging' developing during the life of the car. Lamination to foam also produces the attractive appearance of deep sew lines, see Fig. 4.29 and 4.30. The scrim on the back of the foam helps control stretchability of the laminate, especially with knitted fabrics, and it also improves sewing seam strength, which is likely to be important for lighter weight woven fabrics. In addition, it acts as a 'slide aid' during sewing and making up of the car seat, because polyurethane foam tends to 'grip' surfaces. The scrim is generally knitted nylon or polyester in suitable constructions to produce the stretch control required, e.g. queenscord, to reduce excessive stretch. If the scrim is required only as a 'slide aid', a non-woven fabric may be used.

The design and colour of the fabric must match the appearance and colour of the rest of the car interior and, indeed, must also complement the exterior design and capture the mood and fashion prevailing in the market in which the car is being sold. The importance of the design of the car

4.30 Modern car seats are always attractive with well defined sew lines, and are completely free of creases or 'bagging' owing to the polyurethane foam backing on the fabric. Interior fabric design and construction has become an important factor in attracting customers. Photograph supplied by Caligen Foam Ltd and reproduced with kind permission.

interior has grown very significantly in recent years and there are regional and even local preferences for particular designs and fabric handles and touch properties. In Europe flat woven fabrics, although gradually reducing in volume, still predominate, while in Japan, softer touches from woven pile fabrics and knitted velour are more popular. Woven pile fabrics are especially popular in the American market. Use of leather and synthetic leathers, notably Alcantara (by Toray), is growing and seat covers made entirely from these materials or with textile inserts are believed to make up about 12% of the total market in Western Europe.

4.3.4 Coating of car seat fabric

Woven car seat fabric is sometimes back coated to increase the abrasion resistance and to improve the FR properties of the whole tri-laminate. The back coating resin can be acrylic, polyurethane or SBR based and can be coated on, knife over air as a straight resin or knife over roller or table as a foamed resin. Add-ons vary from about 10 to 60 gm^2 or more, but the tendency is to put on as little as possible for reasons of cost and also so that the seat cover laminate is as light as possible. In some cases resin add-on may be even higher to control the penetration of spilt drink into the seat cushion. The resins must be soft and flexible and have minimum effect on the fabric drape. Excessive add-on or penetration can cause the fabric laminate to feel harsh or stiff or 'crack' when draped around a concave or convex curve. Excessive use of surfactant in the resin formulation must be avoided because 'tide marks' or ring marks may appear on the surface if spotted with water. Woven velvet fabrics must be back coated to lock in the pile. This can be carried out by direct knife coating using either thickened or foamed resin, or by a back-licking process using a resin of relatively low viscosity.

The FR requirement is generally a burn rate from a horizontal burn test based on the Federal Motor Vehicle Safety Standard FMVSS302. Technically this is not a particularly difficult test to pass, but in practice it may not be easy to do this at a particular price structure and fabric and laminate weight. FR chemicals, usually based on the antimony/bromine synergy, are added to the coating at high concentrations and this, now heavily filled, resin may give problems in actual coating, especially if housekeeping is not of the highest order.

Occasionally, seat fabric is froth coated on to the face to improve abrasion resistance. This process is more of a finishing treatment, because the coating is not visible and merely coats the individual threads without forming a film. Great care is needed in selecting resins or any finish to be applied to the surface of a car seat fabric because of the very demanding conditions it will have to withstand, i.e. high prolonged temperatures, varying relative humidity, and abrasion by car occupants' clothes which may be wet or damp. Problems which may arise include white or sticky deposits appearing on the fabric surface, or stains, especially 'tide marks' which can appear after wetting. In fact, apart from the soil release fluorocarbon type, very few finishes are applied to fabrics for of these reasons.

Fabric to be used on door casings is sometimes coated with PVC latex to enable it to be HF welded to other materials used in door construction. A certain amount of stiffening may be tolerated for door casing fabric, but the other considerations that there should be no effect on appearance or staining (marks) because of the coating process are still important.

4.3.5 Lamination of car seat fabric

Car seat fabric is generally laminated to polyurethane foam by flame lamination, which is a quick, economical way of producing the triple laminate in a single process. A gas flame impinges directly on to the surface of moving polyurethane foam under controlled conditions. This process melts the foam surface, and the molten foam is immediately covered with one of the fabrics to which it is being laminated. The same process takes place on the other side of the foam, and thus three substrates are fed into the machine and a single product emerges.

For the fabric to be made up into car seat covers, the tri-laminate must be dimensionally stable without any curl and must lie flat on the table for panel cutting. This is not easy because all three materials are capable of stretching during lamination, especially the foam and scrim fabric. Handling of the materials, which must be uniform in their properties, needs to be carried out with extreme care and the lamination machine must be designed to do this at commercial speeds of up to 30 m per minute and higher.

The modern global automotive industry is so competitive that only the absolute minimum amount of foam thickness may be lost during the lamination process for the operation to be commercially feasible. Thickness is lost by a combination of foam burnt off and foam compressed during the nipping process in lamination. To compensate for this, the input foam must be thicker than specified by the customer for the finished product. This adds to the material cost of the process, and to have the least amount of foam burnt off while still producing a satisfactory bond and fabric of the correct quality, at the required speed, is a measure of efficiency. Deciding the various combinations of the gap settings between the lamination rollers, and the air/gas ratio which determines flame temperature and lamination speed is a highly skilled art – based on scientific principles. The actual machine settings need to change with the type of foam being laminated, i.e. different foam density, FR grade, polyurethane foam type (polyester polyurethane or polyether polyurethane – see below), and also with the type of fabric being laminated.

Foam density can be anything from 28 km^3 to about 70 km^3 depending on the end use required by the OEM. Density will influence the compressibility of the foam and FR chemicals influence melting and flow characteristics. Polyether polyurethane foam, required by some OEMs, in the unmodified state cannot be flame laminated. Producers of polyurethane foam such as Caligen Foam have produced varieties which can be flame laminated, but the melting characteristics are significantly different from those of polyester polyurethane foam. Polyether polyurethane is more hydrolysis resistant than polyester polyurethane and is used in those parts of the world which have high ambient temperature and humidity.

In addition to being dimensionally stable, the laminated fabric must pass a variety of tests to simulate many years of actual use and also to assess its suitability for further processing. The bond quality is assessed by peel bond tests carried out after exposure to water and certain solvents, and after heat ageing tests, which can be as severe as several days at temperatures as high as 120 °C. In general the bond should initially be in the order of seven newtons per 5 cm width, although some OEMs require even higher than this.

The flame lamination process, while being very quick and economical, requires a certain amount of polyurethane foam to be burnt off which produces potentially toxic fumes. These must be satisfactorily treated before release to atmosphere, and this abatement process must be included in the processing costs of flame lamination. Fumes can be treated by 'scrubbing' or by carbon adsorption. The process is controlled by environmental regulations in most parts of the world. A number of flame laminators have been reported to have been closed down or been required to convert to cleaner methods of lamination. However, many large scale laminators operating in the automotive area continue to flame laminate with fumes being treated before release to atmosphere. It appears that when very large volumes of fabric are being processed, the fixed cost of installation of fume control equipment can be absorbed into the overall processing costs. In addition, until relatively recently, alternative methods of lamination either cost more or were incapable of producing the high standards of excellent peel bond and flexible handle required by the automotive industry for seat fabric. Powder adhesives generally require quite high add-ons, and great care is needed to produce laminates which do not 'crack' when folded or when draped around a convex or concave curve. The most promising alternative, at the time of writing is believed to be moisture cure polyurethane adhesives, either by slot die extruding or application from gravure roller. More stringent air quality regulations for flame lamination may make these alternatives more economically feasible in the future. Hot melt powders, films and webs, however, have been used for many years for items which are more rigid and require less flexibility, such as door casings, headliners and other areas of the car.

Flame lamination is a versatile method of lamination and can be used to laminate a variety of different substrates. Polyurethane foam about 1–2 mm thick, sometimes referred to as 'mini foam', can be used purely as the adhesive to join, for example, nonwoven material to fabric or films to fabric. The mini foam is virtually all burnt off during the lamination process and a flexible product is obtained, because the foam joins the two materials in a discontinuous manner without penetration.

4.3.6 Other seat coverings

Until the late 1970s, most car seats were covered in real leather, or 'leather cloths' produced from PVC coated fabric. Both straight PVC and 'blown' PVC varieties were used, the latter having a softer touch. However, these coated fabrics did not 'breathe' and were uncomfortable to sit upon, especially in hot weather. Attempts were made to improve this by producing a 'tape yarn' by slitting PVC into thin strips and weaving it.

Natural leather has always been used for car upholstery and is still regarded as the ultimate in luxury. Even leather is usually laminated to polyurethane foam to produce a soft touch to the material, and there is usually a fabric scrim on the back to aid processing. Natural leather is sometimes lacquered with a polyurethane resin to improve abrasion resistance, but this reduces breathability of the leather. Leather processing has been modified in recent years to reduce environmentally unfriendly effluent.

Despite its luxury appeal, natural leather is difficult to work with because hides come in different shapes and sizes and their quality varies widely. Several hides have to be examined very carefully to select sufficient matching leather for use in the same car. There is also a shortage of hides, and this may intensify as people in developed countries eat less red meat. At the same time, the demand for leather is increasing and rising standards of living mean that more people can afford it. There have been many attempts over the years to produce artificial leather and many products have appeared on the market, some of which have been withdrawn, e.g. Corfam (Du Pont). At the time of writing the most successful man-made products, two grained leathers and eight suedes, are all Japanese made. The companies involved anticipate a significant increase in demand. Toray, who produce Alcantara (in Europe together with Matsui), predicted in 1997 an increase from the 16 million m^2 in 1995 to 25 million m^2 by 2005, with a significant proportion going into European cars.[81,82] Currently nine car makers are using a total of about 1 million m^2. Kuraray are entering the European market with their Amaretta brand.[83] At present very little man-made leather appears to be used in the USA, but this is expected to change. Man-made products have the important advantages of availability in roll form, uniformity of quality and lighter weight. The first two factors allow more efficient production planning and the third is of special importance in the automotive and transportation industries.

4.3.7 Headliner structure

The modern car headliner is a multiple laminate of up to seven or more components all joined together, see Fig. 4.31. Each layer is there for a specific purpose, either for aesthetics, or to provide sound insulation and vibration damping, or to provide rigidity to the whole structure. Much research work has been carried out, especially to optimise the sound and vibration damping.[84–86] The centre core is generally a layer of semi-rigid thermo-mouldable polyurethane foam, initially about 15–30 mm thick; alternatively it can be composed of waste fibre (recycled garments) bound with semi-cured phenolic resins. The centre core of polyurethane is bonded to two layers of chopped fibreglass rovings, one on each side. The fibreglass rovings are bound together and embedded in thermoplastic material, i.e. either hot melt adhesive powder or hot melt adhesive film, e.g. Xiro film, or a combination of both.[87] These materials also act as the adhesive when the layers are joined together. Opinion differs as to whether continuous or slit film adhesives contribute the most to noise reduction. The glass roving layers help impart rigidity to the structure and are not always necessary when phenolic resinated waste fibres are used. Attached to the side facing inwards is the decorative material; a nonwoven polyester scrim is usually attached to

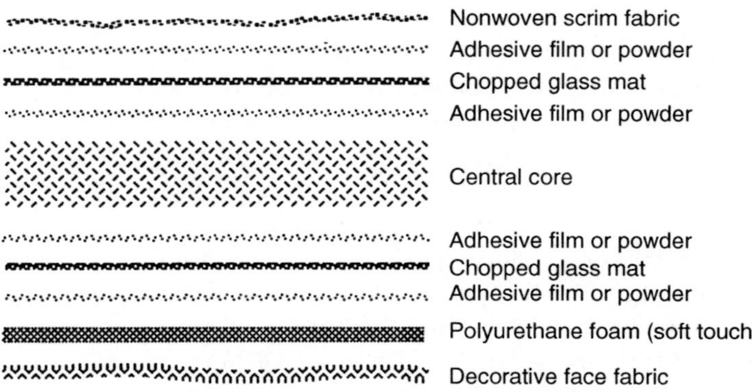

4.31 Typical headliner construction. The central core is semi-rigid polyurethane foam, resinated shoddy waste fibres or some other material. Heat is applied to the assembly in a flat bed laminator to join all the components together. The material is then cut into lengths, heat is applied and the headliner is press moulded to the required shape. During the latter process the assembly is compressed to about one-third of its original thickness. The adhesive film or powder permeates through the chopped glass mat and consolidates it. The chopped glass mat contributes to rigidity and acoustic insulation.

the other side. All layers are joined together by action of the hot melt adhesives in a flat bed laminator, taking care not to damage the aesthetics of the decorative material or to reduce the thickness of the centre core. The correct temperature and pressure must be optimised.

The composite sheet is then moulded to produce the required shape, usually by preheating the assembly with IR heaters just before placing it into an unheated mould where pressure is applied. During this moulding operation the semi-cured phenolic resins are fully cured and the thickness of the central core is reduced by about two thirds. All of the hot melt adhesives have to be selected to meet the heat resistance specifications. Sometimes, relatively inexpensive corrugated cardboard is used as the central core, but this material does not always allow sharp, well-defined lines when moulded. Phenol resinated cotton is relatively low cost and has good formability and acoustic properties, but it is heavy and, if damp, can distort and give off odours.

There are other method of bonding the headliner structure together, such as the Tramivex™ method[85] which involves dipping the polyurethane foam into a bath of liquid chemicals. In other cases spray adhesives are used. The patented 'high calorific transfer medium' (HCTM) process, which makes use of superheated steam to activate hot melt adhesives, is used for headliner construction as well as for door casing and seat making. To ensure rigidity, headliners in larger vehicles generally have more layers, and they may also be thicker in luxury cars for more effective sound proofing. In the USA, alternatives to fibreglass are being explored, because of complaints of dermatitis by workers who handle the material.[88] In addition, because headliners have become more complex and incorporate more items, the fibreglass is more easily damaged during the assembly process; phenolic resinated fibreglass is especially brittle. Fibreglass began to be used because of its exceptional properties in sound absorption. Nonwoven researchers are attempting to replace fibreglass and the centre semi-rigid polyurethane with polyester nonwoven, to achieve a 100% polyester article, which should be easier to recycle.

In the USA, knitted headliners are losing ground to nonwovens which are used extensively in Europe and Japan.[89–91] Nonwoven cover fabrics have the advantages over knitted fabrics of less cost, and of not trying to shrink back during moulding operations, but staying in place allowing deeper draws. The latest developments in headliner nonwovens include the use of fine denier polyesters, three to six denier per filament (dpf), which have good covering ability at low weight. Polyester fibre has a higher melting point than polypropylene, which means that in thermal moulding operations, a higher temperature with shorter processing times can be used. Typical nonwoven headliners are about 200–220 g/m^2.

4.3.8 Other interior coverings

Fabrics to cover door casings, parcel shelves or other areas of the car are sometimes used in a combined lamination and moulding operation in which the plastic, rigid carrying component is formed from molten polymer in a mould with the fabric already in it. This single operation replaces three separate processes, but it requires a barrier material laminated on the back of the face fabric – or the back of the foam to which the face fabric is joined – to prevent molten polymer striking through before it solidifies. This barrier can be a nonwoven fabric, a film, a polyurethane foam of higher than normal density, i.e. say over 50 km^3, or a combination of these materials. Most door casings, however, are still produced by more traditional methods of moulding and lamination, usually with fabric to foam laminates. Both polyurethane and polyolefin foams are used for a soft touch and thermal and sound insulation, see Fig. 4.32.

Small items, such as armrests and headrests, are sometimes made by a 'pour in foam' technique. Panels of fabric laminate are cut and made up into the article cover. Liquid polyurethane foam is poured into this 'bag' which then solidifies to form the finished armrest or headrest. However, it is necessary to provide some kind of barrier to prevent the liquid foam seeping through the fabric before it solidifies. For these small articles, it is sufficient to use a foam of higher than normal density in the fabric

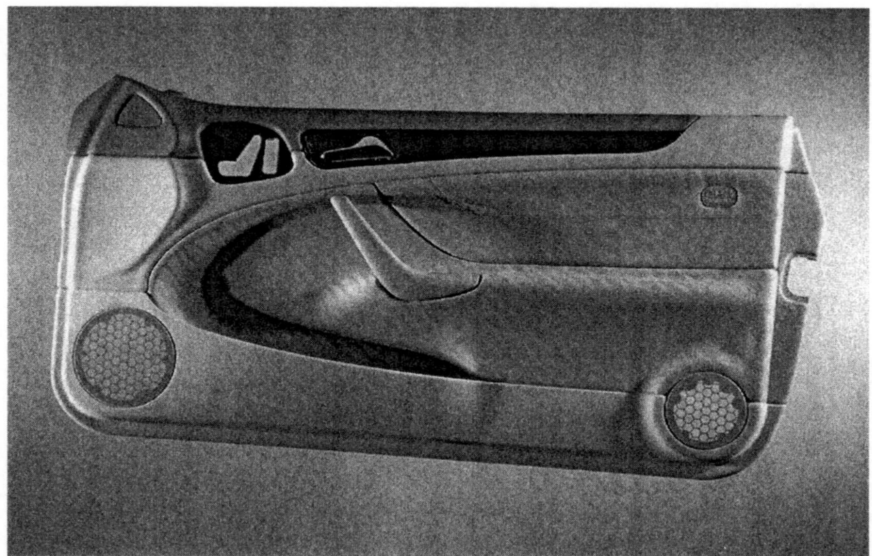

4.32 Car door with well defined curves produced by a moulding technique. This particular door features the 'soft touch' and sound and thermal insulation of polyolefin foam. Photograph supplied by Alveo (Sekisui) and reproduced with kind permission.

laminate. Around 1990, some car seats were produced in this way, but a polyurethane film was required as the barrier because higher pressures of foam developed inside the much larger structure of a car seat cushion or squab. The process was discontinued for large scale seat manufacture, possibly because of seat thermal discomfort, i.e. discomfort in hot weather (see Section 5.6.13) and other reasons.

4.3.9 Airbags and associated products

Airbags have come to be widely used only since the early 1990s, but they have in fact been around since before 1970. Their growth has been quite spectacular, mainly because of legislation in the USA. The federal safety standard FMVSS208 requires all passenger cars sold in the USA to have airbags for both the driver and front seat passenger. A second federal standard FMVSS201 required 10% of cars to be fitted with some type of head protection by May 1999.[92,93] The standard did not specify the means by which this should be achieved, and it could be a type of airbag or some kind of padding in the headliner area of the car. FMVSS201 requires 100% of cars to be fitted with some head protection facility by May 2003. The USA has a higher proportion of sports utility vehicles (SUVs) and light trucks, which have a higher centre of gravity than cars, leading to more roll-over types of accident. Head protection is especially relevant in these circumstances. In Europe BMW introduced a head protection device, an 'inflatable tubular structure' (ITS), and also a side impact airbag in their 7 series which has been extended to their 5 and 3 series of cars. In fact, a BMW concept car featured 12 airbags,[94] although this is probably impractical because if they all went off together, the noise and the sudden increase in pressure within the vehicle could be harmful to human health. Side airbags provide protection in roll-over accidents by shielding the occupants from side window glass and protecting the head.[95,96] Airbags for protection of the knees and legs are being considered. Autoliv have three different head protection systems, the combined Head and Thorax bag (HAT-bag), the ITS and the inflatable curtain (IC).[97] Volvo installed an Autoliv side airbag during June 1994, which was the world's first side airbag, and a short time later, they were also the first OEM to use an IC, which is held inside the headliner and covers the length of the car interior. Typically, because each car manufacturer has its own views on appropriated design for their cars, major producers of airbags make dozens of different types, see Fig. 4.33 and 4.34.

The world market for airbags in 1998 was worth \$5.5 billion and was expected to reach \$7 billion in 1999 – there is still room for considerable growth in both front passenger airbags, especially in Europe and Japan, and side impact airbags, especially in the USA. Acordis Industrial Fabrics believe that worldwide consumption of yarn for airbags will increase from the 1999 level of 50000 tonnes representing 200 million m^2 of fabric and reach 80000 tonnes and 300 million m^2 of fabric by the year 2004. The figures for Western Europe are believed to 18000 tonnes (75 million m^2) in 1999 increasing to 25000 tonnes (115 million m^2) by 2004.[98]

Airbags operate by a triggering device, which sets off explosive chemicals when it senses that an accident at a speed greater than about 35 km/hr is about to happen. This causes the bag to inflate, which cushions the human body and prevents it from hitting a harder object. The airbag inflates and deflates all within a fraction of a second – less than the time to blink an eye. The fabric from which the bag is made must be able to withstand the force of the hot propellant chemicals and, more important, these must not penetrate through the fabric to burn the skin of the car occupant.[99] Polyester is not used for airbags because its thermal properties are not suitable. Compared to nylon 66, about 40% less heat is needed to melt polyester and hence this fabric could allow the penetration of hot gases.[100]

The first airbags were Nylon 66 coated with Neoprene rubber, but in efforts to make bags which are light and thin enough to fold up into a compact pack, use of silicone coatings soon followed. Silicone coatings are applied by knife on air coating, and there are now 100% 'solids' silicone

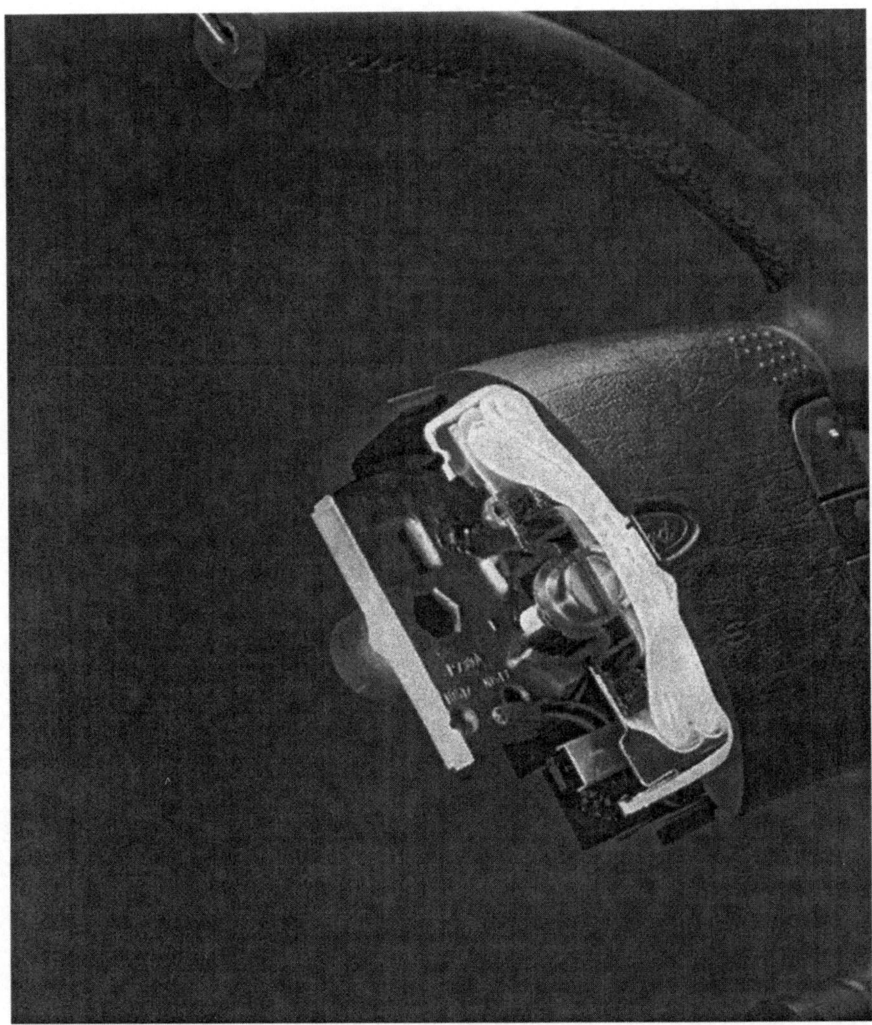

4.33 Cut through of a Ford steering wheel showing airbag in place. Source: Autoliv website and reproduced with kind permission.

fluid resins which contain no solvent to evaporate off. They have the advantages of long term UV light and heat stability, and performance at low add-on which allows smaller and more compact packing.[101,102] Some airbag fabric is coated on both sides, and the cost of coating is more than double the base fabric cost. Needless to say, there is a constant search for savings both in cost and weight. The most recent development involves film lamination for

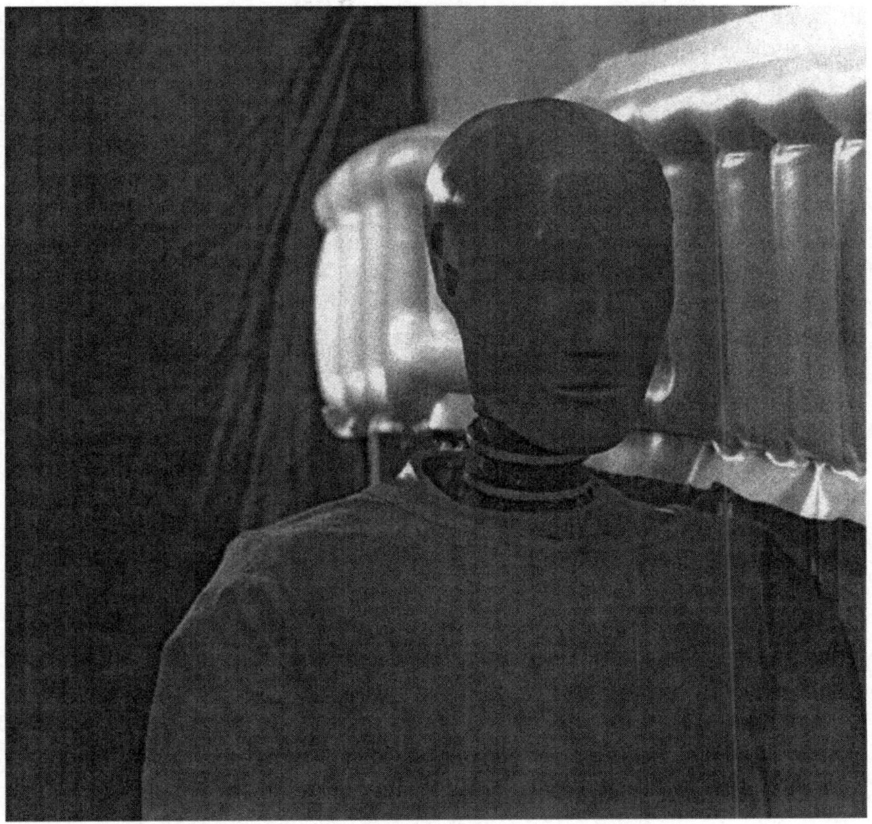

4.34 An inflatable curtain (IC) for head protection.
Source: Autoliv website and reproduced with kind permission.

the ITS, which provides protection in roll-over accidents and which must remain inflated for five to seven seconds.

There has been considerable research and development to improve the deployment, the design and also the production efficiency of airbags. The huge volumes involved – not to mention the life preserving factors – justify the effort being put into airbags and all associated safety devices. To save weight and cost, uncoated fabrics have appeared which make use of fabric construction to control air permeability. These appear to be the preferred option for the future. However, there are advantages and disadvantages for both coated and uncoated fabrics. Coated fabrics do not fray and are easier to cut and sew, and air porosity can be controlled better. Non-coated fabrics are lighter, softer, less bulky and can be recycled more easily. The sizes of air bags vary with the car they are going into and also whether they are to be used for the driver or the passenger. In Europe, driver side airbags are about 35–70 l capacity, while front seat passenger airbags are a little larger, about 60–100 l capacity. Airbags in the USA are generally larger than European ones, because in Europe the airbag is designed for use in conjunction with a seat belt, while in the USA not all drivers use seat belts, relying on the airbag alone for protection.

Airbags are typically woven from high tenacity, multi-filament nylon 66 yarns in the approximate dtex range of 210–840 with 470 being the most widely used in Europe and Japan.[103,104] Fabric weights, uncoated, are about 170 and 220 g/m^2. A small amount of nylon 6 is used; this is claimed to be softer, which minimises skin abrasion, and also to have better packing compactness. Autoliv has developed a 'one-piece weaving', double layer system which produces airbags directly from the loom.[97] Airbag fabric is not dyed, but it needs to be stabilised by heat setting, and scoured to remove impurities which could cause mildew or other problems. The fabric must be strong with high tear strength and high anti-seam slippage, and it needs to have controlled air permeability, usually measured using ASTM D37–75 or DIN 53877.[105] It must be capable of being folded up in a small space for over 10 years or more without deterioration and, in the case of coated fabric, without blocking or sticking together. Coating quality is assessed by the Swiss 'scrub test' and by rigorous 'environmental' ageing procedures, for example treatment for 14 days at, say, 120 °C followed by a period of time at –40 °C and then subjecting the specimen to high humidity. Some tests specify 75% property retention after 4000 hours at 90–120 °C, the equivalent of 10 years UV exposure, and also cold cracking down to –40 °C. A selection of the main test methods is in Table 5.2. DuPont have tested two of their airbags in nylon 66, one coated and one uncoated, which were installed in a car in the 1970s, and found no loss in performance after 16 years. Nylon 4.6, recently introduced by AKZO, has a melting point of 285 °C and should be well suited to airbags, although the extra cost is a disadvantage. Improved thermal performance can be obtained by coating.

In the USA, the National Highway Traffic Safety Administration (NHTSA) estimates that 4750 people are alive today (9th December 1999) because of their airbags.[106] They also estimate that the use of an airbag in combination with a lap/shoulder belt reduces the risk of serious head injury by 81% compared with 60% reduction by the use of a belt alone. While airbags undoubtedly save lives, they are believed to have caused the deaths, in the USA, of an estimated 148 persons in low severity accidents since 1990. These included 42 female drivers, from a total of 56 drivers, 68 children (aged between 1 and 11) and 18 infants. Analysis showed that many of these persons were not wearing a seat belt and were sitting close to the steering wheel or airbag. This has resulted in the development of 'smart' airbags that can sense the size of the seat occupant, or even if the seat is unoccupied, and deploy accordingly. Other designs have been put forward, including an airbag which deploys outwards from the seat belt.[107,108] Research is also directed towards more gentle inflation and less abrasive fabric material. Injuries caused by airbag inflation include eye damage, fractures, bruises and chemical burns caused by penetration of the inflating material through the fabric. Integrated safety systems, especially for children, are being developed. Airbags are not yet compulsory in Europe, and it has been suggested that tests should be carried out to determine the most suitable for use in conjunction with seat belts which are compulsory and which are in widespread use. In recent years there seems to have been some progress in the development of airbag fabric from nonwoven material.[109]

Statistics compiled by the World Health Organisation in 1998 record that world-wide 500 000 persons are killed and 15 million injured in traffic accidents every year. This is expected to increase sharply as car ownership increases in the developing nations and also as the young adult populations grow. Traffic accidents are amongst the main causes of premature death in many of the developed counties; efforts are being made to reduce this by several means including making cars more safe. In the USA there is a whole series of FMVSS aimed at improving car safety.[92] These include:

- FMVSS 201 – head protection in cars and light trucks, 100% compliance by May 2003;
- FMVSS 208 – airbags for both cars and light trucks, 100% compliance by September 1998;
- FMVSS 213 – child restraint systems, 100% compliance by August 1994;
- FMVSS 214 – side impact for passenger cars and light trucks, 100% compliance by September 1998;
- FMVSS 581 – bumper properties.

Side impact accounts for about 25% of all injuries in car accidents; a high proportion of these are serious or fatal, with half involving the head. The US transport department estimates that about a quarter of these annual fatalities (2400) could be prevented if all cars had side impact head protection. Car occupant safety is becoming an electronic 'high tech' industry with the development of sensor systems for the seat belt, to anticipate the type of accident, e.g. roll-over, side impact, etc., and for the airbag.[110] These sensors activate the relevant components to optimise the protection performance. In the USA, new legislation to regulate advanced airbags is expected in the near future. Continental analysis of airbag deployment has drawn attention to the economic cost of airbags deploying unnecessarily.[111] New fabric structure development continues to improve safety devices.[112] Recent developments have been summarised, together with a review of expected NHTSA proposals,[113] and an account of the latest method of one-piece airbag weaving.[114] Airbag technology is summarised in a recent Textile Institute publication.[115] The NHTSA publish periodic reports on airbag related accidents and investigations on their website.[116]

Manufacturers of both polyurethane and polypropylene foam are developing grades of foam with optimum properties for absorption of impact energy. Eventually, every surface in the car interior will incorporate some kind of energy absorbing material, which will not only improve comfort and reduce noise levels, but also contribute to safety. In Europe, legislation is in preparation for side impact safety, and, in the not too distant future, laws are likely to appear for the protection of pedestrians in accidents up to 25 mph (40 km/h). Softer car exteriors, possibly incorporating coated or laminated textiles, may contribute to this new challenge. There has been progress in the development of external airbags, which deploy on the bonnet for pedestrian protection.

4.3.10 Coverings for convertibles

A variety of coated fabrics have been used for this application, including rubberised cotton fabric and PVC coated cotton, nylon and polyester. PVC is still used in some volume production models, but more up market vehicles now use a triple textile laminate with spun dyed acrylic fibres in the top layer. This outer material has to be extremely well engineered to be resistant to UV and other sunlight radiation, rain, frost, ozone, micro-organisms, dirt and traffic fumes as well as car wash chemicals for the life of the car. In addition, it needs to be dimensionally stable under all weather conditions and to have excellent abrasion resistance.[117] Soil resistance is achieved by application of a fluorocarbon. The market leader is Dolan 25, now being replaced with improved Dolan 65, both made from Acordis acrylic fibre.[118] Aesthetics are important as the vehicle top is very visible and its appearance must be integrated into the rest of the vehicle design.

Cotton is now unsuitable for any part of the convertible top because it will not pass tests for weathering and resistance to micro-organisms. Polyester twill woven fabric is used for the inner surface of the triple laminate and the middle layer is rubber. The recently improved Dolan 65 now uses acrylic yarns in both the warp and weft of the top layer. Earlier versions of this product used acrylic only in the weft with polyester in the warp, but the polyester warp hydrolysed in subtropical locations of the world. At the time of writing, only spun dyed acrylic satisfies all the demanding tests; the latest UV stabilised polypropylene fibres were evaluated, but certain processing problems could not be overcome.

The fabric is tested for weather resistance using a test such as DIN 53387 and also specially designed sagging tests at 75 °C with a weight hung on the end of the test strip of material. The percentage of permanent elongation after a 24-hour test period is calculated from measurements before and after testing. These laboratory tests are correlated with outdoor exposure tests in the southern states of America. Society of Automotive Engineers (SAE) tests for exterior materials include J1960 JUN 89 (accelerated exposure using a water cooled xenon-arc), J1961 JUN 94 (accelerated exposure using solar Fresnel-reflective) and outdoor weathering test J1976 FEB 94.

4.3.11 Bonnet (hood) liners

These are generally made from a laminate material, the main function of which is to absorb and dampen engine noise. The main constituent is generally phenolic resinated waste, shoddy fabric or fibreglass, usually laminated on both sides with nonwoven fabric. The covering side facing the engine needs to be resistant to fluids such as oil, fuel, windscreen cleaning fluids and water, and this can be improved by a flurocarbon finish. The effectiveness of the sound absorption has to be balanced by increased weight – generally the thicker the liner, the better the noise insulation. Both polypropylene and polyester nonwovens are used in this application. The adhesives used for lamination have to be resistant to fluids and also to heat, and they must be capable of lasting the life of the vehicle. Polyurethane foams, which have the advantage of lightness, are sometimes used in place of the shoddy material. More recent bonnet liners are being designed entirely in polyester or entirely in polypropylene to facilitate recycling and to remove the fibreglass, a potential skin irritant. Spunbonded nonwovens, such as Lutradur (Freudenberg), are well suited because they allow deep draw, well defined moulding to the required shape.

4.3.12 Wheel arch liners

These external components are a relatively new application for textiles, but are becoming more important in the quest for road noise reduction and increased road safety. As well as significantly cutting down noise, they also reduce spray in wet conditions and protect the bodywork against stone impact and corrosion. An EC directive has been issued to reduce road spray for increased road safety. Performance and properties are covered by DIN 661151 and DIN 61210.[119] Needlepunched polyester and polypropylene coated with specially formulated SBR latex have been found to be suitable materials. They replace PVC and EPDM which are stiffer, harder, less flexible and heavier than the textile alternatives, and not as effective in reducing noise or spray. The textile is more effective because of its porous structure; this disperses water into smaller droplets which helps to reduce road noise. In addition, use of textile material in place of the heavier plastic reduces car weight by about 3 kg. Constructions include polyester monofilament yarn knitted in a spacer fabric construction about 12 mm thick. They are lighter than plastic equivalents and about six are required for each vehicle. Wheel arch liners are also used inside the car to assist with road noise insulation for the occupants.

4.3.13 Car carpets

This was once considered a luxury item, but it is now an essential part of interior trim, not only with a view to aesthetics and sensual comfort but also because of the part it plays in noise and vibration control. Each car contains about 3.5 to 4.5 m^2 of carpet, made by either tufting or needlepunching. There are considerable differences depending on where in the world the car is made.[120] In Western Europe, approximately a third of all cars have carpets tufted mainly from bulked continuous filament (BCF) nylon yarns. The rest, about two-thirds and gradually increasing, is needlepunched, mainly from polyester but also includes increasing amounts of polypropylene. About the same proportion of tufted and needlepunched carpets appear in cars made in Japan, and most of the yarn used in tufting is BCF nylon, with a very small amount of polypropylene. In the USA at the time of writing all car carpets are tufted mainly from BCF nylon. The remainder is staple spun nylon, and solution (spun) dyed fibres are being increasingly used.

The poor compression resilience of polyester prevents it being used in tufted carpets. In recent years, needlepunched carpets have appeared in the USA and demand is expected to grow. An increasing trend worldwide is to produce lighter carpets using finer gauge yarn; this has more covering power towards the lower weights – approximately 12 oz/yd^2 for tufted and 450 g/m^2 for needlepunched or even lower. Tufted carpets are generally more resistant to wear and tear, but needlepunched carpets have better mouldability. In addition, unlike tufted carpets, needlepunched carpets have no tendency to exhibit 'grin through', especially around convex curves.

Tufting is a relatively new process, many times faster than weaving, and tufted car carpets first appeared in the USA in the 1950s.[121] The tufts can either be cut or uncut looped pile, but the vast proportion of tufted car carpets are cut pile; uncut looped pile is used only for special effects. The most popular gauge of both cut pile and loop pile tufting is 1/10 inch. Needlepunching of car carpets is an even younger process than tufting and is an even more rapid and economical process. Needlepunching has been much refined in recent years and is now producing very attractive materials, comparable in quality with tufted carpets. Tufted carpets are standard in up market cars in Europe, e.g. Audi, BMW, Mercedes, Saab and some Rover cars. Tufting of automotive carpets is carried out on a polyester, spunbonded, nonwoven material weighing between 110 and 120 g/m^2, called the primary backing. Both AKZO (with Colbond) and Freudenberg (with Lutradur) have developed special fabrics for this purpose, which also facilitate moulding to the shape of the car interior. The weight of these fabrics has been reduced by using finer fibres, but the stretchability has been increased to allow deeper draw moulding which is especially important for the European and Japanese markets. American cars generally have less contoured floors.

Both tufted and needlepunched carpets require about 70–100 g/m^2 of a binder coating on the back, usually SBR or acrylic latex, to stabilise them and to lock in the fibres. Another layer of a suitable material is then applied to both types of carpet to confer good thermomouldable properties. This is important for process efficiency, and also to produce a good fit which will reduce vibration and maximise noise insulation.[122,123] Polyethylene powder is used on both tufted and needlepunched carpets for this purpose, and about 250–600 g/m^2 is applied by powder scattering and IR heating. The correct thermal characteristics are critical, because no softening must occur at temperatures below 90 °C. However, softening should occur sharply, in the region of about 110–140 °C, which is the temperature at which moulding is carried out. An alternative to polyethylene powder, used mainly on needlefelt carpets, is a layer of thermoformable fibres which is needled on to the back of the needlepunched carpet. These thermoformable fibres must also have the thermal characteristics already mentioned. In older plants, the thermomouldable layer is a further, much heavier coating of SBR latex filled with chalk (calcium carbonate) or barytes (barium sulphate), a heavy material which helps to improve sound absorption, see Table 4.1.

All carpet materials have to be selected for mouldability, good adhesion to the fibre and other substrates, heat stability to withstand further pro-

Table 4.1 Automotive carpet structure. The main elements are: decorative top layer of fibre; thermoplastic material for thermomouldability; and acoustic and vibration damping layers

Layer	Materials used
Top decorative	Tufted BCF nylon or needlepunched polyester or polypropylene back, latex coated with SBR or acrylic latex
Thermoforming	Polyethylene powder, meldable fibres, EVA or a further thick layer of compounded SBR latex
Acoustic	'Heavy layer' of EPDM, shoddy fibres or polyurethane foam

N.B. Luxury cars have more than one acoustic layer. Bitumen is also used in many cars as a further layer next to the metal.

cessing in the car factory, and also durability during the life of the car.[124,125] After the application of the thermomouldable layer, the carpet is thermomoulded to the shape of the car to which it is to be fitted. A good fit is essential for good acoustics control and for ease and efficiency in installation, and it must be dimensionally stable under all conditions of temperature and humidity.[126,127]

Many cars now have a sound absorption barrier layer of barium sulphate filled EVA/EPDM polymer, weighing approximately 2000–7000 g/m^2, which is laminated to the thermomouldable layer before thermoforming. The combination of a heavy layer against a flexible padding layer is beneficial in reducing noise inside the car. Up market cars will also have an extra noise and vibration insulation layer between the carpet and the vehicle floor, see below.

Carpet manufacture is made more complicated by the holes and gaps required for cables and ducting, provision must also be made for the fitting of seats and control consoles. Certain areas of the carpet will have extra heat insulation pads, for example the part at the front facing the engine. As well as being an essential item for comfort by providing thermal and noise insulation and vibration damping, carpets now contribute directly to safety through the use of energy absorbing backing foam. In a similar way to the headliner, the car carpet has become a complex module system in its own right.

In addition to fitted car carpets there is also a sizeable market for secondary carpet mats – known as 'thrown in' or 'option' mats – totalling perhaps 4 million m^2 in Western Europe alone.

4.3.14 Noise control in the car – the contribution of the carpet

Sound is propagated through the air and by vibration of the car body, and there are three basic mechanisms for reducing it: by absorption, by damping and by isolation or insulation.[85] In general, a thick piece of material will absorb more sound than a thinner piece of the same material. There are a number of layers of material and permutations of layers of materials used in noise and vibration damping. Density, air porosity and thickness of the material influence sound absorbency, but the actual frequency of the sound waves is also relevant.[128] Damping can be obtained by putting soft materials next to a harder material such as the metal car body structure. Efforts to isolate engine vibration and noise are made by designing the engine mounts and the car suspension to isolate the engine from the passenger compartment. Engineers are also trying constantly to improve the fit of doors and windows and to eliminate all the bangs, squeaks and rattles in cars.[129–134] Flock covered parts, both plastic and textile, are sometimes used for this purpose.[135,136]

At one time only luxury cars had a noise insulation pad under the carpet but, like many luxury items, this has become or is becoming a standard requirement. Bitumen sheeting is widely used, but this is now supplemented by resinated waste or shoddy fibre. These materials generally have to be fitted in small pieces, which is time consuming and produces an insulation performance which is inferior to that of a continuous layer. In some vehicles this insulation layer is formed directly on the back of the pre-formed carpet itself by back injection moulding using polyurethane foam. A barrier film on the back of the carpet is necessary to prevent liquid foam from penetrating to the carpet surface during moulding. This is considered to be a more satisfactory method giving more consistent results, and the objective is to obtain maximum sound absorption with the lightest possible material. Various types and densities of foam and non-woven fabrics have been evaluated at different vibration frequencies.[128,137]

A method of adding barium sulphate filler to a polyurethane elastomer which is then froth coated directly on to the back of the carpet was developed in the late 1980s. Barium sulphate has a high density and this produces a sound insulating 'heavy layer' in a reduced number of manufacturing steps and with much reduced fogging.[138] Foams developed in the late 1990s by ICI (now Huntsman) allow the individual noise and vibration characteristics of a vehicle to be selectively damped by tailoring the foam properties to suit the particular vehicle. Large sections of floor covering can be precisely moulded to the shape of the vehicle's floorpan to enhance acoustic absorption as well as contributing to assembly efficiency.[139]

Engineers working in research departments at carpet manufacturers and OEMs as well as in universities are applying much effort to the reduction of noise and vibration. Collins and Aikman have a test dummy in their Michigan acoustic laboratory, called Oscar, which is equipped with specially designed ears to locate the origin of noises.[122] This laboratory is one of the most up-to-date in the world for the investigation and elimination of noise, vibration and harshness (NVH). Noise frequencies and levels are characteristic of each specific car model, and the materials and designs have to be individually optimised.[85,130–132,134] The carpet, headliner and parcel shelf are probably the three major items by which interior noise can be controlled. However, all the components in a car will influence noise levels, and it is difficult to separate individual contributions. The best tests are done on the car as a whole, on the road – where the ultimate tests will be carried out by the customer. Noise contributes to driver fatigue, a major cause of accidents, and reducing it is a valuable contribution to road safety.

Road noise is now considered to be a form of environmental pollution. The EU and national governments are applying pressure to OEMs to reduce *external* car noise levels by 3dB, from 74 to 71dB, which means by a factor of 50% because noise is measured on a logarithmic scale.[140] Textiles are also contributing to overcoming this problem, as was shown earlier in Section 4.3.12.

4.3.15 Drive belts

Automotive drive belts require high performance under extremely demanding conditions over a period of several years. They are small items but essential to the reliability of the vehicle. High tenacity polyester and aramid yarns are now in widespread use for these products. The textile requirements are low extensibility, good vibration resistance and damping qualities, high strength at the lowest weight and resistance to oils, fuels, oxidation and moisture. Neoprene is probably the most used rubber coating, because of its overall good oil, chemical and heat resistance properties. Acrylonitrile–butadiene (good oil resistance) and butyl rubber, which has good strength-to-weight ratio and excellent heat and chemical resistance, are also used. The rubber coated textile belts have replaced chain drives in cars because they are quieter, lighter in weight, do not require lubrication, and are more flexible, allowing a more compact design using smaller diameter pulleys and thus saving space under the car bonnet. V-belts are made from cords of high tenacity polyester or aramid embedded in specially formulated rubber, usually chloroprene (Neoprene), and covered with rubber coated fabric. The V-belt shape is designed to produce maximum frictional grip.

4.4 Marine applications

4.4.1 Inflatable craft

Inflatable craft have many advantages over rigid boats and have become widely used since around 1960. They are used as life boats and rescue craft, as freight carrying vessels and as pleasure craft, as well as having several military applications. When they are not being used or when they need to be transported to a different site, they can be deflated, folded and packed into a relatively small space. They are made from individual buoyancy tubes or several different compartments, so that, even if a particular section is damaged, the whole craft will still float and be capable of supporting weight.[141–143] There are national standards and minimum performance specifications for the coated fabric and the craft itself. These standards do not actually specify which materials should be used. A number of different coatings are used, polychloroprene, Hypalon polyurethane and PVC.

Nylon fabric about 145 g/m^2, woven from typically 470 dtex yarn, is used in the lighter craft; heavier yarns, typically 940 d/tex, are used for larger boats; and even heavier yarns of 1880 dtex woven in a 2 × 2 twill may be used in even larger boats. Good tear strength is important to prevent propagation of any damage. Polyester's higher yarn modulus gives it some advantages over nylon, but generally there are more disadvantages. It is usually more difficult to bond rubber coatings to polyester, and polyester can be degraded by certain compounding ingredients. In addition, polyester is heavier than nylon. If cost allows, aramid fibres may be used, thus saving a significant amount of weight. An aramid fabric weighing 85 g/m^2 can be used in place of a nylon fabric weighing 170 g/m^2.[131–133]

Carbon dioxide is used to inflate life rafts, which are designed to be self inflating. Inflated pressure is about 15 kPa. Baffles are used to disperse the gas as it emerges from the cylinder, but the chilling effect could embrittle the fabric coating if it is not formulated well out of a material with a low glass transition temperature such as natural rubber. One problem with the making up of inflatable craft is production of durable, air tight seams. This is a time consuming process, but it can be made simpler if a thermoplastic material, such as polyurethane or PVC, is used. PVC, however, has its limitations, but is used in more economical craft.

4.4.2 Marine safety apparatus

Coated woven nylon is used for life rafts, buoyancy tubes, canopies and life jackets, and personal flotation devices (PFD), see Fig. 4.35 and 4.36. Polymer coatings include butyl rubber, natural rubber, polyurethane and polychloroprene, and total weights of the coated fabric are between 230 g/m^2 and 690 g/m^2. Quality tests include hydrostatic head to ensure waterproofness (carried out after ageing and after flexing), coating adhesion, breaking load and tearing strength. Life jackets are generally woven nylon coated with butyl or polychloroprene rubber to give total coated fabric weights of between 230 and 290 g/m^2. Tests for these items include assessment after immersion in water for 24 hours. The specifications and performance standards for life jackets and life rafts are subject to government regulations. The Underwriters Laboratory (USA) issue specifications, e.g. UL 1123, Marine Buoyant Devices, and UL 1180, Recreational Inflatables, which are available for purchase. Specifications are also issued by the military in various countries.

Aircraft survival equipment such as life jackets, life rafts and escape chutes is generally made from woven nylon coated with polyurethane or a synthetic rubber. PVC is avoided because of toxic gases should it catch fire. Life jackets, made from polyurethane coated fabric, about 250 g/m^2 total weight, are individually checked at regular intervals for serviceability. 'Slide-raft' escape chutes inflate automatically and become life rafts when

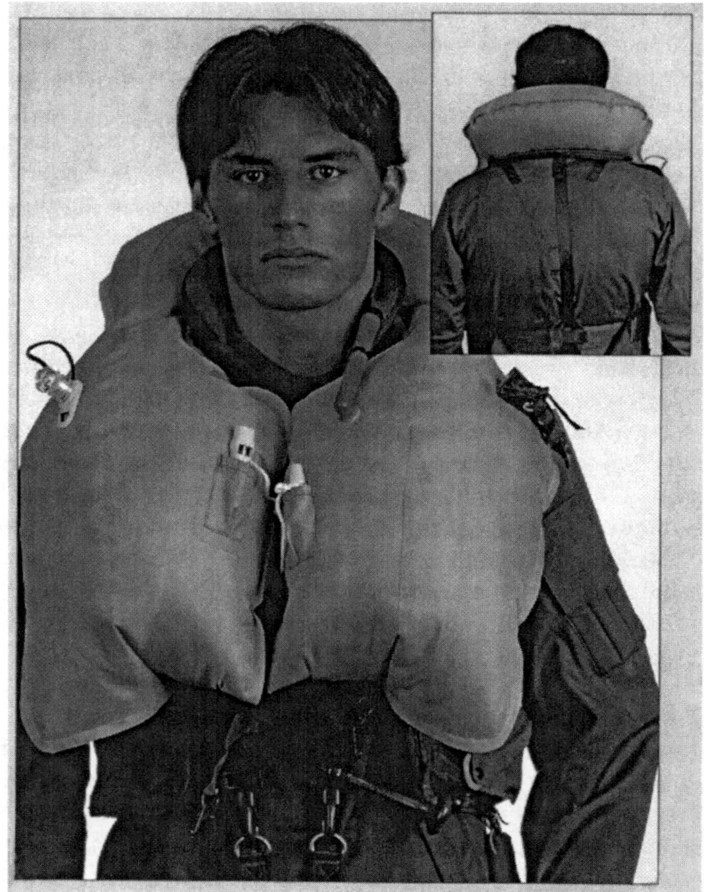

4.35 MK.30T Series Lightweight Lifepreserver by Beaufort Air-Sea Equipment Ltd is produced from high performance FR fabrics. Photograph supplied by manufacturer and reproduced with kind permission.

detached from the aircraft. Weight and space saving are factors to consider for all aircraft equipment.

4.4.3 Oil booms

These articles are designed to contain accidental oil spillages in rivers and estuaries. They are usually produced from woven nylon base fabric of about 175 g/m^2 coated with Hypalon, polychoroprene, PVC or PVC/nitrile rubber blends. The booms do not normally have to withstand severe conditions and

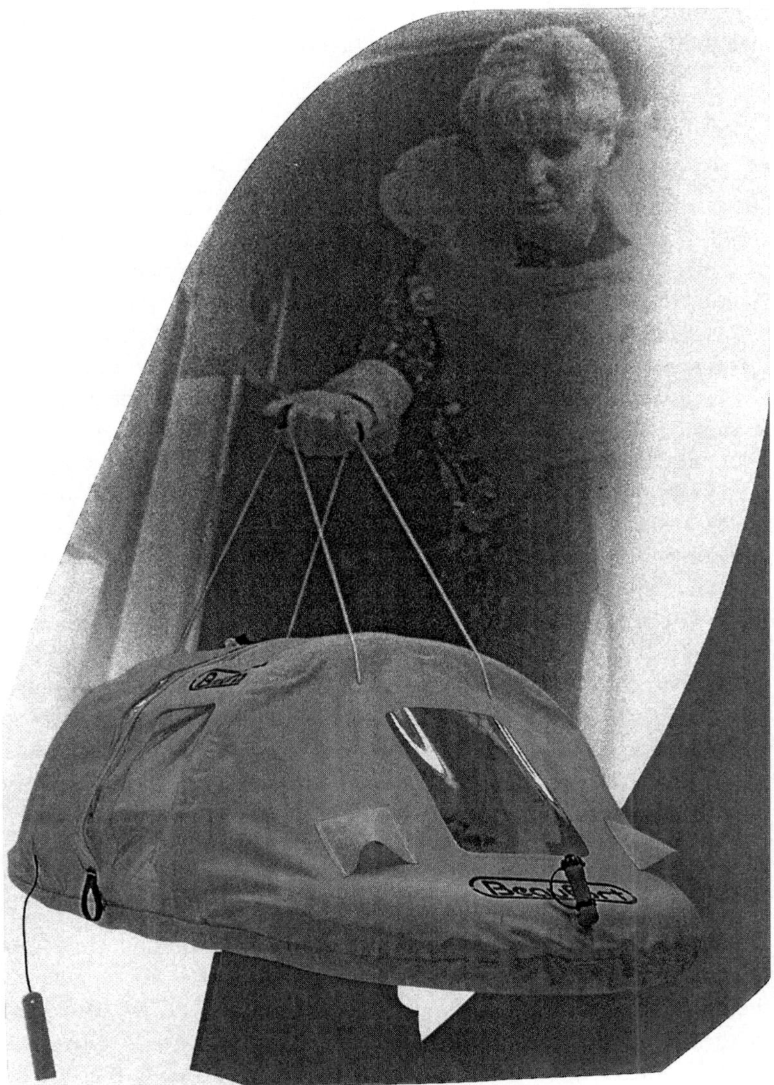

4.36 Baby's Floating Survival Cot MK4 by Beafort Air-Sea Equipment Ltd will accommodate babies of ages up to 18 months and provides a high level of protection against exposure and drowning. It features a separately inflated floor for thermal insulation and buoyancy, is produced from high visibility and FR materials and the structure has self righting properties. Photograph supplied by the manufacturers and reproduced with kind permission.

are inflated to a fairly low pressure, but they do have to be oil resistant. A two-compartment structure is generally used so that the boom is still serviceable if one compartment is damaged.

4.4.4 Hovercraft skirts

Hovercrafts were invented in Britain in the mid-1950s and, from the mid-1960s until quite recently, October 2000, they provided a cross-channel ferry service, crossing in 25 minutes in good weather. However, their use is likely to continue in other parts of the world, and they are built to withstand Arctic conditions of –60 °C. The Life Boat Service in the UK are reported to be buying a fleet of about 12, because hovercraft can be deployed on terrain such as marshes where boats cannot go. The skirt material is a nylon fabric coated with a polychloroprene/ natural rubber blend or natural rubber/polybutadiene and compounded for oil resistance. PVC blended with nitrile rubber was evaluated but was not as satisfactory. The nylon cords, e.g. 940 dtex/2 and 940 dtex/3/3, from which the fabrics are woven are highly twisted to impart fatigue resistance in order to withstand the rapid and continual flexing in use. Fibre to rubber adhesion must be of the highest standard, and a suitable RFL priming coat must be compounded. Nylon is the best overall fibre for this application. Polyester yarns are less affected by water, but in actual use coated polyester fabrics did not last as long as nylon materials.[141] This is believed to be due to a poorer polyester/rubber bond compared with nylon/rubber. Cotton and rayon absorb too much water and are not generally strong enough for this application. Aramids, which are stronger than both nylon and polyester, have been tested, but they broke down quite rapidly because of low fatigue resistance.

4.4.5 Sails

The first sails were made from panels of hides or woven natural fibre fabric sewn together. When synthetic fabrics appeared they were used because of their advantages of greater strength and resistance to micro-organisms and mildew, minimal water absorption and less distortion.[144] Leisure sailing in the UK dates from the late 1700s in Devon, but during the last half of the twentieth century became the sport of larger numbers of people and, with increasing disposable incomes, it is set to grow even further. Sails required for competitive racing are different from simple cruise sailing, and sail making is a highly skilled occupation. The sail has to be tailored to the vessel and to the purpose required; sail design is a crucial factor for optimum performance and efficiency.

The main requirements for sail cloth are: light weight, dimensional stability (low creep and minimum distortion), puncture resistance, high tear strength, high seam strength, low porosity to wind (i.e. good cover), low water absorbency, good resistance to microbes and UV degradation and smoothness; all these factors contribute to a long service life. Sails which absorb water add weight to the upper parts of the vessel reducing efficiency, while a sail with a rough surface will cause frictional drag. Bias stretch is a limiting factor in sail cloths, and this can be reduce by calendering and by impregnating or coating with a resin. It was usual to 'break-in' new sails made from cotton and other natural fibres, i.e. to allow them slowly to stretch before actual use. This is less important for sails made from synthetic fibres and even less important for laminated sail as will be seen below. Sails should be easy to handle and easily stowed away.

As with protective clothing, the seam is a point of weakness, and under high stress the sewing holes can become larger and thus influence sail porosity. Under prolonged high stress, fabrics can distort causing the sail to lose its shape. Polyester is better than nylon in this respect, but nylon is better than polyester for spinnaker sails which require a certain amount of stretch and elasticity to allow them to take their shape. Some manufacturers have developed techniques to glue rather than sew seams; they claim that these seams reduce seam distortion allowing sails to maintain their shape longer and that in addition, there are no stitching holes.

Some modern racing sails are produced from polyester film laminated to rip-stop woven polyester or nylon fabric. Laminated sails generally have excellent shape holding properties with very little distortion or stretch, because the polyester film has a high modulus of elasticity in all directions. The film, being only thousandths of an inch (100 micrometres) thick, is significantly lighter in weight than a fabric and offers advantages in weight saving. The sail laminate must be bonded with an adhesive which can withstand sea water and UV degradation. Sail laminates can comprise two layers of film laminated to each side of the rip-stop fabric or the other way round, i.e. two layers of rip-stop fabric laminated to each side of a layer of film. More recent laminates are believed to consist of a single layer of film and fabric laminated together. Random-laid nonwoven fabrics have been used; Kevlar and even carbon fibres are also sometimes used for high strength with low weight – but at high cost. Kevlar, however, has poor UV stability, losing strength perhaps twice as fast as polyester. However, in racing, when performance at low weight is all important, the Kevlar, having up to nine times the modulus in grams per denier than high tenacity polyester, is the preferred material. Kevlar can actually lose up to 50% of its strength, but still be stronger than polyester. Total weights of laminates vary from 5 to 6 oz (170 g/m^2 to 204 g/m^2) to as much as 7 oz (238 g/m^2) or higher.[145–147]

Ultra modern racing sails comprise film laminated to threads of yarn laid on the bias of the sail to produce an article of high dimensional stability and very light weight. Spectra, the ultra high tenacity polyethylene yarn from Allied Signal (now Huneywell), is even stronger than Kevlar and is used in some sails. It is also lighter in weight than Kevlar with a density of approximately 0.9 g/cm^3. An even more recent fibre called Vectran has been introduced by Celanese in the USA. This liquid crystal fibre has excellent UV resistance, a modulus similar to Kevlar 29, and virtually zero creep, enabling it to retain its shape and laminate well to films and fabrics. Spectra, Vectran and carbon fibre[148] all have good UV degradation resistance compared with nylons, polyesters and aramids.

Laminated sails do not last as long as fabric sails, but they are invaluable when speed and performance are the critical factors. When sails are cleaned, any detergent or cleaning agent must be rinsed away because they may affect laminated bond strength or contribute to UV degradation of the materials. Hot melt adhesive films are generally used, and these factors must be taken into consideration when the adhesive is selected. Development work continues to improve lightness, dimensional stability and overall durability of sails. Six catamarans, believed to be the fastest yachts ever built, set off from Barcelona on a round the world race on 31 December 2000. The sail cloth of each vessel had a maximum area of over 11 600 ft^2 (1078 m^2) and cost around £150 000 ($210 000) each.[149]

4.5 Buildings and architecture

4.5.1 Permanent structures

Fabrics and animal hides have been used for structures for human habitation or use probably since prehistoric times. Nomads took their home with them as they moved from place to place. In modern times, with the availability of synthetic fibres and plastics, the scope for producing inhabitable structures from fabric has been greatly increased. The advantage is portability, although many are now constructed as permanent buildings or parts of permanent buildings. They vary from temporary marquees for exhibitions and shows to sports arenas and airport buildings. The best known structure constructed from coated fabric at the time of writing is the Millennium Dome at Greenwich, London. This is claimed to be covering the largest enclosed space on Earth of 80 000 m^2 or 20 acres (870 000 square feet), see Fig. 4.37. PTFE coated, glass fibre fabric, about 1 mm in thickness, was used in the construction, which is claimed to have a minimum life of 25 years. It is reported that the cost of the PTFE coated glass fibre roof was £14 million compared to an original costing of £6.1 million which would have used PVC coated polyester fabric.[150]

Papers have been published on coated fabric used in buildings.[151–158] Small marquees sometimes comprise a metal framework covered by inexpensive polypropylene extruded tape yarn, coated with polypropylene to produce

4.37 The Millennium Dome, Greenwich, London is made from Teflon coated glass fabric and is only 1 mm thick.

a coated material – although it resembles a piece of plastic sheet more than a fabric. This is probably the least expensive material available, but it is adequate for the purpose and is light weight and easily erected, taken down and transported. PVC coated nylon or polyester have been used, but in more recent years, the more advanced PTFE polymers have come to be used more despite the much higher price. Glass fibre, despite its higher price and heavier weight, is also used in preference to nylon and polyester. PVC coated fabrics have a life expectancy of about 15–20 years, but PTFE coated glass fibre lasts much longer because both components are very inert to most chemicals and have a high resistance to UV radiation. Buildings made from the more advanced materials are considered 'permanent'. PTFE and glass fibre both have inherent excellent FR properties, but PVC is likely to require FR chemicals. A polyurethane or acrylic lacquer is usually applied to the top surface of PVC to improve both resistance to UV degradation and cleanability.

All coated materials need to be joined with watertight seams, and an advantage of a particular material is the ability to be repaired *in situ*. The comfort of the people inside needs to be considered carefully. As well as giving protection from wind and rain, these buildings can act as greenhouses and become unbearably hot in sunny weather. To overcome this, the coating is formulated to control the amount of light and heat radiation entering the structure. This depends on the human activity inside and the purpose of the building, and also on the part of the world in which it is being erected. A typical covering may absorb about 14% of solar energy, i.e. daylight and heat, transmit about 13% and reflect about 73%. Other technical requirements are abrasion resistance, puncture resistance, high tear strength, and also resistance to the combined and prolonged effects of UV and light radiation and widely varying temperature and humidity. In hot climates, the absorption of solar energy may result in savings in air conditioning, while in less sunny climates, the 'greenhouse' effect could result in savings in heating. Dimensional stability under all these atmospheric conditions and the ability to withstand high winds are essential. The base fabric should, therefore, have a high modulus and low extensibility and be temperature stable. The coated material must also be rot proof and not vulnerable to damage by insects. Self cleaning properties are an advantage – if not a requirement.

Coated fabrics are used in structures with traditional shapes, but many are used in modern advanced 'tension' forms, e.g. pointed conical to 'horn shaped' that need metal framework and cords. Others are 'air supported' structures with no framework, but an air pump must run continuously to keep the structure 'inflated' and to replace air lost through doors and small leakages. In fact, the air pressure required to achieve this is only approximately 0.3% more than the air outside. People walking in and out do not notice the difference, but air-lock doors are generally required to reduce the amount of air lost. In some designs, the roof sections are made from double layer material to produce an enclosed air layer which acts as both thermal and acoustic insulation. Circulating warm air within the enclosed area can also facilitate the melting of snow.

The advantages of buildings constructed from coated fabric include ease and speed of erection – especially in emergencies. Less costly foundations are required, and the weight of coated fabric has been estimated as about one-thirteenth of the weight of bricks and mortar. A fabric envelope roof allows a large area unobstructed by pillars, making it ideal for shows and public meetings. However, joining finite widths of coated fabric with seams that are 'invisible' is a challenge to fabric engineers and designers. Building from coated fabric requires the collaboration of building engineers, architects and the textile technologist. The American ASTM Standard D 4851-97 details test methods for coated and laminated fabrics for architectural use, including PTFE and PVC on polyester and glass fabrics. This comprehensive document contains test methods such as coating adhesion, resistance to weathering and tear strength, which may be applicable to coated fabrics for other applications.

An interesting development area is believed to be development of covering fabrics with sufficient light transmission to allow natural grass to grow.

There are even reports of exploiting the 'greenhouse' effect obtained in these structures to reduce the need for fossil fuels for heating in northern latitudes. The new PTFE fibre with enhanced tensile strength could offer some benefits in building structures and should facilitate eventual disposal or recycling.

4.5.2 Tents

There are a wide variety of different types, from small back-packing tents, to large marquees and military tents of different sizes and purposes.[159] Small tents are generally not coated, and the only coated material is likely to be the ground sheet which is usually PVC or some rubber coating on nylon polyester fabric. Tent fabrics need to be strong enough to withstand high winds, resistant to rotting and microbes and, most important, resistant to sunlight and UV degradation. Polyester is better than nylon for UV degradation resistance, which can be improved by additives applied from the dyebath and by careful dye selection. Acrylic fibres are best for UV degradation resistance but are generally not strong enough. Wax finished cotton produced by padding is still used, but lighter tents are closely woven nylon or polyester with a silicone water-repellent finish which also improves the tear strength. Light weight coated tents are generally made from polyurethane coated fabric, the larger ones from PVC or Neoprene.

4.5.3 Awnings

The Oxford English Dictionary definition of an awning is a 'sheet of canvas, etc. used as shelter against sun or rain, especially on ship's deck'. Awnings are also used over shop windows, in domestic gardens, and at open air shows and exhibitions. There seem to be many more manufacturers and suppliers in the USA compared to the UK, presumably because of the sunnier weather and generally higher standard of living. Cruise ships use considerable numbers of awnings, and this is perhaps a growth area with opportunities for innovation with more novel designs. Since 1985, cruising holidays have increased by a factor of seven amongst UK residents, and several large cruise ships, some well over 100000 tons, have been or are being built. A variety of base fabrics, some in rip-stop constructions, and polymers appear to be used, including nylon, polyester, polypropylene and acrylic as base fabrics and PVC, polypropylene, and acrylic polymers as the coating. Acrylic lacquers are used in some cases to improve UV and sunlight resistance, and PVDC lacquers are known to be used to confer heat weldability. Spun dyed acrylic fibres are especially popular because of the bright shades available together with the highest light fastness and UV degradation resistance.

The main requirements of awnings are weather and sunlight durability, resistance to insects and microbes and adequate physical strength for the purpose. Aesthetics, a wide variety of colours and design, are likely to be important and fading, discoloration or dimensional distortion should not occur. Anti-soil properties, ease of cleaning, FR properties and screening to UV radiation against skin cancers are other possible requirements. When used in public places and especially in transportation or marine applications, FR requirements are likely to be stringent. The growth in cruising holidays and the overall increase in leisure time out of doors should present growth opportunities for these products.

4.5.4 Coated roofing materials

The roof is continuously exposed to the elements and roofing materials must have excellent resistance to UV and light radiation, ozone and general weathering, heat, moisture, and rotting. They should also be tough and have FR properties, some resistance to oils and solvents and ideally some water vapour permeability. The base fabrics should be non-wicking (see tarpaulins), resistant to microbes and dimensionally stable. Traditional built up sheet roofing comprises layers of felts impregnated with and bonded together with bituminous materials.[160]

Coated fabric roofing material is generally produced from coated woven polyester or nonwoven glass fibre fabrics and the coating materials are EPDM, Neoprene and Hypalon, CPE (chlorinated polyethylene) and PVC. EPDM and other products could possibly be combined with the reinforcing fabric on the calender just after sheet formation.

4.6 Household products

4.6.1 Introduction

Coatings used in furnishings should be tested for skin irritation, which can be caused by the release of formaldehyde. This substance is probably more likely to be associated with fabric finishes, especially certain FR finishes, which are outside the scope of this book. However, fabric which has already been treated with a finish may, under the action of further heat during a coating process, release odours or fumes which may be irritants. In the modern eco-conscious world, there is a trend towards testing articles more for possible heath hazards, and the standards issued by the Oeko-Tex Association members, at present active in 12 European countries, are increasingly being used. See Chapter 6 for further information. Materials used in the home must be free of potentially harmful substances, especially with their prolonged exposure to older people and young children. There are a number of articles on coated household products.[161–165]

4.6.2 Furniture upholstery

One of the main requirements for furniture upholstery is FR properties. In addition, the coated fabric must be tested for any effect on colour and light fastness, staining or water marking if spotted with water, and of course the drapability of the fabric must not be unduly changed. Domestic furnishings in the UK are subject to the Furniture and Furnishings (Fire) (Safety) Regulations of 1988/89. All visible parts of upholstery must pass the cigarette ignitability test (BS 5852; Part 1: 1979) and all visible cover fabrics (with some exceptions) must pass the butane 'match' flame test in BS 5852: Part 1: 1979. In addition, all covers or loose covers which have been chemically finished or coated must be treated with a water-soak test detailed in BS 5651:1978 before the ignitability test is carried out. The test methods are specified for most domestic furniture items, including furniture and loose furniture coverings, mattresses, cushions and seat pads, but not curtains and carpets. Curtains for domestic use are not covered by these regulations, but contract curtains are subject to BS 5867: Part 2; 1980. Whether carpets are 'furnishings' or part of the structure of the building is apparently still being debated, but there are FR tests relating to carpet ignition, BS 4790 (hot nut test) and BS 6307 (methenamine tablet). There are excellent recent summaries of FR requirements for UK textiles.[166,167]

Many different fibres in various yarns and fabric constructions are used in furniture coverings to produce attractive patterns and aesthetic designs. A large proportion of the fabric is back coated with FR coatings, to meet legislative requirements and to help improve seam slippage. Coatings are generally acrylic resins, although SBR is occasionally used for economy. Non-coated furnishings are generally made FR in some other way, e.g. by Proban (Albright & Wilson) or Pyrovatex (Ciba) treatments.

4.6.3 Mattress ticking

Traditional mattress ticking is usually cotton fabric closely woven with coloured stripes, although plain cloths are also made. More modern mattress fabric is produced from a polyester/polypropylene blend, the polypropylene yarn providing an aesthetic shiny effect. A back coating improves the seam slippage and, if suitably formulated with FR chemicals, allows any legislation requirements to be met. In some cases, a fungicide is added to the coating recipe at the request of certain customers such as health care institutions. Sometimes, a coating is applied to the back of the fabric to control permeability and prevent any feathers or fibres in the mattress filling penetrating through to the exterior. Application of a coating, especially a crushed foam acrylic one, may allow a lower cost base fabric to be used.

4.6.4 Duvet and pillowcases

These are made from closely woven cotton or polyester/cotton fabric covering. This is sometimes calendered to close up the fibre structure further in order to give more cover and prevent fibre or feather filling from penetrating through, see Fig. 4.38. As with mattress ticking, it is possible to coat a less closely woven – and hence less expensive – fabric with a crushed foam acrylic coating to achieve the same degree of porosity and containment of filler. The resins used should be able to withstand washing without significant reduction in performance. In recent years the general public have become educated about dust mites which some believe may be contributing to asthma and other medical conditions. This has led to the development of barriers against these small creatures for use on mattresses

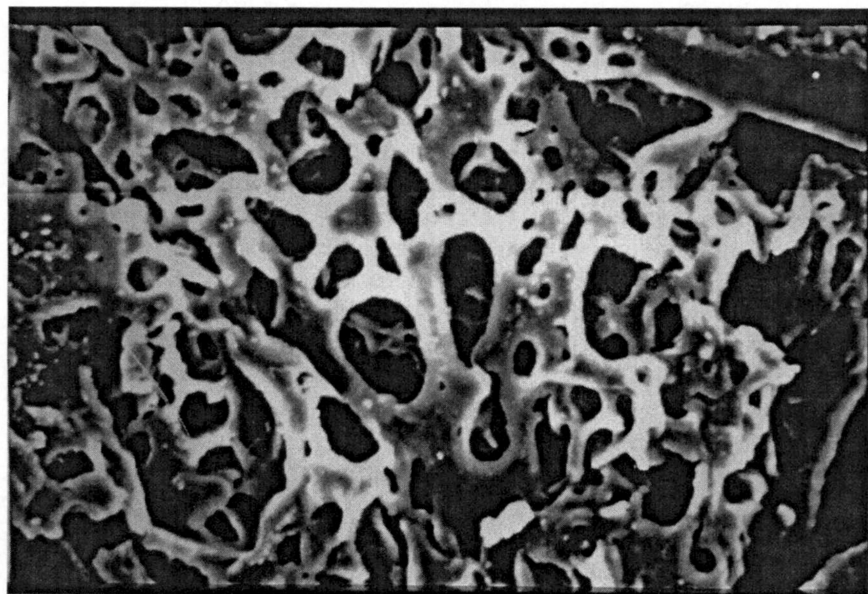

4.38 Crushed foam coating suitable for a duvet or pillowcase as a barrier to filling penetration. The coating may allow savings in fabric construction. Photograph supplied by Noveon and reproduced with kind permission.

and pillows. Some of the products offered for sale are transfer coated polyurethane on light weight knitted nylon or polyester.

4.6.5 Shower curtains

This product went through a period of sharp growth in the early 1980s as households installed showers. The performance requirements are 'rain resistant' standard rather than waterproof level (see Section 4.1.2). Some customers may require only a spray rating while others may specify testing using the Bundesmann or WIRA apparatus. Aesthetics and the correct handle – crisp but not stiff – are of the highest importance. Woven fabric shower curtains should not be too stiff, should drape correctly without curling and should not discolour in use or after washing. Shower curtains are prone to mildew, especially the bottom areas which are exposed to soap and water. An anti-microbial finish helps, but is sometimes not entirely satisfactory because the mildew is not located on the actual fabric itself. Soap residues attach themselves to the fabric and the mildew grows on the soap – regular washing is necessary to prevent this. The coated fabric should, therefore, be washable at 40 °C and able to withstand ironing. Shower curtains from coated fabric, generally 100% woven polyester, are considered to be more up market than PVC sheet which does not drape as well, invariably has plasticiser odours and which could eventually stiffen and crack.

4.6.6 Curtain linings and black out curtains

The production of these articles has already been described in Section 3.6.2. Three layers of crushed foam coating are necessary when light coloured curtains are coated: a base white, the black-out layer and a top layer of white pigmented resin. The coating formulation must be washable or dry cleanable. Some manufacturers incorporate micro glass spheres or other micro spheres into the coating and claim increased thermal insulation. This technique has also been used in the production of roller blinds and louvre blinds – with the same claims of saving heat energy by providing better thermal insulation, see below. Some curtain linings are produced with aluminium flake in the surface coating layer, which is also claimed to provide some thermal insulation by reflection of heat radiation. Black-out curtains are of interest to contract customers and hotels, especially near airports, rather than to domestic consumers.

4.6.7 Roller blinds

Woven fabrics are mainly used for this product, but it is possible to use heavier weight, warp knitted fabrics with a close construction. Medium stiff resins are used, usually with fillers for opacity, and FR agents are usually required, especially for contract markets. Fluorocarbon, soil-repellent finishes are also usually applied to the base fabric face side either before or after coating. Durability and dimensional stability and no distortion in use under all likely conditions of temperature, relative humidity, UV radiation and sunlight are the other performance requirements. Cleanability, stain resistant properties and anti-glare as well as overall attractive appearance are the other properties consumers require.

4.6.8 Vertical louvre blinds

General requirements are FR, stain repellence, drape, i.e. to hang straight vertically without twisting or spiralling, no glare in sunlight, the correct amount of opacity, and durability without discolouring in strong UV containing sunlight. The durability tests should include heat ageing and resistance to varying relative humidity. Filled resins, for opacity, are applied to both sides of the fabric, taking care that that the material lies flat when it is cut into strips. The blinds must not twist or spiral in use, even when the length exceeds six feet (1.5 m). Louvre blinds can be made from natural fibres or synthetic fibres, but if synthetic fibres are used, the fabric must be stabilised by heat setting at least to the temperature used for drying off water during coating. Coating prevents fraying of edges both in making up and during use.

4.6.9 Aprons

Generally these are made from coated PVC. Apart from properties such as flexibility, drapeability, and waterproofness, they have to be resistant to food oils and greases such as fish oil. A common variety are made from clear PVC coated over printed cotton or cotton/polyester using knife on air or calender on a knife on air base.

4.6.10 Wipe-clean tablecloths

A very large number of coated materials can be used for this end use. The coating most frequently used is PVC, applied to a variety of different fabrics, generally woven polyester or polyester/cotton, by knife coating or by calendering. The appropriate plastisol should be used for materials coming into contact with food, and it must have some degree of temperature resistance. Water-based resins, acrylics, etc. are reported to be increasingly used for this end use, because they produce an article with more textile properties and aesthetics and also allow a move away from PVC.

4.6.11 Synthetic wash leather

This material can be produced from a nonwoven base fabric, preferably cellulosic based, or from a raised cotton fabric such as flannelette. The resin applied may be a rubber latex or be acrylic based and is applied by a foam coating process. The resin should have good water resistance, some solvent resistance and good abrasion properties. Some degree of hydrophilic nature is necessary for water holding properties. Modern domestic washing materials contain microfibres for better wipeability – an indication of how the 'high tech' culture extends these days even to cleaning cloths!

4.6.12 Baby pants fabric

During the 1980s the production of waterproof baby pants which could be machine-washed at 60 °C several dozen times and still retain their waterproof properties was a reasonable volume business. The coated fabric was produced from light weight, woven nylon coated with specially formulated polyurethane or water-based acrylic resins. It had a very soft handle and was also printable. The product, however, lost popularity with the introduction of disposable nappies, but the time may be right for a re-introduction of washable products as reduction in household waste is encouraged. There may also be an outlet in third world countries whose populations cannot yet afford disposable products. Some baby diapers are made from foam laminated fabric, see Fig. 4.39.

4.6.13 Ironing board covers

These articles can be produced by incorporating aluminium powder into a crushed foam resin formulation and coating it on to the surface of the fabric. On crushing, a smooth, heat conductive and heat reflective surface is formed on the surface of the fabric. The chosen resin must be heat resistant, preferably up to 200 °C, be completely non-blocking, have excellent abrasion properties and also be resistant to moisture under high temperatures. In general, the aluminium powder should be added to the mix just before use to prevent the formation of hydrogen gas from the interaction of the aluminium with the compound. The finished coated fabric must be completely flat and wrinkle free, and must of course retain these properties throughout its serviceable life. Fabrics with reflective surfaces may have other applications, such as thermal curtain linings and heat shields. However, the actual heat reflected back is quite low compared to an aluminised film or mirror. The finer the aluminium powder, the better the reflective surface. Cotton laminated to polyurethane foam about 2 mm thick

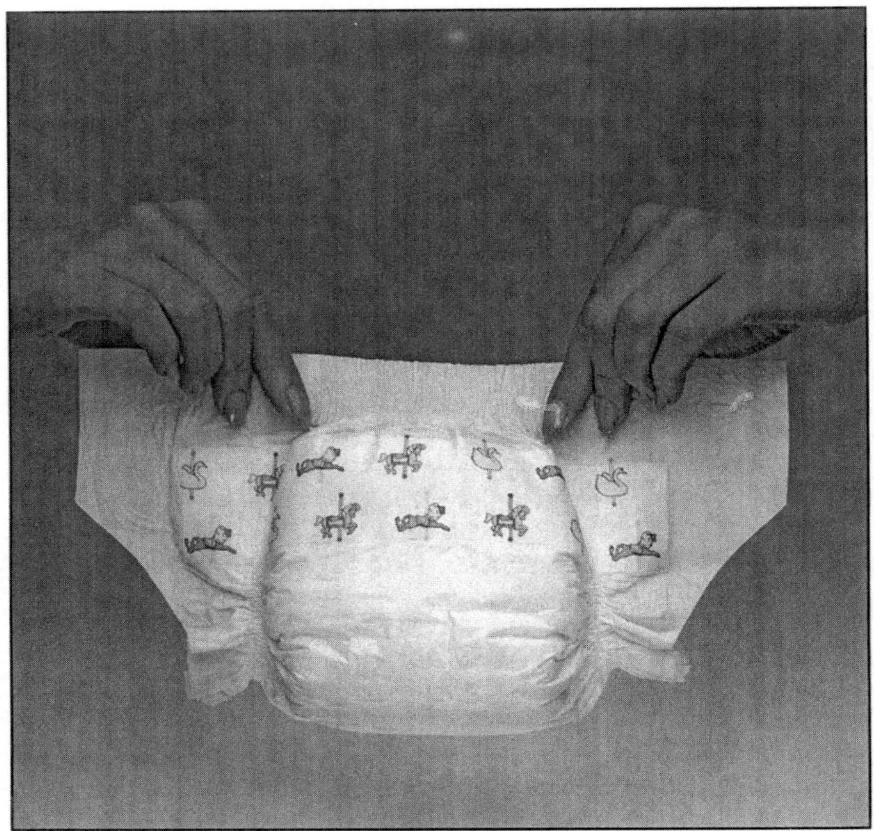

4.39 Foam backed fabrics have many current and potential applications. Baby diapers are produced from fabric laminated to polyurethane foam. Elasticity and tensile strength combine with softness for baby's delicate skin. Photograph supplied by Caligen Foam and reproduced with kind permission.

is also used as ironing board cover material. The benefits claimed are smoother and faster ironing and non-slip properties. Both aluminised surfaces and printed fabrics are available.

4.6.14 Carpets

This is one of the largest markets for water-based resins, and SBR (43 million dry pounds weight used in the USA in 1995) is the most used because of its versatility, availability and economical price. Coatings are required to lock in the pile and to assist with thermal, shock and acoustic insulation and also water resistance.[168–171] The coating, usually applied to a height of 2–6 mm (just over a sixteenth to a quarter of an inch), must be sufficiently elastic to avoid creases being formed when the carpet is rolled up for transportation. The resin must be formulated for good adhesion to lock in the tufts, and it must have good dimensional stability, be resistant to water, light and heat ageing and reduce edge fray. There has been much development, especially in recent years, to test for complete vulcanisation of the rubber coating. If this is not accomplished properly, the carpet will not only have poor resiliency, but may also give off offensive odours.[170,171] EVA is also used because it has good adhesion to fibres, is inexpensive and has less tendency to produce odours. Waterproof carpet backings are sometimes required for hospitals and entrance halls and other areas where there may be wet contact.

Contract markets for offices and hotels generally require some anti-static properties, which are achieved by various techniques. Finishes can be applied to the carpet surface, carbon black can be added to the resin mix or conductive fibres such as Negastat or Resistat (BASF) can be used in the carpet pile. The latter method is permanent and cannot wear off, but is probably the most expensive option. Conductive fibres are extremely expensive, but only a very small proportion is required to produce some measure of anti-static performance. The widespread use of electronic equipment, such as computers, makes anti-static properties more necessary. Routine tests on carpet assess colour fastness, anti-soiling, dimensional stability when wet, pile height, etc., but the tests which are relevant to the coating technologist include pile pull out, bond strength between the backing materials and FR.

In recent years, the question of indoor air quality has become more of an issue, and emissions and odours from carpets have become a critical item. Disposal of carpets has also become a problem which is being tackled by fibre producers and others. Landfill, previously the most used method for carpet disposal, is now restricted, see Chapter 6, and recycling methods are being devised. It appears that nylon 6 is more easily recycled than nylon 66 and this may eventually influence the choice of which fibre to use. Research into renewable natural materials is being conducted, and Interface Inc. of Atlanta expect to produce large volumes of carpet from polymers such as poly lactic acid derived from corn and other plants.[172] These aspects are discussed further in Chapter 6.

The largest carpet producing country in the world is the USA, and in Europe the UK, Germany and Belgium are the major producers. Carpets are used significantly less in hotter countries. Carpets for contract use invariably have high FR safety standards, especially in the case of public transport applications such as trains, planes and marine vessels, and Neoprene latex is widely used. Flame retardancy requirements have already been discussed in Chapter 2, and they are particularly stringent on trains and aircraft.

4.7 Medical uses

4.7.1 Protective clothing

Since the discovery of AIDS, it is essential that medical practitioners, nurses and others do not come into contact with the patient's blood.[173,174] Surgeons' gowns must be impermeable to microbes and blood, but also need to be comfortable in all the ways discussed earlier. At present, in the USA, most medical protective clothing, maybe as much as 80%, is made from nonwoven materials and used only once. The leading materials are Sontarna made by DuPont and nonwoven laminates produced by Kimberly-Clark. The outer surface of the nonwoven is treated with fluorocarbon for liquid repellency. In situations where a total barrier is required, the nonwoven is laminated to a waterproof film, and in some cases waterproof breathable films such as Gore-Tex are used. Because of its critical importance, this laminate needs to be very carefully tested for barrier properties to microbes, and special tests have been and are being devised for this purpose. See Section 5.8 in Chapter 5. In the UK and Europe, the situation is different and a high proportion of protective clothing is reusable. It is produced from tightly woven cotton and polyester/cotton fabrics which are laundered at high temperatures, e.g. 74 °C, and sterilised.

Fabrics for medical textiles must also be completely non-linting, and in some cases must have FR properties because of the use of lasers and because they may be used in atmospheres rich in oxygen. Work continues to develop economical materials which can be washed and sterilised and still retain their barrier properties. Disposable medical protective clothing must add to the problem of waste that needs to be disposed of in an environmentally friendly way. There are likely to be opportunities for the development of washable and sterilisable medical products in the future, and perhaps this process has already started. However, they must be capable of withstanding multiple launderings and sterilisation treatments while retaining their protective properties. An EC funded project, BIOBAR, is evaluating test methods for protection against biological hazards.

4.7.2 Mattress covers

Polyurethane, transfer coated, raised knitted nylon fabrics are used as mattress covers for incontinent patients. The material must be washable and capable of withstanding repeated sterilising and disinfectant treatments and still being water resistant. A test to simulate suitability is a ten minute boil in 4% caustic soda after which the material should still have a hydrostatic head of 80 cm. Other tests include treatment with a variety of solvents and dry cleaning. Some degree of breathability is desirable for patient comfort and to reduce the incidence of bed sores, but this is not a requirement of BS 5455, which is needed for contract to the British National Health Service. In some cases, the bed is lined with an unsupported film of polyurethane under the sheet. The film, an example of which is Platilon (by ATO), is waterproof and some grades have some measure of breathability.

4.7.3 Health care items

A growth area is incontinence garments and devices. Reports advise that 5% of the US population, 13 million, people are incontinent and this is likely to grow as the aged population increases. Analysts are expecting 5–7% annual growth, as the 'baby boomers', i.e. those people born just after the Second World War, begin to age. The retail market in the USA for incontinence garments reached $594 million and sales of $2.1 billion are expected in the institutional and retail market together by 2005. The world market is estimated to be worth $5.8 billion.[175]

At present most of these garments are disposable, but with the future emphasis on reduction and recycling and the problems associated with disposal by landfill and incineration, there may well be a requirement for washable and reusable articles. The same may be eventually be true for related items such as baby diapers or nappies (see Section 4.6.12). A very recent report details an improved and cheaper method (Reifenhauser) of producing breathable waterproof film/non-woven laminates by a combined film extrusion and lamination process.[176] This method incorporates coating by an electrostatic process rather than by using mechanical pressure. The laminate material produced could have a variety of applications, including clothing and industrial coverings.

4.8 Military uses

4.8.1 General comments

The military are bulk users of coated fabric for protective clothing, including specialist protective clothing for all three services, for coverings and temporary structures, and for survival apparatus such as inflatables and life jackets. There are also some more specialist uses such as camouflage and clothing to protect against chemical warfare agents, and some quite novel applications such as inflatable decoy aircraft and decoy tanks. The latter products are used to deceive the enemy and make them use up expensive ammunition on virtually worthless imitations, and they also give the impression that the force is larger than it actually is. The ever-improving detection capabilities of military equipment necessitate ever-improving countermeasures making use of different areas of the electromagnetic spectrum, e.g. visible light, IR, UV radar. Many camouflage materials used by the military are coated fabrics, and the coating allows the incorporation of specialist fillers and pigments to perform various functions of absorption and reflection. More information on military uses is available from a number of sources, including the US Army soldier systems website.[177–181]

4.8.2 Protective clothing

Military waterproof protective clothing – known as 'foul weather' clothing – is produced to the highest standards of performance. Protection from the cold and wet is of paramount importance in the field. It is a well known fact that there are as many casualties from sickness as there are from enemy action. The soldier subjects any material to much more 'punishment' than the walker, rambler or average industrial worker. The coated fabric apparel is subject to prolonged exposure to rain and wind and also to wet ground and grass as the soldier crawls on his hands and feet. The elbows and knees of the garment are subject to very high pressure, and water can leak in at these points if the material is not produced to a high standard of waterproofness. The UK MOD require a hydrostatic head of over 70 kPa (700 cm of water) after washing and after ageing and flexing. The US Army once quoted 25 psi hydrostatic water entry pressure to prevent leaking under all conditions – this is equivalent to 1750 cm head of water of 175 kPa. For comfort and maximum efficiency, the material must also breathe as much as possible, but at the time of writing, it is too expensive to equip every soldier with Goretex and it is used only by specialist forces. The rustling noise produced by garments made from coated and laminated fabric is a security issue in stealth operations. Foul weather garments are generally produced from material printed with a 'disruptive' pattern to reduce visibility in the field.

The nuclear, biological and chemical warfare suit material (NBC) is a laminate, which incorporates activated carbon and other material for adsorbing toxic agents. The garment issued to UK forces has an outer layer of nylon and modified acrylic (for FR properties) in a twill construction. The precise design varies from nation to nation.

Soldier's clothing must be as versatile as possible, because they may not able be to change in the field and, during a particular operation, the weather may change significantly. In addition to wind, rain and cold, the soldier needs protection against thorns, solar radiation and heat stress. Other factors to consider are the weight of garments and the restrictions they place on movements. Like other research institutions, the MOD are investigating 'intelligent' and highly sophisticated materials and garment designs which are capable of reacting to the weather and to the soldier's skin temperature.

4.9 Other materials involving coating and laminating

4.9.1 Sliver knit products

Pile fabrics can be produced directly using circular knitting machines by feeding in slivers (continuous bundles of fibre) which are incorporated into a backing fabric as cut tufts of material.[182] Virtually any fibre or blend of fibre can be used as the sliver although, because the purpose is to produce a soft fleecy fabric, acrylic or modified acrylics are the most widely used. Sliver knit fabrics need to be stabilised by heat setting, and the tufts need to be locked in by back coating usually with acrylic resin. This can be applied by spraying, back roller kissing or by back coating with a foamed resin. The resin needs to be soft and flexible but non-blocking and to have good adhesion to the fibres. Washability or drycleanability may be required for certain products and cross-linking resins should be used. Acrylic fibres should be processed at lower temperatures than polyester or nylon, and this should be taken into consideration during resin drying and curing; otherwise there may be risk of fibre stiffening or discoloration.

Sliver knits appear in a wide variety of products, including apparel and shoes as thermal liners, blankets, baby products, rugs and soft toys. Using higher deniers, sliver knit fabrics can be processed to produce attractive simulated animal fur for outerwear.

4.9.2 Natural leather

Natural leather has been used since ancient times as a covering material for clothing and tents, and as a flexible, reasonably waterproof material for making shoes and articles for water storage such as buckets. Leather has a combination of properties which make it the preferred material for shoes: waterproofness, breathability, softness and the ability to 'mould' to the shape of the wearer's feet – leather shoes 'wear in' and become more comfortable. The modulus of leather is low for the first 10–15% of elongation, and this makes it very suitable for the shaping or lasting process used in the manufacture of shoes. Processed leather drapes well, can be very soft and has aesthetic and fashionable value. It also has functional properties and is used for making bags and luggage, items which sometimes also need to have an attractive appearance and an element of fashion to match clothing. Natural leather is sometimes coated with polyurethane resin to improve abrasion resistance. This reduces water permeability in articles where it is relevant, for example in car seats or apparel.

Natural leather is a network of fibrous bundles, which are themselves made up from finer bundles, which in turn are composed of even finer bundles and so on right down to the collagen protein molecule itself.[183,184] The fibres are finest on the grain side of the skin. Suede is leather which has been napped. Because of the popularity of leather and its general shortage at particular times, for example just after the Second World War, many attempts have been made over the years to produce artificial forms. Leather is a by-product of the food industry – cattle are not raised for their hides – and a shortage is forcast for the future for two reasons. One reason is that less meat is being eaten in some developed countries, and the other is that more people can afford to buy leather goods which are regarded as luxury items, thus creating more demand for leather.

Automotive natural leather is frequently laminated to foam which is then backed with a scrim fabric in the same way as fabric seat covering before being made into a seat cover. The leather is also usually lacquered with a polyurethane resin on the face side to improve abrasion resistance, but this is believed to reduce breatheability. In recent years, leather processing has undergone certain changes to comply with environmental laws.[185] Leather is universally regarded as the ultimate in seat luxury, but it is expensive and a shortage is forecast in the future, partly because of the increased volumes used in car manufacture, and partly for the reasons explained above. At the same time, leather is also being used more in design combinations with textiles. The shortage of leather, together with the increased preference for leather designs, is an opportunity for expansion in man-made leather products – which require textile base materials. The odour of natural leather, generally regarded as part of the overall luxurious image in a car interior, is believed to be disliked by some Japanese customers.

4.9.3 Man-made leather and suede

Transfer coated polyurethane fabrics and PVC coated fabrics, especially the 'blown' types made by chemical blowing agents and by mechanical foaming, sometimes referred to as 'leathercloths', have been available for some time but do not breathe. However, researchers in Europe, the USA and Japan have attempted over the years to produce artificial leathers from polyurethane. The objectives have been to produce a material which is comfortable, breathes and resembles natural leather more than the basic 'leathercloth'. The technique used is coagulation of polyurethane resins to produce a pore-like structure, which will produce the soft touch and breathability of natural leather. At the time of writing the most successful man-made products, two grained leathers and eight suedes, are entirely

Japanese made, and the companies involved have production expansion plans.[81] The estimate made in 1997 (see Section 4.3.6) is already looking very conservative. The base materials are generally nonwovens using microfibres in polyester, which constitute 68% of the weight, the remainder being polyurethane resin. For use as car seat covers, the man-made suede is polyurethane foam backed with a scrim fabric in the usual way. The best known is Alcantara,[82] made in Italy since 1975 by a Toray/Enichem joint venture (now Toray/Mitsui) and initially used mainly in Italian cars. A second production line for the material is being built. Kuraray, one of the pioneers of man-made leather, has recently entered the European automotive market with their Amaretta product.[83] Man-made products have the important advantages over natural leather of availability in roll form, light weight, uniformity of quality, uniformity of thickness and other physical properties which allow more efficient production planning and minimisation of waste.

Because Alcantara and other successful man-made suedes are produced by a solvent coagulation process requiring expensive plant and environmental controls, attempts have been made to develop more environmentally friendly aqueous-based methods.[186] One key factor in achieving the quality of man-made suede is believed to be due to the microfibres in the base fabric, and the very latest products use ultra fine filaments of 0.001–0.003 dtex. Special techniques have been developed to produce these ultra fine filaments, an example of which is the 'islands in the sea' technology developed by Kuraray for their Clarino material. Two polymers of different solubility are mixed for yarn spinning and, after spinning, solvent extraction produces filaments as fine as 0.0001 tex. However, the polyurethane polymer must have the right properties and there is considerable skill required in the final sueding operation.

In 1994, Enichem launched a new artificial leather called Lorica.[187] This has several advantages over natural leather which include better elongation, tear strength, mouldability and high frequency weldability. Lorica is made from polyamide microfibres and polyurethane and is available in a variety of colours. Car seat covering is one of the applications being explored.

4.10 References

1. Smith J, 'The comfort of clothing', *Textiles*, Vol 15 No 1, 23–7.
2. Monkman JRP, 'Water repellency of textile materials', *Insight*, Vol 2 No 4 Dec 1979, Quality Control Services Ltd, Wembley, 1–3.
3. Factsheet No 6, Feb 1983, Channel Four Television Consumer Programme, 5.
4. Norris CA, 'The testing of water repellent finishes on textiles', in *Waterproofing and Water-Repellency* (Editor Molliet JL), Elsevier, Amsterdam 1963, 270.

5. Ashton S, 'Technical review; breathable waterproof fabrics', *High Magazine*, Sept 1985, 52–5.
6. Keighley JH, 'Breathability of outer fabrics', *Survival '85 Symposium*, Leeds University 28 Mar 1985.
7. Keighley JH, 'Breathable fabrics and comfort in clothing', *JCF*, Vol 15 Oct 1985, 89–104.
8. Sellers G, 'Introducing John Keighley – the man who tests your gear', *Climber and Rambler'*, Mar 1984, 58–9.
9. Sellers G, 'Breathable waterproofs under the microscope', *Climber and Rambler*, Nov 1985, 49–53.
10. Umbach KH, 'Physiological problems of breathable garments, *Survival '85 Symposium*, Leeds University 28 Mar 1985.
11. Marxmeier H, 'Membranes for all-weather clothing effects of outside wetting on permeability to water vapour, *Chemiefasern Textilindustrie*, Vol 36 No 88 Jul/Aug 1986, 575–6.
12. Anon, 'Magic membrane for all-weather clothing' (Sympatex), *Textile Month*, Jan 1995, 20–22.
13. Anon, 'Impact 87 Conference Coverage – Part 1', *Nonwovens World*, May–Jun 1987, 85–8.
14. Shishoo RL, 'Recent developments in breathable high functional fabrics', *Nonwovens World*, Feb 1987, 89–94.
15. Lomax GR, 'Coated fabrics: Part 1 – Lightweight breathable fabrics', *Textiles*, Vol 14 No 1, Spring 1985, 2–8.
16. Anon, 'Waterproof, breathable membrane system' (Permatex), *Textile Month*, May 1995, 48.
17. Holmes DA, 'Waterproof breathable fabrics', Chapter 12 in *Handbook of Industrial Textiles* (Editors Horrocks AR and Anand SC), Woodhead, Cambridge 2000, 282–315.
18. Overington Y-H and Croskell R, 'Standards for breathable fabrics', *International Dyer*, May 2001, 23–7.
19. Croskell R and Overington Y-H, 'Determining water vapour permeability of clothing materials', *TTi*, Apr 2001, 9–13.
20. Spijkers JCW, 'Development trends in workwear with Sympatex memzbranes', *IMMFC*, Dornbirn 17–19 Sept 1997.
21. Weder M (EMPA), 'Performance of rainwear materials with respect to protection, physiology, durability and ecology', *JCF*, Vol 27 Oct 1997, 146–68.
22. Weder M, 'Performance of breathable rainwear material with respect to protection, physiology, durability and ecology', *IMMFC*, Dornbirn 17–19 Sept 1997.
23. Gretton J, 'Condensation in clothing systems', *World Sports Activewear*, Spring 1999, 38–43.
24. Watts IC, 'Moisture interaction', *Textile Horizons*, Oct 1994, 44–9.
25. Anon, 'Eta woven from long staple cotton', *World Sports Activewear*, Vol 3 No 1 Spring 1997, 15.
26. Norris CA 'The testing of water-repellent finishes on textiles', in *Waterproofing and Water-repellency* (Editor Moilliet JC), Elsevier, Amsterdam 1963, 271. Original paper Baxter S and Cassie ADB, *Journal of Textile Institute*, 1945, 36 T67.

27. Raheel M, 'Protective clothing; an overview', in *Protective Clothing Systems and Materials* (Editor Raheel M), Marcel Dekker, New York 1994, 1–23.
28. Shaw A, Cohen E and Wicke H, 'Personal protective equipment for the agricultural worker', *Industrial Fabrics Product Review*, Jun 2000, 48–54.
29. Anon, 'The implementation of PPE – a reminder', *BTTG Independent*, Aug 2001, 13.
30. Masri M, 'The European Directive on personal protection equipment', *TTi*, Jul 1992, 17–20.
31. Slater K, 'Comfort or protection; the clothing dilemma', *Fifth International Symposium on Performance of Protective Clothing*, San Francisco 25–27 January 1994. Papers published by ASTM as *Performance of Protective Clothing; Fifth Volume (STP 1237)*, West Conshohocken PA, USA 1996, 488.
32. Stull JO, 'A review of the ASTM F 1001 battery of chemicals and its effect on the chemical protective clothing industry', *Performance of Protective Clothing; Fifth Volume (STP 1237)*, ASTM Publications, West Conshohocken PA, USA 1996.
33. Stull JO, 'Assessment of chemical barrier properties', in *Modern Textiles Characterization Methods* (Editor Raheel M), Marcel Dekker, New York 1996.
34. Adanur S, 'Safety and protective textiles', in *Wellington Sears Handbook of Industrial Textiles* (Editor Adanur S), Technomic, Lancaster PA, USA 1995, 415–74.
35. Shaw A, Cohen E and Wicke H, 'Personal protection for agricultural workers', *Industrial Fabrics Product Review*, Jun 2000, 48–54.
36. Truong Q and Wilusz W, 'Material design for chemical and biological protective clothing', *IMMFC*, Dornbirn 17–19 Sept 1997.
37. Brown PL, 'Assessment of the protective properties of textiles against microorganisms', in *Modern Textile Characterisation Methods* (Editor Raheel M), Marcel Dekker, New York 1996.
38. Wadsworth L and Allen HC Jnr, 'Development of highly breathable and effective blood/viral barrier laminates of microporous films, staple fibres and nonwovens', *JCF*, Vol 28 Jul 1998, 12–28.
39. Butler N, 'Modern textiles provide greater protection for London's firefighters', *TTi*, Oct 1998, 13–16.
40. Holmes DA, 'Textiles for survival', in *Handbook of Technical Textiles* (Editors Horrocks AR and Anand SC), Woodhead, Cambridge 2000, 483–9.
41. www.iaff.org/iaff/health_Safety/safety_alert_2.html (International Association of Fire Fighters) International Association of Fire Fighters' website www.iaff.org/ (15/12/1999).
42. Sager B, 'Reflective inks offer protection and fashion', *TTi*, Oct 1999, 20–22.
43. Adanur S (Editor), *Wellington Sears Handbook of Industrial Textiles*, Technomic, Lancaster PA, USA 1995, 464–9.
44. Taylor PJ, *Latham's Technology of Clothing*, Third Edition, Blackwell, Oxford 2000, 212–20.
45. Viikna A, 'The adhesive properties of interlinings', *Textiles Magazine*, Issue 3, 2000, 8–12.
46. Hoechst Information Leaflet, *Trevira in focus – Complan*, 12687/98e.
47. Bajaj P and Sengupta AK, 'Tarpaulins and covers', *Textile Progress*, Vol 14 No 1, The Textile Institute, Manchester 1982, 25–6.

48. Eichert U, 'Weather resistance of coated fabrics for automotive industry', *IMMFC*, Dornbirn 17–19 Sept 1997.
49. Dartman T and Shishoo R, 'Prediction of performance of coated fabrics', *Technische Textilen*, Nov 1995, E43–E46.
50. Cass J, 'Revisiting vinyl coated cotton', *Industrial Fabrics Products Review*, Aug 2000, 24–8.
51. Smith R, 'Top graphics; truckwise', *Industrial Fabrics Products Review*, Aug 1999, 52–7.
52. Peterson S, 'Trucks; changing stripes', *Industrial Fabrics Products Review*, Mar 2001, 60–65.
53. Jagielski K (Fabrene Corporation), 'Extrusion coated and laminated woven polyolefin fabrics – an overview of the material and markets', *5th International Conference on Textile Coating and Laminating*, The Fort MacGruber Inn, Williamsburg VA 13–14 Nov 1995, Technomic, Lancaster PA, USA.
54. Intertape Polymer Group Newsletter, April 2001.
55. Klarr R, 'Big bags', '*Industrial Fabrics Products Review*, Feb 1997, 40–42.
56. Hoechst Technical Information Leaflet, *Resistance of bulk goods containers made from coated Trevira high tenacity woven fabric to various types of product*, F 0301 Jul 1977.
57. Floyd KL, 'Coated textiles as fluid containers', *Textiles*, Vol 6 No 3 Oct 1977, 78–83.
58. Thomas E, 'Coated materials for specialised end uses', *Progress in Textile Coating and Laminating*, BTTG Conference, Chester 2–3 Jul 1990, BTTG, Manchester.
59. Silvey DH and Rugman G (BF Goodrich), 'Belting', *Kirk Othmer, Encyclopaedia of Polymer Science and Engineering*, Third Edition, Vol 2, John Wiley, New York 1985, 193–201.
60. Murphy G, 'Conveyer belting', in *Textile reinforcements of elastomers* (Editors Wake WC and Wotton DB), Applied Science Publishers, London 1982, 133–76.
61. Butterworth P, 'Endless woven Tygaflor PTFE coated belts', *Developments in Coating and Laminating*, Shirley Institute Conference, Manchester 27 Mar 1981, Shirley Institute Publication S41, BTTG, Manchester.
62. Sperati CA (DuPont), in *Handbook of Plastic Material and Technology* (Editor Rubin II), John Wiley, New York 1990, 125.
63. http://www.hotship.com/
64. Wallace S (Cameron Balloons), private communication to author on 27 Dec 2000.
65. http://www.cameronballoons.com/1998/fabrics_98.htm
66. Anon, 'Firms bask in balloon success', *PRW*, 26 Mar 1999, 8.
67. Griffith C, 'Near space balloons – NASA's new workhorses', *Industrial Fabrics Products Review*, Oct 1999, 38–44.
68. Anon, 'Ultimate flight – around the world in 18 days by balloon', *World Sports Activewear*, Dec 1995/Jan 1996, 48–53.
69. Boyes R, 'A new dawn for the airship', *The Times*, Feb 6 1999, 18.
70. Woodkead M, 'The airship rises from the ashes', *The Sunday Times*, Jun 4 2000, 16.
71. CargoLifter Team, private communication to author on 5 Jun 2001.
72. www.polyfibron.com (Polyfibron Technologies – textile blankets).

73. Suits LD, 'Geotextiles', *Kirk Othmer, Encyclopaedia of Chemical Technology*, Fourth Edition, Vol 1, John Wiley, New York 1994, 503–11.
74. Fluet JR, 'Coated Fabrics in geotextile and geomembrane applications', *JCF*, Vol 14 Jul 1984, 53–64.
75. Kowalski M, 'Automotive fabrics to date', *Textiles*, Vol 20 No 2 1991, 10–13.
76. Fung W, 'Present and future technical requirements of automotive fabric', *Industrial, Technical and High Performance Textiles, World Textile Congress*, Huddersfield University 15–16 Jul 1998.
77. Horsforth A, 'One big technical headache-on wheels', *JSDC*, May/Jun 1992, 108, 243–6.
78. McCullum JB (Ford), 'Engineering requirements for automotive textiles', in *Automotive Textiles* (Editor Ravnitzky M), PT-51 SAE, SAE Inc, Warrendale PA, USA 1955, 117–22.
79. Parsons MA (Rover), 'Fabric requirements for automotive use', *Autotech*, Seminar 9, NEC Birmingham 1991.
80. Kowalski M, 'Automotive fabric presentation', *Autotech*, Seminar 9, NEC Birmingham 1991.
81. Anon, 'Man-made leather grows in product and demand', *JTN*, Jun 1997, 66–73.
82. Borri C, 'Comfort for car interiors – Alcantara – features and advantages of a unique product', *IMMFC*, Dornbirn 22–24 Sept 1993.
83. Tanaka N and Tanaka J (Kuraray), 'New man-made leather for automobiles', *IMMFC*, Dornbirn 15–17 Sept 1999.
84. Pikula D, Koesis MJ and Brandon RH (Dow), 'Acoustic evaluation of automotive headliner composites with various adhesive systems', in *Automotive Textiles* (Editor Ravnitsky M), PT-51 SAE, SAE Inc, Warrendale PA, USA 1995, 173–83.
85. Souders SL, Doerer RP and Scott TE, 'Engineering optimisation and tuning of vehicle interiors sound insulation', in *Automotive Textiles* (Editor Ravnitsky M), PT-51 SAE, SAE Inc, Warrendale PA, USA 1995, 185–203.
86. Westrick RW and Grey JW, 'What's new in automotive headliners', in *Automotive Textiles* (Editor Ravnitsky M), PT-51 SAE, Warrendale PA, USA 1995, 123–6.
87. Fichtenthal NH, 'Meeting tomorrow's economic and environmental requirements', *Progress in Textile Coating and Laminating*, BTTG Conference, Chester 2–3 July 1990, BTTG, Manchester.
88. Pfortner P (Freudenberg), 'Nonwoven applications for automotive interiors today and tomorrow', *Inter Auto*, Amsterdam 13–14 Oct 1998, organised by *Inside Automotives International*, Michigan.
89. Daniels J, 'Nonwoven headliners – a European perspective', *Inside Automotives International*, Jan/Feb 1995.
90. Gardner C, 'Headliners, rising to the occasion', *Inside Automotives International*, Dec 1994 19–23.
91. Lebovitz R, 'Nonwovens offer new interior possibilities', *Automotive & Transportation Interiors*, Sept 1997, 18–19.
92. Braunstein J, '2001 and beyond, a safety odessey', *Automotive & Transportation Interiors*, Apr 1996, 27–34.
93. Thompson J, 'Restraining orders', *Automotive Interiors International*, Feb 1998, 16–20.
94. Mound H, 'BMW gives a boost to hyperinflation', *The Times*, 12 Apr 1997.

95. Anon, 'Eye on environment', *Inside Automotives International*, Oct 1998, 14–19.
96. Crighton KN, 'Tubular side airbag technology takes another shape', *Automotive & Transportation Interiors*, Apr 1995, 16.
97. Sonderstrom P, 'Side impact airbags, the next step', *Inside Automotives International*, May/Jun 1996, 12–15.
98. Anon, 'Western Europe airbag yarns', *Chemical Fibers International*, Vol 51 Feb 2001, 6.
99. Davidson A, 'Growing opportunities for airbags', *TTi*, May 1992, 10–12.
100. Smith TL, 'Tough stuff', *Automotive & Transportation Interiors*, Aug 1996, 30–32.
101. Bohin F and Ladreyt M, 'Silicone elastomers for airbag coatings', *Automotive Interiors International*, Winter 1996/7, 66–71.
102. Kerr RC, 'A new generation of textile substrates and coatings for automotive airbag fabric', *6th International Conference on Coating and Laminating*, Dusseldorf 4–6 Nov 1996, Technomic, Lancaster PA, USA.
103. DuPont Automotive TI leaflets H-48030 and H 48032 (USA).
104. Siejack V (AKZO), 'New Yarns for lighterweight airbag fabrics', *IMMFC*, Dornbirn 17–19 Sept 1997.
105. Barnes JA, Partridge JF and Mukhopadhyay S, 'Air permeability of nylon 66 airbag fabrics', *Yarn and Fibre Conference*, Textile Institute, Manchester 2–3 Dec 1996.
106. Insurance Institute for Highway Safety (Arlington VA, USA), safety facts, airbag statistics, website http://www.hwysafety.org/
107. Ross HR (AlliedSignal), 'New future trends in airbag fabrics', *IMMFC*, Dornbirn 17–19 Sept 1997. See also *A Technical Discussion on Airbag Fabrics*, StayGard™ nylon 6, Technical Information Brochure, AlliedSignal 1993.
108. Smith TL, 'Airbags and seat belts; fabrics role in safety restraint systems', *Automotive & Transportation Interiors*, Dec 1995, 53–4.
109. Lennox-Kerr P, 'Stichbonded airbags' (3M patent), *Nonwovens International*, July 1999, 37–8. See also US Patent 5826905.
110. Braunstein J, 'Occupant safety/electronic supplement', *Automotive & Transportation Interiors*, Jun 1999, E1–E10.
111. Wolff H, 'Problems with airbags – unneeded airbag firing is expensive and dangerous', *IMMFC*, Dornbirn 15–17 Sept 1999.
112. Mowry G and Head A, 'Braided inflatable tubular structure technology in crash safety', *IMMFC*, Dornbirn 15–17 Sept 1999.
113. Braunstein J, 'A steady march forward', *Automotive & Transportation Interiors*, Dec 1999, 32–41.
114. Braunstein J, 'Autoliv introduces one-piece weaving to North America', *Automotive & Transportation Interiors*, Dec 1999, 34.
115. Mukhopadhyay SK and Partridge JF, *Automotive Textiles*, Textile Progress, Vol 29 No 1/2, Textile Institute, Manchester 1999, 68–87.
116. National Highway Traffic Safety Administration website www.nhtsa.dot.gov
117. Wohlgemuth J and Nordhoff R, 'Properties required of convertible top material as exemplified by the Mercedes-Benz CKL Cabriolet', *Plastics and Textiles in Automotive Engineering VDI Conference*, Mannheim 25–6 Mar 1998.
118. Walkenhorst W, 'Dolan outdoor for soft top textiles', *IMMFC*, Dornbirn 17–19 Sept 1997.

119. Eisele D, 'Outer textile linings for cars – an innovation', *Technical Textiles*, Vol 42 Apr 1999, E28–E29.
120. Cheek M, 'Automotive carpets and fibres; an international perspective', *Textiles in Automotive Conference*, Greenville SC, USA Oct 1991.
121. Cox JH, 'Tufted carpet for auto use', in *Automotive Textiles* (Editor M Ravnitsky), SAE PT-51, SAE Inc, Warrendale PA, USA 1995, 145–50.
122. Creasy L, 'The great cover-up', *Automotive & Transportation Interiors*, Aug 1997, 16–22.
123. Zimmermann M (Rieter), 'Textiles for motor car interior fibres', *Technical Textiles*, Vol 42 Apr 1999, E27.
124. *Mouldable Needlepunched Nonwovens for Auto Applications*, BASF Technical Information Sheet TI/ED 1382, Nov 1988.
125. *Binders for Automotive Interiors*, Synthomer Technical Information, Jul 1996.
126. Laser J, 'Moulded automotive carpets – their influence on the interior noise', *Technical Textiles*, Vol 41 Feb 1998, E4.
127. Schurian A, 'Computer-supported development of integrated carpet for cars', *IMMFC*, Dornbirn 15–17 Sept 1999.
128. Saha P and Baker RN, 'Sound adsorption study for auto carpet materials', in *Automotive Textiles* (Editor Ravnitsky M), SAE PT-51, SAE Inc, Warrendale PA, USA 1995, 199–203.
129. Schuster D, 'New concepts for car trim parts to improve noise reduction', *IMMFC*, Dornbirn 15–17 Sept 1999.
130. Braunstein J, 'The search for silence proves a never-ending quest', *Automotive & Transportation Interiors*, Oct 1995, 32–5.
131. Markel A, 'Sound judgement', *Inside Automotives International*, May 1998, 14–17.
132. Murphy J, 'The science of sound', *European Automotive Design*, Summer 1997, 30–33.
133. Anon, 'Feel the noise', *Automotive Interiors International*, Oct 1998, 48–50.
134. Anon, 'Silent partners', *Automotive Interiors International*, Dec 1998, 32–8.
135. Smith TL, 'Taking flock technology a step further', *Automotive & Transportation Interiors*, Feb 1996.
136. Lebovitz R, 'Beyond flocking – new fibre technology battles the rattles', *Automotive & Transportation Interiors*, Dec 1995.
137. Ozsanlav V, 'Specific applications for jute/synthetic blends', *Industrial, Technical and High Performance Textiles, World Textile Congress*, Huddersfield University 15–16 Jul 1998.
138. Berthevas PR, Fanget A and Gatouillat G, 'The development of a sound insulation package for car floor coverings using a combination of polyurethane technologies', *JCF*, Vol 18 Oct 1988, 124–41.
139. Anon, 'PU foams give selective sound damping', *BPR*, Sept 1997, 45.
140. Anon, 'EU applies pressure to reduce noise', *PRW*, 30 Apr 1999.
141. Gardiner ER, 'Marine applications', in *Textile Reinforcement of Elastomers* (Editors Wake WC and Wootton DB), Applied Science Publishers, London 1982, 197–223.
142. Thomas EJ, 'Coated material for specialised end uses', *Progress in Textiles Coating and Laminating,* BTTG Conference, Chester 2–3 July 1990, BTTG, Manchester.

143. Sowden E (Avon), 'Fabric design factors in the production of inflatable craft', *JCF*, Vol 13 April 1984, 250–57.
144. Howard-Williams J, *Sails*, Sixth Edition, Adlard Coles, London 1988, 39–69.
145. www.ullman-sails.co.uk (Ullman Sails UK, Southampton).
146. www.doylesails.com (Doyle Sailmakers, Marblehead MA USA).
147. Anon, 'Weaving into the wind', *World Sports Activewear*, Dec 1995/Jan 1996, 16–19.
148. Anon, 'Carbon sails to the front', *TTi*, May 1992, 28–31.
149. Wheatley K, 'Roar of the big cats', *The Sunday Times*, 31 Dec 2000, 12.
150. Mottram R, 'News feature', *PRW*, 3 Nov 2000, 10.
151. Bajaj P and Sengupta AK, 'Coated fabrics in civil engineering', *Textile Progress*, Vol 14 No 1, The Textile Institute, Manchester 1982, 17–20.
152. Adanur S, 'Applications of coated fabrics in building structures' in *Wellington Sears Handbook of Industrial Textiles* (Editor Adanur S), Technomic, Lancaster PA, USA 1995, 205–15.
153. Sallis R, 'High performance coated fabric structural materials', *JCF*, Vol 14 Jul 1984, 36–45.
154. Forster B, 'The engineered use of coated fabrics in long span roofs', *JCF*, Vol 15 Jul 1985, 25–39.
155. Cumberbirch RJE, 'Textiles, engineering and architecture', *Textiles*, Vol 16 No 2 1987.
156. Effenberger JA and Dery M (Chemfab Corporation), 'Membranes for permanent fabric architecture', *5th International Conference on Textile Coating and Laminating*, The Fort MacGruder Inn, Williamsberg VA, USA 13–14 Nov 1995, Technomic, Lancaster PA, USA.
157. Anon, 'Tensioned textile structures – new thinking, new aesthetics and new architecture', *TTi*, Jul/Aug 1999, 21–4.
158. Butler N, 'IFAI award winners', *TTi*, Jan/Feb 2000, 11–21.
159. Doherty G and Higginbotham RS, 'Fabric for tents', *Textiles*, Vol 8 No 1 Feb 1979, 7–11.
160. Adanur S, 'Architectural and construction textiles', *Wellington Sears Handbook of Industrial Fabrics* (Editor Adanur S), Technomic, Lancaster PA, USA 1995 222–3.
161. Morley D, 'Coated fabric for upholstery', *JCF*, Vol 14 Jul 1984, 46–52.
162. Avril MW, 'Non-commercial applications of coatings', *JCF*, Vol 16 April 1987, 225–38.
163. Hill D, 'Polyester for safer public buildings', *TTi*, Jul/Aug 1998, 25–7.
164. Harris D, 'Non apparel coating', *JCF*, Vol 24 Jan 1995, 210–29. Also in *AATCC Symposium, Coating and Lamination* Danvers MA, USA 3–4 April 1995, AATCC, Research Triangle Park NC, USA.
165. Wilkinson CL (BF Goodrich), 'Backcoating for performance', *6th International Conference on Textile Coating and Laminating*, Dusseldorf 4–6 Nov 1996, Technomic, Lancaster PA, USA.
166. Eaton PM, 'Flame retardancy testing for textiles', *Review Progress Coloration*, Vol 30 2000, 51–62.
167. Bagnall J, 'Testing the reaction of textiles to fire', *Textiles Magazine*, Issue 4 1995, 12–17.
168. Porter D, 'Latex in carpets', in *Polymer Latices and their Applications* (Editor Calvent KO), Applied Science Publishers, London 1982, 71–91.

169. Stamper K, 'An overview of carpet laminates', *JCF*, Vol 25 Apr 1996, 257–67.
170. Bez W, 'Progress of emission reduction in backcoating of tufted carpets', *JCF*, Vol 24 Oct 1994, 162–73.
171. Pottel H, Albrecht E and Simoens D (Centexbel), 'Foam layer curing', *Textile Horizons International*, Vol 13 No 2 Apr 1993, 23–5.
172. Anon, 'New products – carpets from corn', *Textile Asia*, Jul 2000, 114–15.
173. Adanur S, 'Medical textiles', *Wellington Sears Handhook of Industrial Fabrics* (Editor Adanur S) Technomic, Lancaster PA, USA 348–52.
174. Raheel M, *Protective Clothing, Systems and Materials*, Marcel Dekker, New York 1994, 18–19.
175. Blitz K, 'Soaking up sales in adult incontinence', *Nonwovens Industry*, Mar 2001, 83–6.
176. Anon, 'Nonwoven backing gives new strength to breathable films', *BPR*, Jan 2001, 20–21.
177. Squire J and Gasper N, 'A review of military textiles and clothing for personal protection', *Industrial, Technical and High Performance Textiles*, World Textile Congress, Huddersfield University 15–16 Jul 1998.
178. Scott RA, 'Textiles in defence', in *Handbook of Technical Textiles* (Editors Horrocks AR and Anand SC), Woodhead, Cambridge 2000, 425–60.
179. Adanur S, 'Military and defence textiles', in *Wellington Sears Handbook of Industrial Textiles* (Editor Adanur S), Technomic, Lancaster PA, USA 1995, 359–80.
180. Musgrove J, 'Some military uses of coated and laminated materials', *Progress in Textiles Coating and Laminating,* BTTG Conference, Chester 2–3 Jul 1990, BTTG, Manchester.
181. www.natick.army.mil/products (US Army Soldier Systems Center).
182. Quay E, 'Sliver knitting', *Textiles Magazine*, Issue 4 1996, 5–12.
183. Hoki K (Kuraray), 'Leather – like materials', *Kirk Othmer, Encyclopaedia of Chemical Technology*, Fourth Edition, Vol 15, John Wiley New York 1994, 177–92.
184. Civardi FP and Hutter GF (Inmont), 'Leatherlike materials', *Kirk Othmer, Encyclopaedia of Chemical Technology*, Third Edition, Vol, 14, John Wiley New York 1981, 231–49.
185. Muirehead JAM, 'Automotive leather and the environment', *Inter Auto 98*, Amsterdam 13–15 Oct 1998.
186. Hemmrich J, Fikkert J and van der Berg M (Stahl), 'Porous structured foams resulting from aggregate modification in polyurethane dispersions by means of isothermal foam coagulation', *JCF*, Vol, Apr 1993, 268–78.
187. Bagnoli G and Polette G, 'The revolutionary high-tech leather, Lorica for automotive interiors', *IMMFC*, Dornbirn 22–4 Sept 1993.

4.11 Further reading

4.11.1 Comfort and protective clothing

1. Bajaj P and Sengupta AK, *Protective Clothing*, Textile Progress Series, Vol 22 Nos 2/3/4, The Textile Institute Manchester 1992.

2. Cooper CC, 'Textiles as protection against extreme wintry weather, fabrics and clothing', *Textiles*, Vol 8 3 Oct 1979, 72–83.
3. Curiskis J and Pailthorpe M, 'Apparel textiles and sun protection', *Textiles Magazine*, Issue 4 1996, 13–17.
4. Fourt L and Hollies NRS, *Clothing, Comfort and Function*, Marcel Dekker, New York 1970.
5. Geisow M, 'Keeping warm and dry on a mountain', *New Scientist*, 14 Aug 1980, 535–7.
6. Holdstock P, 'Outdoor clothing for protection against UV radiation', *TTi*, Apr 2000, 7–8.
7. Holmes DA, 'Waterproof breathable fabrics', in *Handbook of Technical Textiles* (Editors Horrocks AR and Anand SC), Woodhead, Cambridge 2000, 282–315.
8. Holmes DA, 'Textiles for survival', in *Handbook of Technical Textiles*, as above, 461–89.
9. *Index '99 Incontinence and Hygiene Sessions*, Geneva 27–30 Apr 1999, EDANA, Brussels.
10. *International Symposiums on Performance of Protective Clothing*, papers published by ASTM, West Conshohocken PA, USA 1990 to date.
11. Keighley JH, 'Breathing fabrics and comfort in clothing', *JCF*, Vol 15 Oct 1985, 89–104.
12. Keighley JH and Steele G, 'An assessment of the functional and design requirements of clothing used to protect the human body', *Clothing Research Journal*, Vol 8 No 1 1980, 28–45.
13. Li Y, *The Science of Clothing Comfort*, Textile Progress Series, Vol 31 No 112, The Textile Institute, Manchester 2001.
14. Lomax R, 'The design of waterproof water vapour permeable fabrics', *JCF*, Vol 15 Jul 1985, 40–52.
15. Masri M, 'The European Directive on personal protective equipment', *TTi*, Jul 1992, 17–20.
16. Mitchell D, 'Making high visibility safety garments that will be worn', *TTi*, Jul 1992, 24–7.
17. Overington Y-H and Croskell R, 'Standards for breathable fabrics', *International Dyer*, May 2001, 23–7.
18. Painter CJ, 'Waterproof, breathable laminates – a perspective from film to marketplace', *JCF*, Vol 26 Oct 1996, 107–30.
19. Raheel M (Editor), *Protective Clothing Systems and Materials*, Marcel Dekker, New York 1994.
20. Slater K, *Comfort Properties of Textiles*, Textile Progress Series, Vol 9 No 4, The Textile Institute, Manchester 1977.
21. *Textiles for Comfort International Seminar*, Manchester 15–17 Jun 1971, Shirley Institute, BTTG, Manchester.
22. Umbach KH, *Biophysical Evaluation of Protective Clothing by Use of Laboratory Measurements and Predictive Models*, Hohenstein Institute Technical Information Report.
23. Van Roey M, 'Water resistant breathable fabrics', *JCF*, Vol 21 Jul 1991, 20–31.

4.11.2 General

1. Adanur S (Editor), *Wellington Sears Handbook of Industrial Textiles*, Technomic, Lancaster PA, USA 1995.
2. Blow CM, *Rubber Technology and Manufacture*, Newnes-Butterworth (Plastics and Rubber Institute), London 1978.
3. Crew AH and Araharosen H *et al*, *Carpets*; *Back to Front*, Textile Progress Series, Vol 19/3, The Textile Institute, Manchester 1987.
4. Evans CW, *Hose Technology*, Second Edition, Applied Science Publishers, London 1979.
5. Fung W and Hardcastle M, *Textiles in Automotive Engineering*, Woodhead, Cambridge 2001.
6. Holker JR, *Bonded Fabrics*, Merrow, Watford, Herts 1975.
7. Horrocks AR and Anand SC, *Handbook of Technical Textiles*, Woodhead, Cambridge 2000.
8. Howard-Williams J, *Sails*, Sixth Edition, Adlard Coles, London 1988.
9. Keller M, 'Making waves', *Industrial Fabrics Products Review*, May 2001, 44–8.
10. Lennox-Kerr P, *Flexible Textile Composites*, Textile Trade Press, Manchester in association with WRC Publishing Co., Atlanta GA, USA 1973. (Based on papers of a symposium at Bradford University)
11. Moilliet JL (Editor), *Water-proofing and Water-repellency*, Elsevier, Amsterdam 1963.
12. Mukhopadhyay SK and Partridge JF, *Automotive Textiles*, Textile Progress Series, Vol 29, (1/2) Manchester, The Textile Institute, 1999.
13. Mukhopadhyay SK, 'Automotive Textiles – 1', *Textiles Magazine*, Issue 3 2000, 5–7.
14. O'Shea M, *Interior Furnishings*, Textile Progress Series, Vol 11 No 1, The Textile Institute, 1981.
15. Ravnitsky M (Editor), *Automotive Textiles*, SAE PT-51, SAE Inc, Warrendale, PA, USA 1995.
16. Rose H, *Fusible Interlinings, Origin and Technology*, Highgate Publications, London 1979.
17. Smith LP, *The Language of Rubber*, Butterworth Heineman (in association with DuPont), Oxford 1993.
18. Taylor PJ, *Latham's Technology of Clothing*, Third Edition, Blackwell, Oxford 2000.
19. Wake WC and Wootton DB, *Textile Reinforcement of Elastomers*, Applied Science Publishers, London 1982.

5 Testing, product evaluation and quality

5.1 Introduction

Product testing is carried out for a number of reasons – the main one being to ensure complete customer satisfaction, thus making it very likely that repeat orders will follow. This is understandable, but monitoring of production during the manufacturing process is also important to ascertain if the product is suitable for the next stage in the production sequence. Value is being added all the time, and the sooner a defect is noticed the better; for example, it is wasteful to coat a fabric if colour is off-shade or it has construction faults which will lead to its eventual rejection. Testing is also carried out for safety reasons, such as FR properties for domestic furniture upholstery – labels must be displayed providing FR information. Testing survival equipment such as seat belts and life jackets is so important that it is subject to government legislation. In recent years testing for environmental friendliness and the concept of eco testing have been introduced; however, formats and procedures have not yet been standardised for every product, and so far eco testing is voluntary. It is, however, likely to grow in importance as consumers become more environmentally conscious, and especially if the proposals to offer VAT concessions on 'green' products become a reality. Eco testing is discussed further in Chapter 6.

Testing attempts to simulate the conditions of wear and tear over several years in either a single, or a series of short laboratory tests. This is not easy or straightforward, and testing must be carried out in the shortest possible time so that goods can be released for the next stage in the production process or to the customer as soon as possible after manufacture. Warehouse storage space is usually limited, and goods must be delivered to on time both for customer satisfaction and for efficient cash flow. Timing is important in the case of fashion items, and even more critical in industries where JIT manufacturing procedures are in operation, such as the automobile industry. Despite rigorous testing, however, goods can still fail prematurely in use, and much research and development is conducted to devise quicker and more reliable ways of testing materials and products. Of course determining the reasons for product failure and investigation of customer complaints are frequently the starting point for new and improved products.

There are national standards such as British Standards (BS) and German Standards (DIN), and now Euro Standards (EN) and also International Standards (ISO). Attempts are being made to standardise and rationalise these various test methods. When this has been achieved for a particular test it is renamed, e.g. for BS1000, this test is now designated BSENISO1000 and in Germany the same test would be designated DINENISO1000. In addition to these national and international standards, technical institutions and large purchasers such as the chain stores, the MOD and car manufacturers (OEMs) may also have their own standards and test methods. This causes confusion and does not only present test laboratories with additional work; it also means they have to equip themselves with all the appropriate test machinery, the purchase and running costs of which can be quite substantial.

Test methods must be realistic and must not involve factors that are not encountered in real life situations. It is easy enough to devise tests which cause rapid breakdown, but the test conditions should bear some relation to conditions of actual use. For example, in UV and light fastness testing, use of a machine which emits higher energy UV light will result in accelerated fading and separate good samples from poorer ones, but the high energy radiation could also cause damage which would not happen in actual use. Light fastness testing is complicated by the fact that conditions are different all over the world, depending to a large extent on latitude, weather and cloud cover. The additional effects of air pollution are more difficult to assess and are the subject of current research.

Certain tests carried out on certain materials, and especially clothing, are related to comfort, which has assumed much more importance in recent years. Human beings cannot function efficiently if they are not comfortable, and if a person is operating machinery or driving a car, comfort becomes a factor determining safety. However, comfort has many different aspects as summarised by one writer, 'Comfort is a complex matter, with physical, physiological, and psychological factors interrelated in an unpredictable combination which constantly undergoes variation.'[1] Comfort of clothing, in particular waterproof protective clothing and industrial protective clothing, is a well researched area and is still the subject of many textile research projects. A high standard of protection against the elements (and against chemical agents or microbes in medical outlets) combined with thermal comfort could not be obtained until the early 1970s when Gore-Tex – the first material to be both water resistant and significantly moisture permeable – was first introduced. This was followed by a whole series of waterproof and breathable materials of varying quality and gave rise to new methods of assessing breathability. The rise in living standards and increased leisure time has led to substantial demands for breathable waterproof protective clothing for walkers, sailors, golfers and other sports activities. In addition to this, employers are now required to supply their workers with protective clothing, which needs to be as comfortable as possible or more accurately as comfortable as cost allows. The medical profession is now more aware of the risks facing doctors, nurses and dentists from contact with blood from patients, and this has led to gloves being worn more frequently and the redesign of protective clothing to make it more impermeable to bodily fluids. Each product has to be tested to a particular standard of performance and, especially with the last mentioned area, new, more effective and more easily carried out tests are being developed.

When fabric volumes are high and schedules tight, it is neither physically nor economically possible to test every single roll or piece of material. The frequency of testing depends on the nature of the process and is decided after consultation with the customer. There are statistical methods to decide the number of samples which need to be taken from a particular population to obtain a representative result.[2] Results are plotted on a statistical process control (SPC) chart with maximum and minimum control limits while the process is underway. Thus the past history of a process, and any present trends in results, can be seen at a glance. Customers are informed of the results, especially if they are not precisely within the specification, so allowance can be made in the next process if necessary.

General checks for the correct construction, weight and dimensions are simple tests to carry out, but are likely to be just as important as more sophisticated tests. A piece of fabric which has passed all other tests, but which is maybe 1 cm too narrow for panel cutting, could be just as unuseable as if it had failed more technical tests by a large margin. The most relevant tests carried out on coated and laminated fabrics include: coating or lamination adhesion (peel bond), waterproofness, tear strength and durability tests which include abrasion, ageing and cleaning. Tear strength is tested after coating because the tear strength of the base fabric can be reduced by resin penetration.

5.2 Quality assurance

Quality assurance (QA) includes all factors which are relevant to quality and customer satisfaction, and has grown out of simple quality control. It goes from the earliest stages of product design, product development, purchase and monitoring of raw materials through to manufacturing, testing and inspection of the finished product. Quality assurance also involves contact with the customer, from the early stages of product design to meetings after delivery, to ensure customer satisfaction has been achieved. The QA department ensures that every member of the workforce and each member of staff is trained to regard quality as their duty and not just that of the quality department. This approach to quality management was set down in BS 5750 and CEN 2900 and further developed into ISO 9000, the standard for a quality management system. This standard is based on the concept of quality set down in ISO 8402 that refers to the 'totality of features and characteristics of a product or service that bear on its ability to satisfy stated or implied needs.' This phrase has given rise to the concept of 'total quality management'.[3]

The ISO 9000 standard consists of guidelines for choosing the appropriate system relevant to a particular company. QS 9001 is the model for QA in design, development, production, installation and servicing functions. Quality assurance systems require detailed and comprehensive specification and documentation of the product being made, the materials used, all production procedures and the machinery used, operative training, test methods, test standards, tolerances allowed and clear records of test results, and regular review and audit procedures. Only materials which have passed a certain standard can proceed to the next stage of manufacture. Test results are put on to a chart and statistical analysis is carried out The objective of the monitoring is to get things right first time by identifying potential problems and allowing corrective action to be taken before unsatisfactory products are made. The philosophy is prevention before, rather than detection and correction after. The automotive industry and others aspire to 'zero defects', i.e. giving the customer exactly what has been agreed in the sales contract. This may be for example one marked fault in every ten yards of fabric.

The automotive industry pioneered mass production with the invention by Henry Ford of the production line. The pioneering process has continued with the introduction of Japanese 'lean production', JIT and team working manufacturing methods. In mid-1994 a series of QA systems was formulated by the three largest American car makers, General Motors, Ford and Chrysler, with in-put from others including the American truck manufacturers. This system, called QS 9000, includes procedures tailored to the particular requirements of the automotive industry, and was revised in March 1998 with contributions from the European car makers. Many car makers (original equipment manufacturers or OEMs) worldwide now require their suppliers to be accredited with this system of operating and generally buy only from a supplier who has been inspected and approved. The company is also subject to random checks to ensure that it continues to comply with the requirements of QS 9000.

It is now a legal requirement under the Trade Descriptions Act to show the fibre content of textile products offered for sale. A textile product is defined as an article that contains 80% or more by weight of textile fibres, and when the textile content is stated it must be accurate to within 3% by weight. Information on washing and care is not compulsory, but is desirable. Information usually appears on sewn in labels but it can appear on the packaging.

5.3 General test considerations

5.3.1 Introduction

The standard laboratory conditions of temperature and relative humidity (RH) for textiles are generally 21 °C ± 1 °C and 65% ± 5% RH. It is interesting that the corresponding conditions for plastics are the same for temperature, but RH is 50% ± 5%, and conditions also vary for specific composites and some other materials. Test apparatus should be calibrated before use, and checked by an external certifying body at least annually and a certificate issued. Test procedures are described in detail in the various British or American (or other national or European) standards and there are articles[4] and several textbooks on the general subject of testing. Some are listed in Section 5.12, Further reading; the books by Merkel and by Saville are especially useful, and some modern specialist methods are discussed in the book edited by Raheel.

Test standards are gradually being raised as the customer demands better value for money and competition becomes even more intense. This is especially true of the automotive industry where cars being produced now are expected to last longer than before and maintain high resale values. The car interior must be in as good a condition as possible after years of use, and abrasion, light and UV degradation resistance are especially critical. Anti-soiling properties and effective cleanability are also becoming more important.

Manufacturing methods are changing, and new tests are being introduced to allow rapid and consistent operation of the new techniques. Again this is particularly relevant in the automobile industry because the component makers themselves, the Tier 1 suppliers (i.e. direct suppliers to the OEMs), have begun to set their own test methods and standards. The chain stores, the MOD and other major purchasers have for many years set their own specific test methods and standards for fabrics and materials.

The problem with so many test methods is that test laboratories must be equipped with two, three or even more different types of apparatus to measure essentially the same property. For example, there are three main testing methods for fabric abrasion in the automobile industry: Martindale, Schopper and Taber. All the time, effort and expense incurred by the multitude of different methods could be directed towards a more constructive purpose. There is some pressure from the United Nations, national governments and professional bodies to harmonise test methods and standards in general, not just in the textile and automotive industries.

5.3.2 Colour shade approval

Product lines may involve many hundreds of thousands of metres of dyed fabric and continue for several months. However, dyebatch to dyebatch shade variation is inevitable and difficult to eliminate even within the same dyehouse, let alone in different dyehouses, possibly in different countries. It is, therefore, important that the customer and dyer agree on what is acceptable and what is not. Nowadays, this is usually assessed objectively on a 'pass or reject' basis using quantitative colour information obtained with colour measurement instruments. This procedure removes human error and subjective assessment variations, which used to present so many problems in the past.

Two types of instrument, the spectrophotometer and the tristimulus colorimeter, both measure colour, from which quantities known as the chromaticity co-ordinates are calculated. This information is measured from an agreed standard master shade and also from the submitted test pattern. A colour computer then processes the data from both sets of measurements and calculates the differences between them. These differences can be mathematically processed and represented by a single figure, delta E, which can be used for pass–fail decisions. The size of this figure relates to the magnitude of the colour difference between the submitted pattern and the standard master shade. To be acceptable the submitted shade should be inside an agreed tolerance figure. There was more than one method of processing this information, however, which led to problems. Eventually work was carried out by a team which included colour instrument makers and major purchasers who worked under the auspices of the Society of Dyers and Colourists (SDC). The CMC(2:1) equation was the result, and this has proved to be a significant improvement on the original CIELAB 1976 for acceptability decisions. This equation, adopted by both ISO and CEN, is published as the International Standard for Colour-difference Measurement (BS EN ISO 105-J03:1997).

It is possible to calculate a value for delta E which represents the actual difference perceptible by the human eye, but this is complicated because the human eye is more sensitive to certain colours, not to mention all the other human limitations, including fatigue. The situation is still not completely resolved because not all dyehouses use CMC(2:1). However, the ultimate decisions are made on the basis of *visual* assessment under specified lighting conditions, and negotiation between the dyer and the customer.

The acceptance tolerances vary from industry to industry; slight colour variations may be acceptable in industrial protective clothing, but not for fashion anoraks sold in a chain store. The coating or lamination process in many cases involves heat, and shade changes may occur, especially if the process stops and material is exposed to more heat or to a higher temperature than normal. In addition, the coating may either cause colour changes during the coating process or catalyse shade change or fading later. All of these factors should be investigated before products are released to the customer.

5.3.3 Durability testing

There are many tests carried out to evaluate durability when used by the customer. These tests try to simulate wear and tear caused by actual use, sometimes over long periods of time, using short laboratory tests. For coated and laminated fabrics, one of the prime considerations is bond strength, i.e. the strength of the adhesive bond between the coating or laminated material to the base textile fabric. It is important to measure the initial bond strength after manufacture, but the article may be subject to continual flexing in the case of a garment, exposure to the elements in the case of a tarpaulin and UV light in the case of a car seat cover. The effect of UV light is discussed in Section 5.6. Inside cars the combined effects of UV radiation, heat and high relative humidity are probably the main factors determining durability. There may be lessons to be learnt from these factors with other materials.

Household goods and apparel have the added requirement of durability to repeated washing which is, in many cases, more 'wearing' than actual use. Tests must, therefore, be carried out after simulated washing and cleaning as well as simulated wearing, which in the case of coated fabrics is achieved by flexing and crumpling tests. A standard durability test for all textiles used for apparel and upholstery is abrasion resistance, but there are different methods for doing this. Those used for car seat covers are discussed below. Properties such as hydrostatic head to assess water resistance are invariably re-assessed after flexing, washing and abrasion.

All materials are also subject to 'ageing', which means the combined effects of time and the ambient conditions under which the article spends its working life. Most materials are exposed to oxygen and daylight, and these two agents cause fading of colour and eventual loss of mechanical properties. Heat and humidity usually accelerate the breakdown process, which is generally a combination of oxidation and hydrolysis reactions. Ageing tests for coated fabric (BS 3424 Part 12; 1996) involve storing the article under test in an oven for prolonged periods varying from, say, several hours to one week at elevated temperatures and high relative humidity. Car manufacturers have their own procedures, some of which are more exacting than those in BS 3424. More details are provided in Section 5.6.11. The objective is to simulate the effect of several years 'ageing' during normal ambient conditions within the space of a short period of time. Some materials 'age' because of loss of additive chemicals, such as plasticiser in PVC coatings, which migrate to the surface and evaporate off or are leached out by some fluid. Most chemical and physical reactions proceed faster under the action of heat, and this of course is the reason why the ageing tests are carried out at elevated temperatures.

A further consideration which has become more relevant in recent times is atmospheric pollution. It is not easy to measure or simulate this factor because it can vary so much from place to place. The main pollutant is traffic fumes which can include oxides of nitrogen and carbon, certain hydrocarbons, oxides of sulphur and particles of soot. Industrial pollution can consist of oxides of nitrogen and sulphur – especially close to power stations. These pollutants can give rise to acidic species and ozone which can have damaging effects on materials exposed over long periods, such as tarpaulins and building material. The effect of atmospheric pollution is being studied by some research workers.[5]

5.4 Testing of coated fabrics

5.4.1 Introduction

The 'parent' standard for coated and laminated fabrics in the UK is the BS 3424 with its 25 or so parts. The corresponding standard in the USA is ASTM D751-98 with 90 paragraphs. This section details the tests common to most coated or laminated fabrics. Specific tests relevant to more specialised end uses or major applications of coated or laminated fabrics appear under the appropriate heading. Amongst the main tests applied to coated fabrics are adhesion, which measures the bond strength between the base fabric and the coating (or material to which it is laminated), water penetration resistance, and various methods to assess the wear and ageing of these properties, including flexing and abrasion of the actual coated surface. Other physical tests, such as tear, tensile and bursting strengths, are also relevant as they influence the performance of the coated article, and also because coating may change some of these base fabric properties. Tests such as colourfastness, etc. of the base fabric are usually carried out before coating or lamination. With increased medical knowledge and higher standards of living, society has become more aware of biological infection, and tests have been developed – and continue to be developed – to test for barrier properties against these agents. Table 5.1 lists some of the funda-

Table 5.1 Selected British and related test methods associated with fabric coating and laminating

Property	British Standards	Other related standards
Coated fabrics (general standard)	BS 3424 (25 parts)	ASTM D 751-98 (90 sections)
Adhesion (peel bond)	BS 3424-7:1982 (1996) Method 9	ASTM D 902 AATCC 136 DIN 53357 ASTM D 751-98 (para 46–49)
Abrasion resistance	BS 3424-24:1990 (1996) Methods 27A and 27B BS 5690 (Martindale)	ASTM D 3389-94 (rotary) ASTM D 4966 (Martindale) ASTM D 3884 (Taber) ASTM D 3885 (Flexing, Stoll) DIN 53864/2 (Schopper) DIN 53528 (Frank Hauser)
Accelerated ageing tests	BS 3424-12:1996	ASTM D 751-98 (para 73–80) Auto companies tests SAE tests
Tear strength	BS 3424:1982 (1996) Methods 7A, 7B, 7C	ASTM D 1424 (Elmendorf)
Fusion of PVC/state of cure of rubber	BS 3424-22:1983 (1996) Method 25	
Elongation and tension set (stretch and set)	BS 3424-21:1993 (1999) Method 24	
Flexing resistance	BS 3424-9 (crumple)	
Low temperature resistance	BS 3424-8:1983 (1996) Methods 10A, 10B, 10C	ASTM D 751-98 (para 62–66)
Dimensional stability to water	BS 3424-17:1987 (1996) Method 20	
Wicking/lateral leakage	BS 3424-18:1986 (1996)	
Air permeability	BS 4443 pt 6 (method 16) BS 6538 pt 3 (Gurley) BSEN ISO 9237: 1995	ASTM D 737-75
Water resistance – spray rating	BSEN 24920:1992 (BS 3702)	AATCC 22-1989
Water resistance – rain tests	BSEN 29865 (Bundesmann) BS 5066:1974 1993 (WIRA)	AATCC 35-1994 AATCC 42 (impact)

Table 5.1 (Continued)

Property	British Standards	Other related standards
Water penetration resistance (waterproofness)	BSEN 20811:1992 (BS 2823:1982)	ISO 811-1981 ASTM D 3393-91 (1997)
Hydrostatic head test	BS 3424-26:1990 Methods 29A, 29B, 29C, 29D	ASTM D 751 (para 37)
	BSEN 20811 (dense fabrics)	AATCC 127-1989
Water vapour permeability (breathability)	BS 3424-34:1992 (1999) BS 7209:1990 BS 3177 (desiccant, packaging)	ASTM E 96-95 (procedures A, B, BW, C, D, E) DIN 53122 (desiccant, packing) CGSB4-GP-2 Method 49 (Turl) EN 31092 (sweating hotplate) ISO 11092 (sweating hotplate) ASTM F 1868-98 (sweating hotplate) Gore Cup
Blocking resistance		ASTM D 751-98 (para 81–85)

N.B. These are a selection of the more important and general test methods and standards. The BSI and ASTM Annual Books of Standards are recommended for product development. Automotive companies and large chain stores will have their own test methods and specifications.

mental tests applied to coated fabrics. Table 5.2 lists some of the standards and test methods applied to specific products.

5.4.2 Adhesion (peel bond)

This is carried out by measuring the force necessary to pull the coating, film or attached material away from the main base fabric. Specimens 5 cm wide (or 2.5 cm wide) are prepared, and the coating or attached material is peeled away manually to a distance sufficiently far to allow the two components to be clamped into the jaws of a universal tester, Fig. 5.1. The jaws are computer controlled and separate at a specified speed, and the force needed to separate the components is recorded. From this result the 'peel bond' per 5 cm width (or 2.5 cm width) is determined, usually by calculating the average of the test results of five specimens taken from the warp direction

Table 5.2 Selected British and related standards associated with products or properties

Product	British Standard	Other related standards
Water penetration resistant clothing Waterproof protective clothing 'Foul weather clothing'	BS 3546 (coated fabric for use) BS 3546-2:1993 (non-water vapour permeable fabrics) BS 3546-4:1996 (water vapour permeable fabrics) BS 3546-5:1995 (immersion suits)	(ASTM D 751-98) MOD specifications ASTM D 2724 (laminated) EN 343:1996
Architectural uses		ASTM D 4851-97
Mattress covers	BS 5455:1977 (polyurethane) BS 5223-4 (hospital)	
Tarpaulins	BS 3408:1992 (1995)	(ASTM D 751-98)
Upholstery	BS 5790 (PVC)	ASTM D 4852 (attached) ASTM D 3690 (PU/PVC)
Venetian blinds	BS 3415:1986	
Buoyancy aids, rescue devices	BSEN 12628:1999	UL 1123
Life jackets and personal buoyancy aids	BSEN 393:1994 to BSEN 396:1994	
Recreational inflatables		UL 1180
Chemical protective clothing	BSEN 368:1993 (resistance to liquid chemicals) BSEN 369:1993 (liquid chemicals breakthrough) BSEN 374:1993 (gloves – resistance to chemicals/ micro-organisms) BSISO 1817 (vulcanised rubber)	ASTM F 739-91 (permeation by liquids/gases – continuous exposure) ASTM D 3132 (solubility of resins and plastics) ASTM F 1001 (battery of chemicals for chemical protective clothing) ASTM F 1383 (permeation by liquids/gases – intermittent exposure) National Fire Protection Association (USA) Standards: NFPA 1991 NFPA 1992 NFPA 1993
High visibility clothing		EN 471

Table 5.2 (*Continued*)

Product	British Standard	Other related standards
Luminescent fabrics		ASTM E 2072 (specification) ASTM E 2073 (test method)
Conveyer belts	BSEN 28094 (adhesion test) BSENISO 252 (adhesion) BS 903 Part A12 (adhesion)	ISO 36 (adhesion of rubber to textile)
Air bags		ASTM D 5428-93A (evaluation of performance) ASTM D 5446-94 (physical properties determination) ASTM D 737-96 (air permeability) ASTM D 5427-93A (ageing) OEM (environmental age tests) 'Scrub' test for adhesion (Swiss test)
Automotive carpets		ASTM D 2646-96 (test for backing fabric) ASTM D-1175 (abrasion resistance) SAE J1530 abrasion and fibre loss AATCC Method 121 soiling
Aero standards	BS Aero F131 (wind socks) BS Aero F136 (rubber coated nylon) BS Aero F139 (aluminised rubber coated nylon) BS Aero F 127 (elastomeric coated nylon)	

N.B. These are a selection of the more important and general test methods and standards. The BSI and ASTM Annual Books of Standards are recommended for product development. Automotive companies and large chain stores will have their own test methods and specifications.

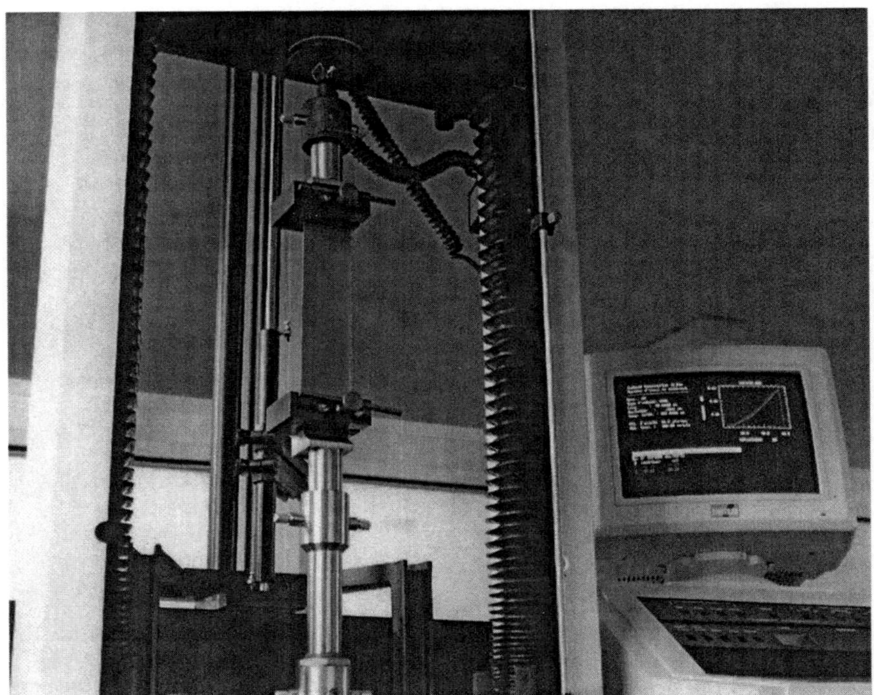

5.1 The universal tester is used not only for fabric tear and tensile strength determinations, but also extensively for coating and lamination adhesion testing (peel bond). The apparatus is also used for stretch and set testing which has become more important for moulding and automated manufacturing operations. Photograph supplied by Bobet Coated Fabric Company and reproduced with kind permission.

and five from the weft direction. The actual test procedure and the standard required (expressed in Newtons per centimetre width of sample) are specified by the customer, but are likely to be based on BS 3424 or ASTM D 751-98. For coating adhesion the level required is generally at least 50N per 5cm width. Peel bonds are also usually carried out with wet samples and somewhat lower figures are generally acceptable, say 35N per 5cm.

When a foam backed laminate is tested, the bond between the foam and the fabric may be stronger than the tear strength of the foam. The result is normally recorded as a 'foam tear' which generally implies more than satisfactory. However, if the foam is very thin and unsupported, i.e. a bi-laminate, the actual tear strength of the thin foam may be less than the required peel bond. If the test result is recorded as a foam tear, this statement is misleading, and it may be necessary to modify the test procedure, for example by sticking a supporting material to the foam (to form a tri-laminate), so that the actual foam to base material bond can be determined. The same considerations apply to thin films laminated to one fabric only.

For coated fabrics where the actual coating itself cannot be peeled away to insert into the jaws of a universal tester, two samples are joined coated face to coated face with a strong adhesive. A short distance at one end is not joined because, when the adhesive has dried and fully bonded, the sample is held in the universal tester using the loose ends in the same way as a laminate, and the adhesion test is done in the same way as before. If the adhesion of a fabric intended for water-resistant clothing is to be tested, and the face side of the material is treated with a silicone water repellent, care must be exercised. The silicone finish can mark off and contaminate the coated sides which are to be joined together to carry out the peel bond test, and this will affect the result. The best practice is to take the sample for peel bond testing before silicone is applied. Fluorocarbons generally reduce peel bonds much less than silicones, and it should be possible to do peel bond tests on fabric which has been fluorocarboned – although pre-testing is best to confirm this.

Peel bond tests on car seat cover laminates, face fabric to foam and foam to scrim, are important to ensure that delamination does not occur in the car during use or during downstream processing. To simulate possible conditions which may be encountered during the life of the car and during subsequent processing, the peel bond test is carried out as received, and also after heat ageing, while wet, and sometimes after treatment with solvents. The new manufacturing technique of direct joining seat cover to the foam squab and cushion may necessitate higher peel bonds, because there are fewer sew lines to help hold the laminate components in place.

5.4.3 Flexing tests

These tests are a further means of assessing coating adhesion by flexing to simulate the mechanical action encountered by garment materials during wear, especially at elbows and knees. The Crumple-flex tester twists fabric samples made into a cylinder through 87° and stretches and crumples them at the same time. The flexed samples are examined for delamination or other damage, and in the case of water-resistant material the water resistance is assessed after a specified number of cycles. The test procedure is in BS-3424-9, see Fig. 5.2. The Schildnecht Flextester is a similar machine, but does not twist the fabric. More samples can be tested at the same time using a Schildnecht machine, but the sample specimens are smaller. Flex requirements specified by the customer may vary considerably with the article

5.2 Crumple-flex testing is important to assess coating adhesion and quality under simulated wearing conditions. A tube of fabric is repeatedly flexed, stretched and compressed and twisted through 87°. Suitable for BS 3424, ISO 8096 and other tests. Photograph supplied by SDL International Ltd and reproduced with kind permission.

being tested, e.g. (Schildnecht) 200 000 flexes for garment material, 500 000 flexes for a tarpaulin.

5.4.4 Abrasion resistance

In the case of a coated fabric, abrasion has two meanings, the abrasion resistance of the coated surface, i.e. the resistance of the resin to damage, and also the abrasion resistance of the face fabric. The face fabric abrasion

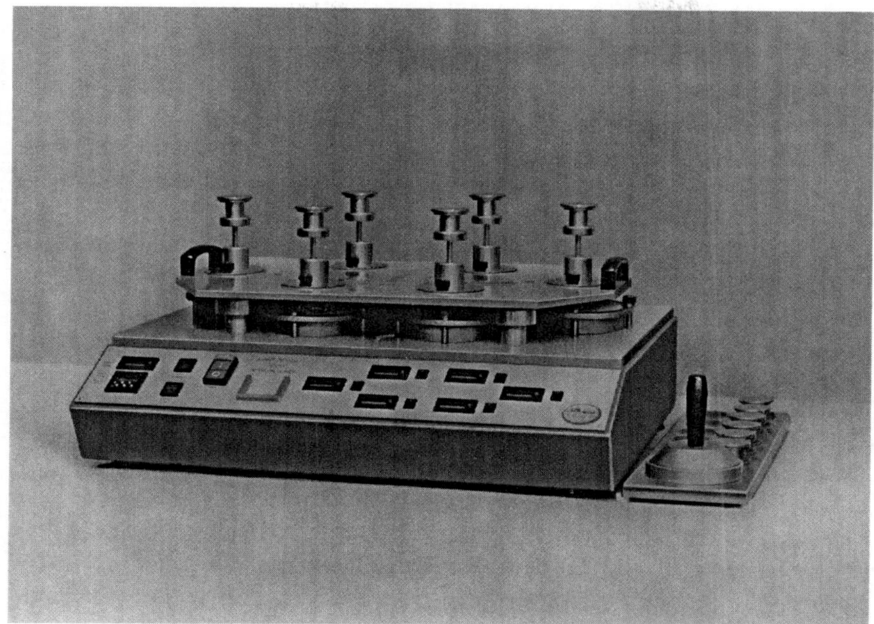

5.3 Martindale abrasion test apparatus. This is only one of several methods in use. The automotive industry also use the Taber and Schopper test methods. Photograph supplied by SDL International and reproduced with kind permission.

resistance of woven fabrics can be improved significantly by coating the back, but of course the material becomes stiffer and less flexible. While this is sometimes unacceptable for apparel fabrics, it is less of a problem with upholstery fabrics or fabrics which are laminated, especially those laminated to a rigid material. Coating fabrics also improves the associated problems of pilling and snagging because the resin helps to lock fibre in place. This aspect of coating and abrasion resistance is discussed in Section 5.6.6, although many of the factors involved are also relevant to domestic furniture.

The abrasion resistance of the coating on a coated fabric could influence the performance of waterproofing, and therefore requires assessment. This is accomplished by use of the Martindale apparatus, see Fig 5.3, placing the coated fabric to be tested on the base of the machine and abrading it with a piece of wool held in the abrading head. This test uses the machine the other way round compared to normal procedure, but it does allow a test sample large enough for waterproof testing after abrading. The material should still be waterproof after say 5000 rubs. There are a number of other abrasion tests, such as the Taber, the Schopper, etc, some of which measure the amount of material abraded away from the test sample by weight reduction.

5.4.5 Waterproofness (water penetration resistance)

Resistance to penetration by water, the ability to provide protection against water or rain, is measured by the hydrostatic head apparatus that essentially attempts to force water through the fabric, the pressure being measured by a column of water, see Fig. 4.11. Details of the test are given in BS 3424 Part 26; 1986 which mentions three procedures, Methods 29A, 29B and 29C. The new procedure for 29A is the method now most widely used and is generally in accordance with BS 2823; 1982, the hydrostatic head test. Method 29A uses a fabric sample of 100 cm^2 area (11.3 cm diameter), and water pressure is applied to one side at the rate of increase of 10 cm per minute until 'the first damp patch' appears. Because this procedure can be very time consuming, many laboratories increase the pressure to the standard level required within one minute and, in addition, may follow the procedure described in BS 2823; 1982, i.e. pressure is increased until the third growing drop appears. Work carried out by Wenzel as far back as 1936 showed that results recording the third drop were more reproducible than those recording the first.[6] The average result of five individual test specimens is taken as the hydrostatic head. Standards have risen since this time, and up market manufacturers may reject an entire batch of material if a single test result fails to meet the minimum. The reason is presumably because a garment needs only a single weak spot to leak water. It should be appreciated that different results may be obtained with different rates of increase in water pressure. The standards do require that the rate of pressure increase be recorded with the test results. Older laboratories may still possess apparatus which uses test fabric samples 5 cm in diameter, as stated in Method 29C. The intention of BS 3424 Part 226; 1986 was to phase this out by the early 1990s.

A height of 100 cm is generally regarded as the absolute minimum for a fabric to be termed waterproof or water penetration resistant. Many quality manufacturers, however, require more than this, up to and exceeding 200 cm. In practice, the higher the better, because if the initial hydrostatic head was 250 cm and during use over six months, or after washing, the performance is reduced by, say, 50%, the article is still water penetration resistant – and so the higher the initial hydrostatic head, in general, the higher the durability. It is usual to test the hydrostatic head after flexing and after abrasion testing; some customers accept a hydrostatic head of 70 cm after these tests, others require considerably more. The term 'waterproof' is discouraged in this test because it implies that the 'waterproofness' is equivalent to the hydraulic bursting strength of the material, and also that the material is completely impermeable to water. This term appears still to be used in the USA; ASTM D 3393-91 (1997) is the Standard Specification for Coated Fabrics – Waterproofness. However, the term is avoided in ASTM

D 751-98 ('hydrostatic resistance' being used). The unit kPa is now being used in place of cm (head) of water; 1 cm 'head' of water pressure is equivalent to 0.098 kPa (1 inch 'head' = 0.25 kPa). BS3546, the standard for water-resistant coated fabric for clothing, states that the term 'water penetration resistant' should not be applied to any coated fabric with a water penetration pressure of less than 10 kPa 'as received', i.e. before any flexing or ageing test. As stated above, however, 'waterproofness' is used in the trade and is therefore used in this book.

Many performance specifications for waterproof protective clothing include requirements of hydrostatic head after simulated wear tests such as flexing, abrasion testing and cleaning. An allowance is normally made for these tests – the hydrostatic head is not expected to be the same as it was 'when received'. Abrasion tests on coated fabric have already been described above. Different customers require differing standards of performance and different tests. The MOD specifications are amongst the most demanding, requiring 700 cm hydrostatic head after the durability tests.

5.4.6 Fabric handle, drape and stiffness

Coating fabric causes stiffening by penetration of the resin into the fabric structure, and generally the more penetration the more stiffening is caused. This is invariably accompanied by a reduction in tear strength, and can also affect the appearance and aesthetics of the face side. Under a microscope, 'grin through' can be seen. In general, the more penetration is controlled, the less the stiffening effect on fabric handle. Sometimes a fluorocarbon finish is applied to a fabric before coating to help control resin penetration. However, the less penetration, the lower the resin adhesion, and in some cases a balance or 'trade off' needs to be established. The stiffness of coated woven nylon for waterproof clothing has always been something of a problem, although anorak fabric is significantly softer than it was in the 1980s. The MOD find the rustle of garments a problem – disruptive print uniforms are designed to conceal the presence of soldiers, but the noise caused by garment rustle may negate this to a certain extent!

Fabric stiffness can be assessed by a bending length tester of the type designed by the Shirley Institute (now BTTG). The Cusic drape tester is a quick method for assessing the drape of light weight material, but it is possibly not suitable for some coated fabrics. Some efforts have been made to quantify surface touch using the Kawabata system, which has had some success in the clothing industry.[7–11] The influence of different finishes on fabric handle has been researched and quantified. Research on fabric handle of apparel and household textiles has revealed national preferences for touch properties and softness. For domestic and contract upholstery, and automotive textiles where the fabric is fixed to a seat or rigid component such as a wall, door casing or headliner, the only relevant mechanical properties out of the five specified by Kawabata are compression and tactile surface touch properties. The other Kawabata properties relating to drape and flexing do not apply in these cases. Kawabata investigations were done on car seat fabric by a research institute, but do not appear to have been followed up by an OEM or seat maker.

5.4.7 Fabric strength – tear and tensile

Good tear strength and resistance to tear propagation is a fundamental property of all fabrics, and is central to good durability and performance. Car seat fabrics are expected to last the life of the vehicle, and torn seat fabric would deter prospective purchasers and significantly reduce the resale value of used cars. The tearing strength of a fabric is influenced by the smoothness and the construction, as well as by the thickness of the yarns. Coating can have a significant influence on fabric tear strength. When a force is applied, if the construction is rigid and inflexible, the applied force breaks the threads one at a time, and low tear strength is obtained. A fabric coating which penetrates the fabric structure and holds the treads in place will generally cause a lowering of tear strength and also, as has been discussed, fabric stiffening. If the threads move under the tearing force and bunch together, several threads will be broken together, producing a higher tearing strength. Some fabric lubricants will allow this to happen, and it is more likely if yarns have a smooth surface. Care must be exercised, however, because lubricants and finishes can reduce peel bonds if the fabric is to be laminated. Silicone finishes and coatings, in fact, very significantly improve tear strength and puncture resistance – but these are *not recommended* on fabric to be coated or laminated. They may be applied after coating or laminating, but with care because, as has already been mentioned, a finishing process on a coated or laminated fabric could reduce bond strength. Silicones are especially not recommended on automotive fabrics, not only because of their effect on adhesion but also because they can contaminate surfaces to be painted, i.e. the car body, and cause resist marks.

Certain constructions such as twill weaves and basket weaves allow threads to group together more easily and thus tend to have better tear resistance than plain weaves. The yarn and fibre type also determine fabric strength; polyester is stronger than wool, and continuous filament yarns will be stronger than those made from spun staple fibre. The use of 'rip-stop' constructions to improve tear resistance has already been mentioned. Tear strength testing is also sometimes required after exposure to UV light. Polyester fabric tear strength is generally satisfactory after exposure, but that of wool fabrics and especially nylon fabrics may be significantly lower. Yarn lustre has a significant effect because of the effect of titanium dioxide – bright yarns with no titanium dioxide are generally much more resistant to UV light. Acrylic fibres have the best resistance to UV degradation.

There are a number of tear tests in use, single rip, wing tear and Elmdorf tear. The wing tear avoids transfer of tear, whilst the Elmdorf method measures energy loss during the tear process. Care must be exercised in clamping the specimen in the jaws of the test machine, because any slippage could be mistaken for a tear or a failure of the test specimen. Jaws and clamps of different designs are available for testing different types of materials. Straight load or tensile strength tests on strips of fabric are useful for investigating the effect of a material or process change on fabric properties. BS 3424-5; 1982 (1996), Methods 7A, 7B and 7C details three methods for determination of tear strength for coated fabrics. Certain knitted fabrics may distort or unravel, and bursting strength tests are likely to be more reliable (BS 3424-6; 1982 (1996), Method 8B).

5.4.8 Bursting strength

This test is more relevant to a knitted fabric or to nonwoven fabrics where the test load is multi-directional. The fabric is clamped in the machine over a rubber diaphragm, and pressure is applied via water or some other fluid. During the test this stretches the rubber and thus applies a force to the fabric. The exact procedure and diameter of the test specimen vary with customer requirements.

5.4.9 Dimensional stability

These are tests for garment shrinkage on washing or during use by the customer. They also check suitability for downstream processing such as panel cutting and sewing or, in the case of the automotive industry, lamination to door casing. Dimensional stability tests involve measuring the percentage shrinkage after soaking in cold water or after heating in an oven at various temperatures. Effective heat setting during fabric finishing significantly reduces the possibility of shrinkage. Figure 2.1 shows the general relationship between heat shrinkage and heat setting temperature with polyester fabric. However, if fabric has been subject to high tension during the coating process, shrinkage may occur when panels are cut or when the material is relaxed. These tests have assumed more importance recently with newer methods of manufacture and tight production schedules, which do not allow for continual machine adjustments. Lamination must be carried out under conditions of no tension if the products are to pass these tests consistently. This is especially relevant if any of the laminate components are stretchy.

5.4.10 Other general tests

Cold cracking tests carried out at, say, –20 °C are specially relevant to PVC coated fabrics such as tarpaulins, because if not appropriately formulated PVC may crack when cooled to temperatures below 0 °C. Tests for dimensional stability under water immersion are also routinely carried out to simulate washing and generally getting wet by exposure to the elements. Stretch and set tests are important for laminates, especially when one or both components have a significant amount of stretch. Lamination must not be carried out when the materials are not in a relaxed state, because this may result in the laminated material curling and not lying flat on the cutting table. In addition, cut panels may distort or even change dimensions slightly, which can cause serious problems if they are used in an automated process such as those used in the automotive industry for door panels, headrests, etc. These factors are discussed further in Section 5.6.9.

5.5 Thermal comfort of apparel

5.5.1 Introduction

Garments produced from coated or laminated fabrics are worn for protection from the elements, or from some specific agent such as heat, microbes, chemicals, radiation or even molten metal splashes. One important factor common to all garments is comfort. The human body cannot function efficiently for any length of time if it is uncomfortable, and hence comfort is a factor determining safety. In some cases the person is so uncomfortable in a particular garment that he or she will remove it and thus be exposed to the danger against which the garment is designed to protect. The main problem is the inability of body perspiration to evaporate to cool the person who will generally be involved in some form of activity.

Thermal comfort is defined in ASHRAE Standard 55–56 as 'that condition of mind that expresses satisfaction with the thermal environment'. Generally, this means that the person does not know whether he or she would prefer a warmer or a cooler environment. Clothing comfort has already been discussed in Section 4.1.

5.5.2 Energy balance of the human body

As long as a human being is alive, he or she is producing heat. Even when sleeping the human body is producing 80 W of heat; when walking at a leisurely pace, this rises to about 210 W; and it can exceed 700 W when walking up a steep hill. In high activity work or sport it can rise to well over 1000 W. However, human body core temperature must always be constant at about 37 °C, and there are automatic body mechanisms which try to maintain this. If we are too cold, the body shivers in an involuntary action to generate heat by muscular contraction – if we are too hot, sweating occurs which causes cooling of the skin by evaporation of the liquid sweat. Water evaporating has a cooling effect because it takes away its latent heat. Heat is lost from the human body by four principal methods: conduction (physical contact), convection (by air or wind), radiation and evaporation of sweat from the skin. The main mechanism for loss of excess heat is by sweating, and if the human body surroundings reach or exceed 37 °C in temperature (the inside of a car on a sunny day), then this the only way the body can lose heat.

At the other extreme, if body core temperature begins to fall below 37 °C, the person begins to suffer from hypothermia.[12] At 35 °C, the person shivers violently and has poor co-ordination. If the body temperature continues to fall, the person becomes incoherent and confused and falls unconscious when 30 °C is reached, death soon follows if they are not warmed and revived. Wind can increase the heat loss by convection and by evaporation of sweat or water from rain on the skin. The effect of a strong wind at a low temperature, the 'wind chill effect', can have a much greater cooling effect than still air at an even lower temperature. Water conducts heat away about 25 times faster than air at the same temperature, and so apart from the discomfort, wetness substantially increases the loss of body heat. In addition, wet clothes have much reduced insulation properties and lose much of their ability to keep the body warm. Waterproof clothes are therefore not only important in keeping a person comfortable and protected from the elements, but are also necessary for survival in extremely cold conditions. At the other extreme, if the human body generates more heat than it can lose to the surroundings, the person becomes uncomfortable and stressed. This can also induce fatigue and cause the person to function inefficiently and become more accident prone if doing a job of work. However, if a person is wearing waterproof clothing, it is important that as much perspiration as possible is allowed to escape and induce body cooling by evaporation. The rate of liquid perspiration exuded from the skin can be related to the work rate in watts, for example gentle walking with a work rate of 200 W produces an equivalent liquid perspiration rate of 7600 g in 24 hours, and active walking on the level carrying a heavy pack with a work rate of 500 W produces an equivalent perspiration rate of 19 000 g in 24 hours.[13] These figures are of course only general, and will vary according to individual body metabolism and ambient atmospheric conditions.

5.5.3 Measurement of water vapour permeability–'breathability'

Most waterproof materials which do not allow the passage of liquid water do not allow perspiration to pass either, i.e. they do not 'breathe'. Breathability is a more difficult quantity to measure to the satisfaction of all customers and of the general public. One of the main reasons, of course, is that everybody has their own particular metabolic rate of producing perspiration and their own particular 'comfort threshold'. In addition, the rate of production of perspiration depends on the level of activity – the more strenuous the activity, the higher the level of breathability required in the garment material. It should be remembered that even an unclothed body would perspire if working hard. Also an unclothed body at rest would perspire if situated in a room at a temperature higher than body skin temperature. Breathability tests seek to assess the amount of perspiration which can pass through the garment material under controlled laboratory conditions which bear some relation to conditions of actual use.

Non-coated fabrics are generally breathable, but they are not water resistant (i.e. do not have a hydrostatic head of 100 cm of water), but most regular, water resistant, coated fabrics, e.g. PVC, Neoprene, regular polyurethane and acrylics, do not breathe appreciably. Coated fabrics which do not have a good hydrostatic head, for example about 30–40 cm of water, will give more protection from the rain than a non-coated fabric. They cannot, however, be classed as 'waterproof', but they are likely to be appreciably breathable. A truly 'breathable and waterproof' fabric should be 'breathable' and also have a hydrostatic head of at least 100 cm of water. An acceptable level of 'breathablity', however, is a little more difficult to define than waterproofness as will become clear over the next few paragraphs.

There are various methods of measuring breathability, the units of which are grams of moisture transmitted through a square metre of the fabric in a given period of time, usually 24 hours. Other more academic methods measure water vapour resistance, i.e. millimetres of still air or other units – see later in this section. Tests generally fall into three categories: desiccant methods such as BS 3177, ASTM E96-95 Procedures A, C or E, and DIN 53122; 'inverted water' methods such as the Gore Cup and ASTM E96-95 Procedure BW, and finally evaporative methods such as BS 7209 and ASTM E-96-95 Procedures B and D. The desiccant methods are used mainly in the films and packaging industry for measuring relatively low levels of water vapour permeability. There is an extensive range of literature on products, test methods, results and discussion of results.[13–18] Chapter 4 discusses waterproof breathable products.

Evaporative tests are probably both the easiest to carry out and the most easily understood by the layman. They are, however, extremely sensitive to test conditions, experimental procedure and error, and inter-laboratory agreement can be poor even when tests are carried under apparently 'identical' conditions. Essentially, they measure the amount of water evaporated through the fabric under a given set of conditions. A small vessel is partly filled with water and the top sealed by the coated fabric being tested, Fig. 5.4. It is then weighed accurately to at least two decimal places, placed on a hot plate set at the test temperature, left for the prescribed period and then re-weighed. From the weight lost and the surface area of the vessel, it is possible to calculate the amount of water vapour transmitted through the coated fabric per square metre in 24 hours – the breathability of the material. This can be done at different temperatures; Procedure B of ASTM E 96 95 uses a temperature of 23 °C (73.4°F), Procedure D uses a temperature of 32.2 °C (90°F). Presumably this temperature was chosen because it is approximately the inside temperature of a garment. The actual temperature of course depends on a number of factors including the layers of clothing worn in between the skin and the outer garment. The higher the temperature of the water inside the test vessel, the higher will be the figure of breathability obtained, because the 'driving force' is the difference between the temperature of water inside the vessel and the outside ambient temperature. The level of water in the vessel, i.e. the air gap between the sample and the top of the water, is very critical – the smaller it is, the higher will be the figures obtained. A difference in air gap of 5 mm within the range 0–10 cm can double 'breathability' results, see Fig. 5.5. The reason for this is believed to be the resistance to water evaporation of the air in this gap. Needless to say for reproducible results, sample temperature, the ambient laboratory temperature and relative humidity should be as constant as possible. Another factor influencing results is efficiency of sample vessel sealing. Table 5.3 shows some figures obtained on a range of products using an evaporative method with an air gap of one inch (25.4 mm).

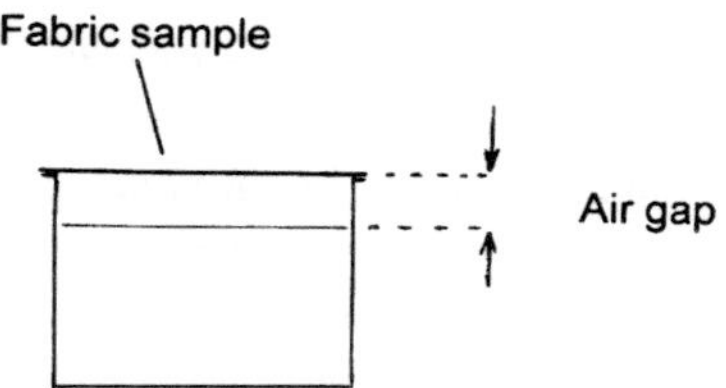

5.4 The evaporative (cup) method of breathability evaluation. The test variables are air gap inside the test vessel, water temperature inside the vessel, and temperature (and relative humidity) of the outside (ambient) atmosphere. The difference between the two temperatures is the 'driving force'. ASTM E 96 Procedure D recommended cup water temperature is 32.2 °C (90 °F). BS 7209 cup temperature is the same as external ambient temperature, 20 ± 2 °C.
N.B. For information on effect of temperature variation on 'breathability' see Ref. 62.
Desiccant methods use the same cup, but with calcium chloride or silca gel inside to reduce humidity.

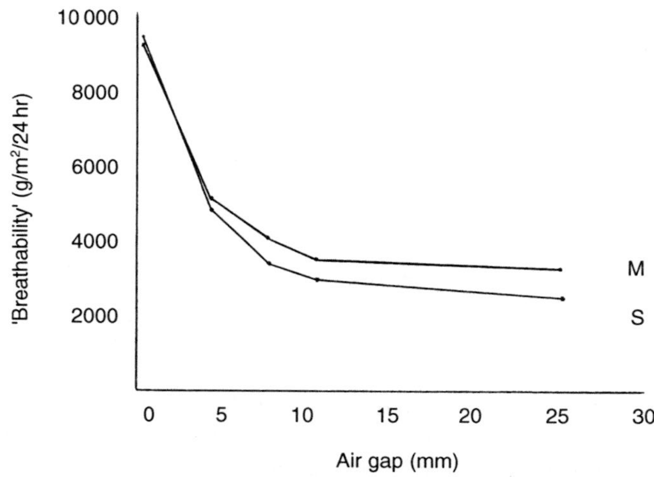

5.5 The effect of the air gap in evaporative or 'cup' tests. This is extremely significant in the 0–8 mm range. These results were obtained using a vessel 3.5 cm wide. The temperature of the water inside the vessel was 37.5 °C and the ambient laboratory conditions were 21 °C and 50–55% relative humidity.
N.B. M = PTFE microporous laminate, S = solid film coating.

When waterproof breathable fabrics were first introduced in the early 1980s, several test methods were used by individual manufacturers, each of whom used methods and test conditions that enhanced the performance of their own product. In addition, even when a particular test was performed by different laboratories using apparently identical conditions, the agreement was poor. As related in Chapter 4, a standard evaporative test procedure was developed under the auspices of the BSI. This was BS 7209, which does not actually heat the water. The test vessels were placed on a turntable which rotated slowly to remove any stationary layer of air which might sit over the surface of the fabric samples and hence influence test results, see Fig. 5.6. The test results were expressed as a percentage of the result obtained with a standard uncoated polyester fabric woven from a mono-filament yarn. This technique improved inter-laboratory agreement, and the results of samples tested by the method detailed in BS 7209 are referred to as the water vapour permeability index (WVPI). Up market, waterproof breathable materials are expected to have a minimum WVPI of 80%, lower grade products a value of at least 50%. As already stated, breathability is meaningless without a high standard of waterproofness, and quality products should also have an initial hydrostatic head of 500 cm of water (50 kPa) while lower grade products should have a minimum of 150 cm (15 kPa).

Table 5.3 Some breathability figures obtained using an evaporative method

Sample	Total fabric weight (g/m^2)	Waterproofness Hydrostatic head (cm of water)	Breathability (g/m^2/24 hr) HDF conditions 37.5 °C	Breathability (g/m^2/24 hr) LDF conditions 21 °C
PTFE 3-layer laminate	200	700+	3600	585
Microporous PU coating (1)	120	560	3300	1009
Microporous PU coating (2)	97	280	3627	942
Microporous PU coating (3)	158	700+	2707	456
Solid PU film hydrophilic coating (1)	150	700+	2900	420
Solid PU film hydrophilic coating (2)	165	700+	2700	266
Solid PU hydrophilic polyester film laminate	140	700+	3890	594
Standard 4 oz polyurethane coating	156	700+	400	163
Closely woven cotton uncoated (commercial product)	202	89	4400	1061
Non-coated woven nylon	123	0	4571	962
Polyethylene film	41	356	143	0
Open test vessel		—	10654	2075

Test conditions:
Test vessel diameter 3.5 cm; air gap 2.54 cm (one inch); relative humidity 50–55%.
HDF = high driving force of 37.5 °C inside test vessel and 21 °C ambient outside.
LDF = low driving force of 21 °C both inside and outside.
N.B. These samples, all on nylon base fabrics, were commercial products on sale at the time of testing (1986) by Courtaulds Research at Spondon, Derby.

5.6 Shirley Water-Vapour Permeability Tester. The apparatus was developed from the Canadian Government specification CGSB4-GP-2 Method 49 and is suitable for BS 7209 and BS 3424–34. Photograph supplied by SDL International Ltd and reproduced with kind permission.

Meanwhile more academic research continued, the objective being to produce a laboratory test which correlated with actual outdoor wear trials. The Hohenstein Institute have been successful in this exercise and have developed the Sweating Guarded Hotplate method, which has now been adopted as a European and international standard test method (EN 31092 and ISO 11092), see Fig. 5.7 and 5.8. The main drawbacks to this method are that the apparatus is expensive to buy, that it needs skilled operatives to carry out the test and is probably not suitable as a quick quality control test. Some results obtained using the Sweating Guarded Hotplate method appear in Table 5.4. A Canadian Government test called the Turl Dish is an evaporative method and measures resistance to evaporation of the test fabric together with the air layer inside the air gap. This test is also believed to agree well with actual outdoor wear trials.

Even now, nearly two decades later, the confusion over breathability figures still persists, with various figures quoted by manufacturers many

5.7 The Sweating Guarded Hotplate. Photograph supplied by the Hohenstein Institute (Germany) and reproduced with kind permission.

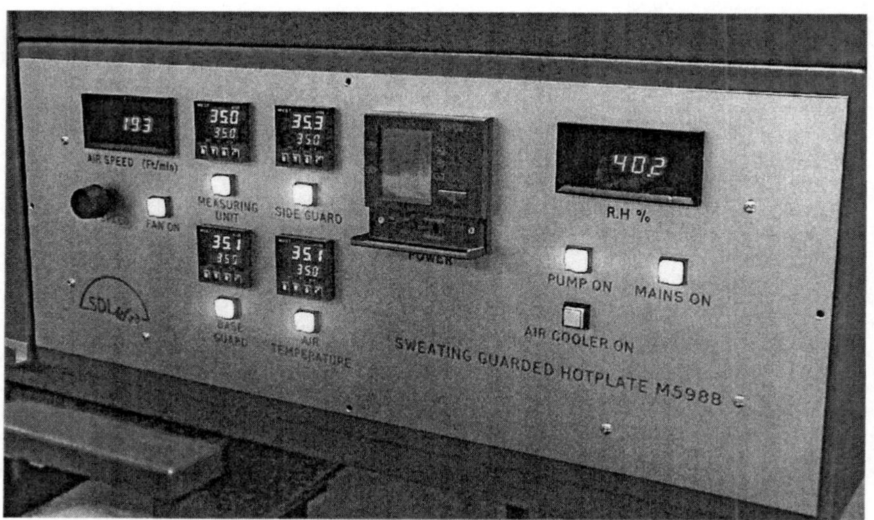

5.8 A commercial Sweating Guarded Hotplate test rig. Photograph supplied by SDL International and reproduced with kind permission.

Table 5.4 Breathability results obtained using the Hohenstein Institute Sweating Guarded Hotplate

	Water vapour resistance Ret (m^2 Pa/W)
1. Non-breathable (standard) polyurethane coating	164.7
2. Woven microfibre (not coated)	1.45
3. Woven polyester microfibre (not coated)	2.30
4. Tri-laminate with hydrophilic polyester membrane	15.49
5. Bi-laminate with hydrophilic polyester membrane	5.12
6. Bi-laminate with microporous PTFE membrane + hydrophilic surface layer	2.58
7. Bi-laminate with microporous PTFE membrane	2.88
8. Hydrophilic polyurethane coating on polyester non-woven	12.05

Notes:
1. Samples were measured by the Hohenstein Institute using their Sweating Guarded Hotplate method which is now an international and European test standard, ISO 11092/EN 31092. The results obtained are valid only for the samples actually tested.
2. The two 'extreme' samples are sample 1, which is standard, non-breathable polyurethane and samples 2 and 3, which are non-coated fabrics woven from microfibres.
3. The higher the water vapour resistance, the lower the breathability.

Source: Dr V T Bartels, 'Survey on the moisture transport properties of foul weather protective textiles at temperatures around and below freezing point', Technical Report AiF 11674, Hohenstein Institute for Clothing Physiology 2001 (in German). Reproduced with kind permission of the author.

of whom do not provide precise details of the test method or conditions of test. The procedures used are sometimes quick in-house methods. However, scientific research into fabric and garment comfort has continued and investigations into comfort in real life dynamic situations and whole clothing assembles are being conducted.[19–21] Thermal comfort depends not only on the ability of materials to allow the transmission of water vapour, but on all other factors which have a bearing on moisture management, e.g. wetting out of textile surfaces, wicking of liquid moisture and the capacity and speed of moisture absorption. Also, as has been mentioned, garment design is relevant too, and some writers have commented that more research should be devoted to this than to the actual fabrics.

5.6 Testing of automotive products

5.6.1 General comments

The automobile industry is the largest user of coated and laminated textiles. In many ways the industry is also the most demanding in terms of both consistent quality and production requirements, because of its highly competitive nature and JIT requirements. The exacting durability requirements of car seat fabric are quite unique for such a large volume material.[22–29] Production must be monitored for two main reasons, first to determine suitability for the next process, and secondly to simulate actual conditions of wear during the life of the car. Simulating actual conditions of use over a period of years with accelerated laboratory tests is neither easy nor straightforward, and each OEM has its own methods of doing this. In addition, climatic conditions around the world vary significantly. Test methods depend on the physical conditions which will apply during the next process, and on where in the car the fabric will be situated. Car seat covers have the highest abrasion resistance requirements, while parcel shelves and dashboards have the highest light fastness and UV degradation resistance requirements. The shade of dyed fabrics must be examined, because two pieces of fabric used together in the same car may have been dyed in different dye lots or even different dye works, and the shades may appear slightly different. In addition, if the fabric is used in conjunction with a coloured plastic foil or dyed leather the colours may again not be exactly the same. Care must be taken to avoid the occurrence of metamerism.

Some properties such as weight and thickness can be monitored automatically by microprocessors. However, the final fabric examination, generally carried out manually, is a slow, relatively costly process, and even experienced examiners miss significant numbers of faults. However, customers now call for 'zero' defects in goods received, i.e. the exact standard agreed in the sales contract. At the time of writing, zero defects allows for one marked fault in typically 10m of fabric. Existing automatic examination systems are not sophisticated enough to cope with the whole multitude of factors which result in second quality material. The technology is probably available, in theory at least, but the cost is prohibitive.

New specialist vehicles such as SUV, 'recreational vehicles' (RV), 'multipurpose vehicles' (MPV), and in the USA pick-up trucks and mobile homes, are creating new requirements associated with the intended use of the vehicle – or the image it creates. An example is the pick-up truck, the 'cowboy's Cadillac', which conjures up the idea of a robust utility vehicle (Ford advertises their product as 'built Ford Tough'), and it is likely to be treated as such! In the USA car leasing has become more widespread and so the 'private life' of the car does not start until it is, say, two years old.

It must not only be in almost showroom condition after this time, but it is also going to have a longer life. The American OEMs are expected to push up the specifications of light and UV degradation resistance and also that of abrasion.[30] In addition, the move towards three-year warranties on new cars by European and American OEMs (to match the Japanese and Koreans) is also likely to lift standards all round. Test methods and standards required by individual OEMs or Tier 1s are generally confidential between themselves and their suppliers. They are, however, usually based on national, international or institutional standards, e.g. BS, DIN, ASTM or SAE. Table 5.5 summarises the main test methods used for seat fabrics and interior trim. Performance standards as well as test methods, however, can vary, and what is acceptable for one OEM may not be acceptable to another. Some attempts have been made to harmonise test methods both in the USA and Europe, which can only be good for the industry as a whole.[31–33]

General items, such as the correct fabric width, weight, thickness, and construction (ends/picks in woven fabrics; courses/wales in knits), must be checked regularly to ensure that they are as specified by the customer. They can, in fact, sometimes give clues to other properties, for example thicker than normal laminates could indicate less burn off or lower pressure during lamination and perhaps lower peel bonds. Fabrics with raised surfaces need to be examined for pile in the correct direction, pile distortion and correct pile height. The fabric design must be examined for regularity, especially warp and weft lines ('bow and skew'), to ensure that they are within agreed tolerances.

5.6.2 Processability quality checks

The main properties required for downstream processing include the following: consistent dimensions of width and thickness; porosity; stretch and set; dimensional stability; elongation; ability to lie flat (not curl); cold water stability; peel bond (lamination adhesion); heat/humidity ageing. Panel cutting, usually on a cutting table, precedes all making up procedures. Many layers of fabric are cut at the same time, and the laminate must lie flat for accurate cutting. There must not be any inherent instability to cause the laminate to distort or alter its shape in any way. If the seat cover laminate has been laminated with any one of the components under tension, it may alter slightly in dimension when unrolled or cut. Large cut panels may stretch under their own weight, especially knitted fabric laminates. This is inevitable, but they must all stretch by a consistent amount within close limitations. Mass production methods and getting the process right first time, every time, requires starting materials with consistent properties. Pour in foam methods require consistent porosity, so foam does not strike through

Table 5.5 Some selected important automotive seat fabric test methods

	British Standard test methods	Selected related test methods
Colour fastness	BS 1006:1990 (1996) Methods of determining colour fastness to about 70 different agencies BS 1006: Grey scales for assessing changes in colour A02 BS 1006: Grey scales for assessing staining A03 BS 1006: BO1 Blue wool scales	AATCC Test method 16 ASTM test methods DIN 54022 (fastness to hot pressing) DIN 54020 (rub fastness)
Light fastness	BS 1006:1990 (1996)	SAE J1885 Mar 92 Water cooled Xenon arc SAE J2212 Nov 93 Air cooled Xenon arc SAE J2229 Feb 93 Outdoor under glass variable angle SAE J2230 Feb 93 Outdoor under glass sun tracking DIN 75202 FAKRA 7/91
Abrasion	BS 5690:1991 (Martindale) NB sometimes tested after UV exposure	SAE J365 Aug 1994 Scuff resistance (Taber) SAE J2509D ASTM D 3884-92 (Taber rotating platform) ASTM D 3885 Flexing abrasion (Stoll) ASTM D 3886 Inflated diaphragm DIN 53863 3/4 Martindale DIN 53863/2 Schopper DIN 53528 Frank Hauser, loss in mass for coated fabrics DIN 53754 Taber
Pilling	BS 5811: pill box (BS 5690: 1991 – Martindale)	ASTM D 3511-82 Brush ASTM D 3512-82 Tumble ASTM D 3514-81 Elastomeric pad
Frosting		AATCC Method 119 screen wire AATCC Method 120 emery
Snagging		SAE J948 Aug 94 (also abrasion of vinyl/leather) ASTM D 5362-93 bean bag ASTM D 3939-93 mace test

Table 5.5 (Continued)

	British Standard test methods	Selected related test methods
Tear strength	BS 4303:1968 (1995) wing tear BS 3424 Pt 5 for coated fabrics BS 4443 Pt 6 Method 15 for foam laminates	ASTM D 2261:96 tongue tear-single rip CRE ASTM D 1117:95 trapezoidal tear DIN 1424-96 Elmdorf tear apparatus DIN 53356 tear propagation
Tensile strength/ breaking and elongation	BS 1932 for yarns and threads BS 3424 Pt 5 for coated fabrics BS 2576 woven fabric/ strip method BS 4443 Pt 6 Method 15 for foam laminates	ASTM D 751 for coated fabrics ASTM D 1578-93 yarns by Skein method ASTM D 5034-95 Grab method DIN 53857 for non-wovens DIN 53571 tensile and elongation
Stretch and set	BS 3424 Pt 21 But BS 3425 Pt 24 still in use	SAE J855 Jan 94 DIN 53853 DIN 53857
Dimensional stability	BS 4736:1996 cold water	SAE J883 Jan 94 cold water SAE J315A DIN 53894
Compression (for foam laminates)	BS 4443 Pt 1 Method 5A stress strain characteristics BS 4443 Pt 1 Method 6A compression set	ASTM D 2406-73 Method B DIN 53572 compression set DIN 53577 stress strain
Air permeability	BS 5636:1978 now BSENISO 9237 BS 4443 Pt 6 Method 16 BS 6538 Pt 3 (Gurley method)	ASTM D 737 DIN 53887
Fogging	BS AU 168:1978	SAE J1756:1994 ASTM D 5393 DIN 75201
Flammability	BS AU 170 1979:1987	FMVSS 302 SAE J369 DIN 75200
Accelerated ageing	BS 3424:1996 Pt 12 coated fabrics BS 4443 Pt 4 Method 11 foam BS 4443 Pt 6 Method 12	ASTM D 2406-73 DIN 53378 Tests of individual auto companies sometimes as pretreatment for further tests e.g. peel bond, effect on appearance. Cycles can range from −40 °C to as high as 120 °C.

For further information see References 22–30 and Further reading 8 and 19.

to the fabric face, and in addition laminate thickness must be within certain limits for some moulding techniques.

Consistency of properties is vital if production is to proceed without continual stops to adjust settings on machines or equipment. With JIT production techniques now in widespread use, everybody's production schedules are linked closely together, and a hold up in one area will cause a hold up everywhere downstream. The ultimate hold up is the OEM assembly line where all parts come together at just the right moment. Certain OEMs impose substantial financial penalties if their production line ('the track') is delayed.

5.6.3 Customer satisfaction quality checks

The more familiar tests associated with fabric include the following: uniformity of shade; regularity of pattern; abrasion resistance; light fastness; wet perspiration dye fastness; dye crocking fastness; tear strength; bursting strength; laminate peel bond; crease recovery; cleanability/soil resistance; 'environmental' tests; flame resistance. Test specifications may seem unrealistically high to the layman, but it is easy to misjudge the wear and tear over several years of daily use and the combined effects of high temperature, varying humidity and UV radiation. Cars produced in, say, the UK must generally be capable of withstanding climatic conditions in any part of the world. In recent years the following factors have grown in importance or become more critical: fogging; odour free; cleanability; anti-static properties.

5.6.4 Colour fastness and crocking

The car seat cover is fixed to the seat and washing fastness is not an issue; however, perspiration dye fastness, cold water leaching and rubbing fastness (tested by crocking) must be checked. Simulated human perspiration liquor is made up, and a test sample of the fabric is wetted out with it and sandwiched between two white undyed pieces of fabric – one is cotton, the other is sometimes wool or can be made from several different fibres (so called multi-fibre test material). The 'sandwich' is placed between glass plates and put into an oven for four hours at 37 °C to simulate body heat. Any staining off of loose dye is assessed using grey scales, of which there are two types. One type is used to assess change of shade (COS) of the dyed fabric, and the other is used to assess the mark off or bleed off of loose dye on to the white undyed pieces of fabric used in the test. Grey scales standards were prepared in accordance with the ISO and are specified in BS 1006 (ISO 105-A02). Rating 5 indicates no change of shade with the COS grey scale and no staining off with the staining grey scale. Rating 4 indicates slight and generally acceptable levels of change of shade and staining off.

Rub fastness, both wet and dry, is assessed using a Crockmeter, a machine with a wooden peg around which is fastened a piece of white cotton fabric. The machine action is to rub the fabric sample ten times with the cotton fabric covered peg, after which any staining off of dye is assessed using grey scales. It is important to check dyed polyester to ensure that the reduction clearing treatment after dyeing has been effective.

5.6.5 Light fastness and UV degradation resistance

This is probably the single most important test and also the most difficult to reproduce; consequently much research has been carried out.[34–41] One of the reasons, of course, is simply that sunlight conditions vary not only according to location in the world, especially latitude, but also at any given place according to the position of the sun in the sky and therefore the time of day. The weather and cloud cover are also relevant factors in addition to variations in actual solar UV radiation, see Fig 5.9. The large amounts of glass in modern cars allow the entry of substantial amounts of sunlight, which heat up the confined space of the car raising the temperature to as high as 130 °C in extreme conditions in the Arizona Desert. On normal summer days in the UK, the temperature of car interior surfaces exceeds 70 °C with ambient exterior temperatures of only 23 °C. As the sun sets, the temperature will fall and this will significantly affect the relative humidity and cause dampness. Some test procedures attempt to reproduce all of these conditions. The daily cycle of heating and cooling could be influencing rate of colour fade and fabric degradation. Some tests, such as the American standard SAE J1885, include a period with the light switched off to simulate this. If the procedure involves the sample becoming wet, the test is best described as a weathering test rather than a light fading test.

Sunlight is a mixture of all the colours of the rainbow plus IR, UV and other radiation. The UV rays are the shortest in wavelength and, having the most energy, are by far the most damaging to fabrics. Whilst much of this radiation is filtered out by car window glass some of the longer UV rays still penetrate. The thickness of the car window glass will have an effect – the thicker the glass, the fewer UV rays will enter the car. Tinted glass also reduces the amount of radiation, including visible light, but there are safety limitations on the degree of tint permissible.

Following investigations over a number of decades, researchers agree that, amongst other factors and combinations of factors, the three most important single ones causing degradation by sunlight are UV radiation, heat and dampness. To obtain test results which will give accurate information on likely performance over several years, of actual use in the

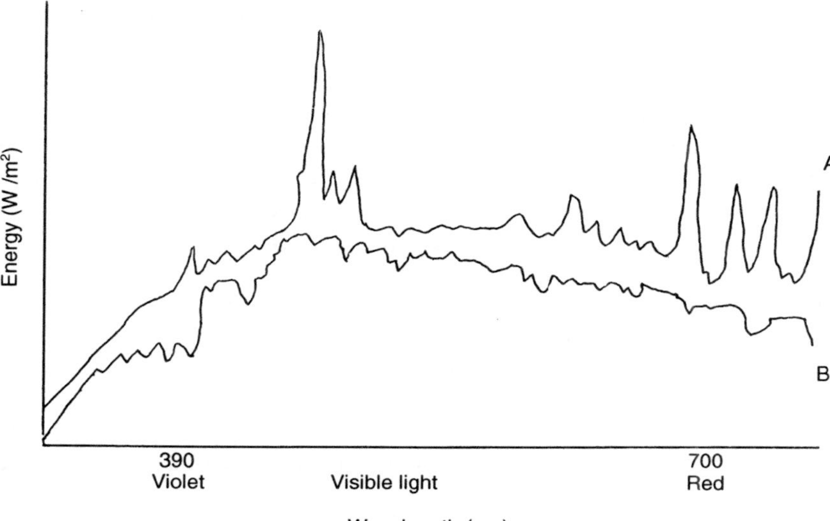

5.9 Spectral energy distribution of daylight compared to that obtained with artificial xenon arc light with filters. Diagram produced from information supplied by Atlas Material Testing Technology BV.
A nanometer (nm) is one millionth of a millimetre. Visible light, made up of the colours violet, blue, green, yellow, orange and red, is in the region of 390–700 nm, with violet (390–430 nm) and red (610–700 nm) at each end. Beyond the range of visible light is UV (30–390 nm) and IR (700–3000 nm).
The shortest wavelengths have the most energy, and the UV is the most damaging radiation to textiles. However, windscreen glass filters out some parts of it.
Sunlight is a mixture of all radiation, but clouds have a filtering effect. The extent to which clouds do this depends on thickness and type of cloud. White light is a mixture of all colours and includes UV and IR. The relative amounts of all the colours and the infrared and ultraviolet, depends on several factors: the position of the observer on the Earth's surface (latitude and whether facing north or south); the time of day (i.e. position of the sun and whether high or low in the sky); the degree of atmospheric pollution; and the amount of cloud cover.
For an accelerated light fading test, it is first necessary to obtain an artificial light source – preferably one which reproduces sunlight as closely as possible. This is neither easy nor straightforward, for the reasons stated. The carbon arc is very rich in UV – much more than in natural sunlight. The best artificial source so far achieved, which closely reproduces natural sunlight, is the xenon arc with filters.
N.B. A = Xenon arc with filters, B = average daylight in Miami.

car, the test machines use these three factors at extreme, but realistic levels, mainly running all together at the same time. It is important that these conditions are comparable to what is observed in actual daylight, because if they are not, misleading information could result. For example, using substantially higher levels of radiation could cause other types of degradation, which would never actually occur under natural conditions. However, it is important that the test is completed in the shortest possible time so that fabric can be released for use as soon as possible after manufacture.

Clear information on the type of test machine and the light source is vital, because the spectral distribution of the light source and the filters used vary from machine to machine. In addition, both lamp and filters have a finite life and deteriorate during use, making it necessary to monitor their performance and to replace them regularly. In some machines the gradual deterioration in lamp efficiency is compensated for by an automatic increase in wattage.

The first lamp developed was the enclosed carbon arc (Atlas in 1920s) which was used in the Fade-Ometer. The spectral distribution of this lamp was very different to sunlight; in particular, UV rays, which are responsible for much of the damage caused by sunlight, were absent. The sunshine carbon arc, which was used in the Weather-Ometer was an improvement. This lamp had a better resemblance to sunlight and did produce accelerated fading, but it contained certain bands of UV radiation not present in natural sunlight, and was judged to be too severe. Furthermore some visible light was absent from its spectrum. Maintenance was expensive because the electrodes of the carbon arc lamps had to be changed daily, and test machines using fluorescent lamps appeared as cheaper alternatives. Whilst fluorescent lamps do give accelerated fading, they are now considered unrealistic because, while their spectrum is rich in UV radiation, other wavelengths are absent.[35]

The latest developments involve the xenon arc lamp, which is at present the best reproduction of natural sunlight commercially available. The first machine of this type, which was introduced during the 1950s by Heraeus, was an air-cooled model. A water-cooled model produced by Atlas followed shortly after, and both types are now in widespread use. However, it is important to specify the method of cooling and which filters are to be used, because the spectral distributions of the two types are not the same. Results of fading tests will be different for the following reasons. The Atlas model has two glass tubes around the xenon lamp, which act both as filters and as part of the cooling apparatus. The spectral distribution of the light is, therefore, the same in all directions. The air-cooled Heraeus model, on the other hand, uses a combination of filters to produce an overall spectral distribution. The carbon arc lamp is still used, but this is declining in favour of the xenon models. Some researchers believe that the whole spectrum of sunlight should be reproduced to give the most accurate accelerated results. For these reasons the OEMs specify the test method and the type of machine. A typical test requirement includes the following information: test machine model and lamp; filter system; humidity; test chamber temperature (ambient inside the apparatus); black panel temperature (temperature of the actual test sample); and exposure time.

The test standard can be specified by the amount of fading or discoloration acceptable, as assessed by grey scales or the wool blue scale after exposure to light for a certain length of time. The grey scales are prepared according to the ISO and BS 1006 in the UK. The wool blue scales were developed by the SDC in conjunction with other relevant organisations, and are based on eight dyes, one for each level of lightfastness rating. Note that the blue scale used in the USA is not the same – it is based on mixtures of two dyes to give the eight levels. With both wool blue scales, each level requires approximately double the amount of energy as the level immediately beneath it to produce the same level of fading. Alternatively the fading or discoloration is assessed after exposure to a measured amount of energy in kilojoules per square metre (kJ/m^2). The American wool blue scale 7 is approximately equivalent to $680\,kJ/m^2$ at 420 nm wavelength of light.

There are additional factors influencing reproducibility. The test substrate itself may not be completely uniform and may have varying amounts of chemical finishes, UV absorbers or other substances on it. In addition, fabric samples could have been produced under varying conditions of scouring, stentering or lamination, etc. Laboratory reproduction of several years exposure to air pollution and traffic fumes, the composition of which will vary widely with location, is especially difficult. These factors may also be playing a part in conjunction with the combined effect of all the other variables, not to mention the surface abrasion and other factors associated with the car occupants sitting on the fabric.

Fibre lustre, or the titanium dioxide delustrant added to the yarn during manufacture, has a very significant effect on UV resistance. Matt yarns, which contain the most delustrant, break down significantly faster than bright yarns. This is thought to be due to the titanium dioxide photosensitising degradation, or, in the case of delustred yarns, because of light being scattered more internally within the fibre filament. Ultra violet degradation is also influenced by the thickness of the filament – the thicker, the better. This is because less radiation will penetrate into the centre of the filament, and the lower specific surface area of the thicker filament reduces the rate of photo-oxidative attack.

When fabric is tested for light fastness and UV degradation, it is important to test it either in the laminated form or with polyurethane foam underneath it. The foam is believed to act as a heat sink, hence more accurately reproducing the conditions actually prevailing inside the car. Different OEMs specify different test conditions, but there are steps to standardise procedures in order to reduce the number of test methods, especially in the USA and Europe. In Germany there has been some successful harmonisation with the Fachnormenausschuss Krafflahzeuge (FAKRA) test procedure, DIN 75202 being widely used, while in the USA, the SAE J 1885 test is widely used. Harmonisation should result in some savings, because at present fabric producers supplying several OEMs must possess every machine necessary and these are expensive to buy and to run.

5.6.6 Abrasion resistance and associated factors

Fabric is normally tested for abrasion in the form in which it will be used in the car, i.e. when laminated to polyurethane foam and back coated if specified. Abrasion results are usually slightly better when the fabric has been laminated to polyurethane foam, compared to tests carried out on the base fabric alone. This is because the foam, joined to the fabric, helps to lock the fibres together in the fabric. When it is necessary to test non-laminated (singles) fabric, for example during development, a small piece of polyurethane foam is placed underneath the fabric being tested in the test holder. Sometimes the foam is attached to the fabric sample with double-sided adhesive tape to simulate lamination. Some test procedures require exposure to light and UV radiation before testing, which significantly reduces the abrasion performance in most cases. The place in the car where the fabric is situated determines the standard of abrasion resistance required, the seat usually requiring the highest; some OEMs specify different standards for the centre seat panels, the bolster (the side and front edges of the seat), and the back of the seat – the bolster requirements are usually the most demanding. Door casing fabric specifications are generally somewhat lower than those for the seat, while those of the headliner are significantly lower.

There are three main test methods for abrasion resistance in use: Martindale (using 12 kPa, 28 oz weight), Schopper and Taber which *very generally* agree with each other – but certainly not always, see Fig. 5.10. The three test methods actually represent different types of abrading motion as well as using different abrading materials. The Schopper machine operates with a reversing circular motion, while the Taber motion is a little more complicated with two circular wheels rotating in opposite directions. Martindale operates in a multi-directional manner, the abrading heads moving in a Lissajous pattern. The Taber uses an abrading wheel made from rubber/aluminium oxide abrasive particles (Calibrase), the Schopper uses fine emery paper, while the Martindale uses a standard grade of woven

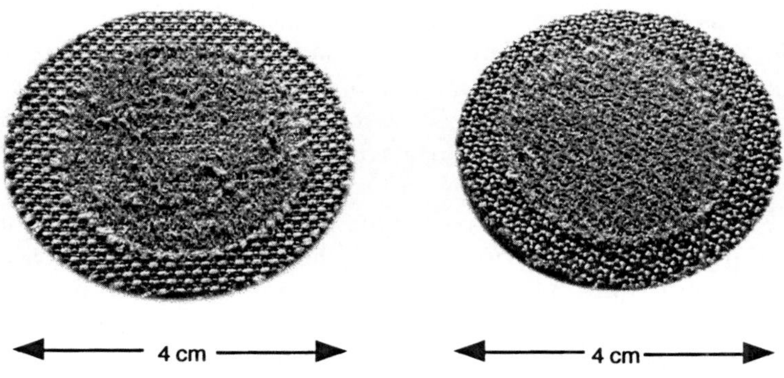

Taber abrasion test [not to scale]

Schopper abrasion test [not to scale]

13 cm

5.10 Abrasion testing of automotive seat cover fabric. The top two were abraded on the Martindale apparatus. The top left sample shows broken threads and some pilling. The right hand sample shows some wear and 'frosting'.
The bottom left sample has been abraded using the Taber apparatus and is satisfactory, while the bottom right, abraded on the Schopper apparatus, is showing signs of wear.
N.B. The photographs are not to scale: the Martindale samples are about 4 cm across, the Taber samples about 13 cm and the Schopper samples about 11 cm.

wool. Many researchers regard the Martindale multi-direction action and use of wool as the abrading material as the most realistic of the three tests, but it does take substantially more time than the other two. To reproduce significant wear in actual use, a minimum of 50 000 Martindale rubs are necessary, taking about 16 hours to complete. In comparison, a similar Taber test only takes 15–30 minutes and a typical Schopper test requires 1–2 hours.

Abrasion resistance is influenced by the fabric construction, yarns used, finishes applied and amount of coating on the back. Yarns of higher dtex/filament generally have better abrasion resistance than yarns made from finer filaments. Highly textured yarns usually have slightly lower abrasion resistance than yarns with a lesser degree of texture. Excessive wet processing, prolonged dyeing or rigorous reduction clearing can all reduce abrasion resistance, and in some cases spun dyed yarns may have better abrasion resistance than yarns of the same shade, which have been aqueous dyed. Fabric construction can have a substantial effect on surface abrasion. Those constructions with long 'floats', or which otherwise provide points for frictional stress, have the poorest abrasion resistance. Fabric finishes can significantly improve abrasion resistance, by acting either as a lubricant in the abrading action, or as a barrier between the fabric and the abrading material. However, they are rarely used on automotive fabrics, because of the risk of fogging and also because they could lead to the development of unsightly or sticky deposits on fabric surfaces over a period of time. This is probably caused by degradation of the chemical by the heat, humidity and light radiation to which car interiors are subjected. In addition drops of water could lead to the appearance of 'tide' marks or discoloration. Certain waxes and silicones in particular must be avoided, because they can affect adhesion of coatings or materials during lamination. They can be applied after coating or lamination, but care is required because the heat or liquid involved in the finishing process can loosen the polymer or adhesive bond. After testing by the prescribed method, the abraded samples are inspected for wear or broken threads. A certain amount of wear is usually acceptable, but most OEMs will not accept a broken thread in car upholstery.

5.6.6.1 Frosting

In some instances the material may not have any significant wear or any broken threads after abrasion testing, but may be whiter in appearance. This condition is referred to as 'frosting' or 'ghosting' and is sometimes associated with fibrillation of the yarns or poor dye penetration. In other cases, especially if a finish has been applied to the face side of the material, the abraded pattern may be glazed and appear shiny. These defects may or may not be acceptable to a customer.

5.6.6.2 Pilling

Also associated with abrasion resistance is pilling, which is the formation of little circular clusters of fibre on the surface of the fabric, produced as a result of the fabric being rubbed against itself or against some other material, see Fig. 5.10. Fibre ends become tangled and twisted together, sometimes with the fibre ends and broken threads of the material it is being rubbed against.[42–45] Unsightly pilling can occur where the seat occupants' clothing is perhaps more to blame than the seat fabric itself. This is sometimes referred to as 'foreign' pilling.[42] Fabrics constructed from spun yarns are significantly more prone to pill than continuous filament, and the problem is probably more pronounced with polyester than wool. Wool is an inherently weaker fibre and pills can break off from the fabric surface before the end of the test cycle. This does not always happen with polyester because of its higher strength, and the pills grow larger, become more conspicuous and unsightly, and unlike wool are present at the end of testing and are assessed. Thus, misleading results which do not reflect actual wear may be obtained.

Pilling can be minimised by chemical finishes, increasing the yarn filament thickness, use of higher twist yarns and by brushing and cropping of the fabric. However, any one of these factors may change the handle and other qualities of the material. Fabric can be tested for pilling using a pill box of the type designed by ICI which consists of twin wooden cubic boxes, each with sides about 25 cm long, the inside walls of which are lined with cork. Fabric samples are wrapped around rubber formers and placed inside the boxes, which are then rotated around a common axis for a measured length of time. The samples are assessed against masters and the degree of pilling assessed on a scale of 1 to 5, the higher the rating the less the pilling. In an alternative test method, the so-called Random Tumble Pilling Tester Method (ASTM D 3512), cotton fibres can, if required, be added as a source of foreign fibre. The Martindale tester is also used to assess pilling by subjecting the test fabric to cycles of, say, 1000 or more rubs and counting the number of pills.

5.6.6.3 Snagging

Snagging occurs when a sharp point or rough surface catches a thread in a knitted or woven fabric. Constructions incorporating long floats are especially prone to this problem. The thread is pulled out of the fabric forming a small loop on the surface, while the thread still in the fabric is stretched and appears as a shiny line – a tight end. The phenomenon of snagging is tested using a Mace snag tester which comprises an array of spiked metal balls. These are abraded against the fabric for a set time, and the degree of snagging is assessed on a scale of 1 to 5.

5.6.7 Fogging

Fogging is the mist-like deposit that forms on car windscreens reducing visibility and which is sometimes difficult to remove, even with soap and water. Fogging is caused by volatile materials vaporising out of *all* interior trim components, such as plastic foils and polyurethane foam, and not just the fabric. Modern polypropylene foils hardly fog at all, but PVC, if not formulated with suitable plasticisers, can fog quite badly. Fabrics, if not stentered well or scoured, can fog very markedly owing to the multiplicity of lubricants applied during dyeing, yarn doubling, warping, weaving, knitting and finishing. Because of the much larger surface area of yarn on the face, velvets or other pile fabrics can fog significantly. Results of chemical analysis of the fogging deposits have been published along with other findings.[46–52]

The fog test is carried out by putting a specified amount of fabric into a beaker, which is covered and sealed by a glass plate, see Fig. 5.11. The light reflectance of the glass plate is measured beforehand. Because the 'non-tin' side of the glass plate must face downwards towards the material being tested, which side is which must be labelled by the supplier. This glass plate is cooled by a metal cooling plate which rests on it, and cooling water is pumped through it at a specified temperature, usually room temperature. The beaker is heated at 90–110 °C for three to six hours, the actual conditions being specified by the OEM. After this time, the light reflectance of the glass plate is re-measured to determine the reduction in reflectance which has been caused by condensation of the volatile materials from the test sample. This is usually expressed as a percentage of the original reflectance, and a good result is generally anything over 90%, although some OEMs may accept less than this value. The three main test parameters, i.e. temperature of sample heating, time and cooling water temperature are specified by the OEM. The apparatus is calibrated by measuring the fogging obtained with DIDP (di-isodecyl phthalate), a plasticiser used in PVC, which should give a fog value of 76–79%, depending on the information supplied by the manufacturer, e.g. Merck. This check should be carried out periodically about once a month, but some test houses include a DIDP standard with every test batch.

Inter-laboratory reproducibility may not be satisfactory, and much discussion has taken place to standardise procedure in order to improve this.

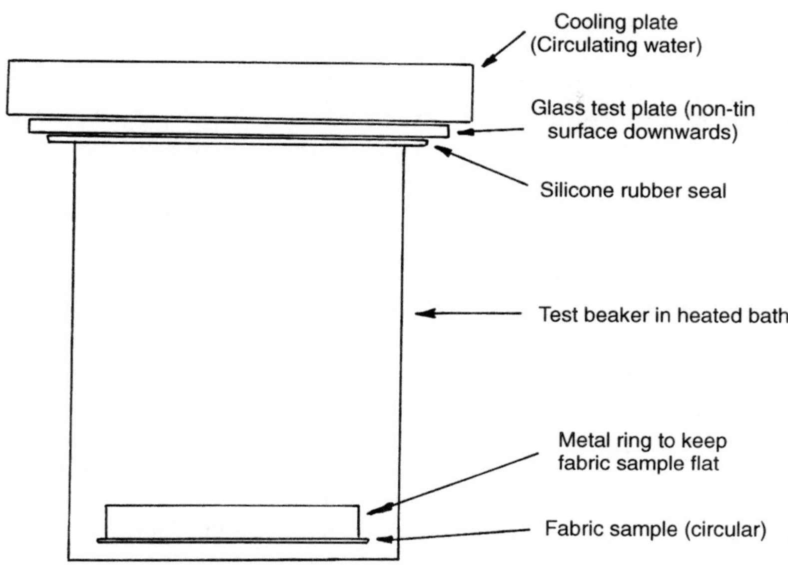

5.11 Automotive fabric fogging testing.

Cleaning of the glass plates is generally believed to be critical. A gravimetric method, which weighs the volatile deposits, has been put forward as more satisfactory, but it relies on a balance capable of weighing to five decimal places. Recent work has focused on the formation of crystals on the glass plate, which may still reduce visibility but give a high reflectance reading. Two test rigs are widely used, those made by Haake and by Hart, but the test specifications are written in such detail that self assembly is possible. Operative procedure is believed to influence the accuracy of the test substantially, and the recommended methods are detailed in the test specifications. A training video has been produced by the Industrial Fabrics Association International (IFAI) for SAE J1756. Fogging has assumed more importance recently and is related to the problem of malodours in new cars, both being associated with volatile materials.

5.6.8 Anti-static properties

The occurrence of static shocks when getting out of cars has been known for some time, and some OEMs require anti-static finishes on their car seat covers.[53–56] The static electricity is generated by the person's clothes rubbing over the polyester car seat cover, especially when he or she stands up to get out of the car. Some individuals are especially prone to static shocks, which are also influenced by clothes and shoes worn and the ambient air conditions inside the car. The polyester seat cover is hydrophobic, i.e. containing very little moisture to conduct away or help dissipate the static electricity. Anti-static properties are easily conferred on to the fabric by application of an anti-static finish by padding, or by foam coating, but these are not permanent and eventually wear off. Their efficiency is measured by surface conductivity methods and they work simply by their hydrophilic nature.

The conductivity meter actually measures surface resistivity in units of ohms per square. A resistivity of about 1×10^{10} ohms per square is considered a reasonable level of anti-static behaviour, but the lower the better. Padding a fabric with a good anti-static agent can easily give a figure of 1×10^{7} ohms per square. The surface conductivity method has its limitations, and some researchers question its suitability for car seats. Some OEMs and research institutions, such BTTG and John Chubb Instrumentation, have developed whole chair tests using human subjects wearing specified clothing. Recently, the possibility of static electricity interfering with electronic equipment controls in the car has been raised, along with the possibility, in extreme cases, of its igniting petrol vapours.[55] There has been renewed interest in development work using conductive yarns such as Negastat and R. Stat. products which can confer permanent anti-static properties. These specialist yarns are very expensive, but only small amounts are required in a fabric to confer some measure of durable anti-static behaviour.

5.6.9 Stretch and set

These properties are important in fabrication of an interior component. After panel cutting, the material may elongate under its own weight and it may then be longer than another piece cut to the same size to which it is being sewn. An oblong shape may need to be folded around into a cylinder shape, say for a headrest, and one edge may be slightly longer than the other. Another fabric panel may require pulling around a seat cushion or squab, and it may not be possible to do this without the use of undue force. Conversely, it could be too loose or too stretchy, and this would cause bagging and creasing during use. These pieces would be rejected because there is no time to trim them to the correct size or to make continual adjustments to the machinery. Stretch and set is determined by testing strips of fabric of specified dimensions at specified rates of extension on a computer controlled universal strength tester.

5.6.10 Soiling and cleanability

These factors are assessed by the application of materials likely to be accidentally spilt on to the surface of car seat fabric. They are applied to test pieces of fabric in the laboratory and then cleaned off using a specified procedure. The degree of soiling left behind is assessed under a standard light source such as CIE (Commission International d'Eclairage) D65 (chosen as closely resembling natural daylight), and either compared to standard patterns or assessed using grey scales. Rating 5 of the grey scale records no noticeable staining, 4 records slight staining, descending down to rating 1, which records very significant staining. The soiling agents include materials such as chocolate, coffee, tea, ice cream, hair-dressing fluid and engine oil. A brand name or chemical type of soiling agent is usually specified for the tests to be reproducible. Anti-soiling seems to be an increasing concern, and there have been recent reports on the subject.[57,58]

The problem of 'linting', or the appearance of white specks of fibrous material on the car seat, is well known and is quite difficult to overcome. These 'lints' are not easy to remove by brushing, and tests have been carried out to minimise the problem using certain soil release agents. Associated with linting is 'minking', the removal of hairs from fur coats by abrasive action of the car seat fabric.

5.6.11 Environment and ageing

These tests try to reproduce several years ageing in the space of one or two weeks in the laboratory in order to pick out poor adhesive bonds, which may allow fabric to lift or 'bridge' in mouldings with sharp angles or corners. Shade change, dimensional stability and peel bonds are also examined after ageing. Typical tests involve exposing the sample to heat for about two weeks at over 100 °C and to relative humidity at 100%. Environmental tests are generally carried out on the component as a whole, e.g. an entire completed door casing. The tests frequently involve a complete cycle of extreme conditions, e.g. exposing the test piece to, say, 24 hours at –40 °C, then to, say, 24 hours at 100 °C and 100% relative humidity.

5.6.12 Air porosity and permeability

For automotive fabric laminates this is usually measured as the volume of air in litres per second required to maintain a specified constant pressure differential across a test specimen of a certain dimension. Test conditions specified by OEMs are generally similar to those in BS 4443: Part 6: 1980 Method 16. High material porosity makes panel cutting of several layers more trouble-free and more accurate, especially when vacuum is used to hold material flat on the cutting table. Porosity also influences seat comfort, see below. It has become more important recently as a measure of laminate suitability for 'foam-in-place' manufacturing methods. When the material is of much lower air porosity, the Gurley Method, which is used widely for packaging materials, is specified.

5.6.13 Car seat comfort

Many factors directly or indirectly influence car seat comfort, and much work has been done.[59–65] Thermal comfort has been discussed in some detail above, and everyone is aware of the thermal discomfort which is experienced in hot weather inside a car. The 'greenhouse effect' is primarily to blame, but the car seat contains thick layers of foam, which is an insulating material and acts like an additional layer of clothing in the restricted area inside a car. If the seat fabric is coated, this is going to restrict the evaporation of perspiration making the seat occupant thermally uncomfortable. Leather and especially PVC seat covers can produce very thermally uncomfortable seats for the same reason. For a short while some car seats were made by laminating a film of polyurethane to the fabric and sewing the car seat cover into a 'bag' into which was poured a liquid which reacted to form the solid foam inside. The film was needed to stop the liquid foam from seeping through the fabric before it reacted and became solid. This led to seats which were thermally uncomfortable because the film was a barrier to the passage of perspiration. If a film of higher permeability to perspiration, such as Gore-Tex or Sympatex, had been used the seats would have been more comfortable, but this was not generally possible because the higher specification material was much more expensive. As has been discussed, there are methods for measuring breathability of films, but more realistic tests are best carried out on the seat as a whole. At present, there are no requirements for water vapour permeability for car seat fabric laminates, although OEMs may test the seat as a whole for thermal comfort during development. Air conditioning in cars, however, is likely to alleviate this problem.

5.6.14 Odours in cars

This has become a problem in recent years, and test methods are being developed.[66–68] At present, some OEMs simply state that the fabric must be free of any unpleasant odours, while Ford have introduced an 'electronic nose' to detect and measure odours. Odours could be related to VOCs and would therefore have health and safety implications. They may in some cases be associated with coatings on fabric or carpets.

5.7 Flamability (FR) testing

5.7.1 General comments

The importance of FR has already been discussed in Chapter 2, and testing articles for accurate information on FR properties is essential. The correct terminology must be used, because sellers of products labelled 'fire-proof', 'flameproof' or even 'low flammability' may leave themselves vulnerable to legal proceedings. Only approved and defined terminology should be used, and reference should be made to the BS or ISO test that establishes the terminology, e.g. 'does not sustain flame' (BS 6336). 'Flame retardant' (FR) has come to be widely used as a useful general term. Because of both the legal and the moral implications, the articles offered for sale must pass the appropriate tests using the procedure set down. Needless to say, laboratory staff should be trained both in carrying out the FR test and in interpreting the results, and commercial test houses are generally validated and accredited in the UK by the National Measurement Accreditation Service (NAMAS). Many papers are available on FR testing and standards in domestic, industrial and transportation situations.[69–78]

5.7.2 FR properties of coated fabrics and laminates

A coated or laminated fabric comprises at least two polymer types, and the FR properties of the combination cannot be predicted using knowledge of the FR behaviour of the individual constituent polymers. The coated or laminated fabric will have its own properties that must be determined by testing. The FR property of the plastics coating will be made more complex by the fabric construction, which may shrink back from an applied flame and not be ignited. Another sample of the same material, but from a different batch, which has been either dyed or stenter finished in a slightly different way, may not shrink back as readily, become ignited and thus display very different FR characteristics. A resin formulation which may pass an FR test on a particular fabric substrate, if it is applied to another fabric which in fact is less flammable than the first, may in the second case burn and fail the test. The reason could be that the supporting fabric of low flammability does not shrink back, or melt and drip away, but instead stays rigid and supports the resin, allowing it to burn. Foam burns more readily than solid material because of the greater surface area available for a given weight.

One researcher has pointed out that the interface of fabric and polymer is, in some ways, an entirely different material to both the fabric and the polymer coating or laminate. If an adhesive has been used, this may be a third different chemical type that could also display different FR properties and produce further modification of the behaviour of the other two components. There will be, in this last mentioned case, two interfaces between dissimilar substances. The factors mentioned above add further to the complexity of FR testing which both research and quality control staff are only too aware of. The FR properties of a coated fabric are thus influenced by the FR nature of the base fabric and by the resin material and its formulation, i.e. FR chemical concentration and type. The FR properties of a coated fabric will also be influenced by coating technique, resin add-on, penetration and the many factors and inter-relating factors which depend on these variables, including fabric handle, drape and flexibility. Mention has already been made in Section 2.8.1 of a statistical study of the effect of coating parameters on FR properties of a cotton fabric coated with a phosphorus FR compound.

5.7.3 FR testing of products

Coated and laminated fabrics which require FR testing include domestic upholstery, curtains and carpets. Any material used in transportation applications or buildings open to the general public is likely to be required to pass stringent FR tests. Flame retardancy requirements are especially important in aircraft, and recent train disasters involving fires in tunnels have highlighted the need for rigorous FR standards for passenger trains. The FR properties of seat covering materials, especially in aircraft, are tested very rigorously because of the foam material used beneath. If ignited, copious potentially toxic fumes may be emitted, but low FR foams are now also used. Fireblocker materials are used as a barrier, and tests are carried out on the whole seat assembly. Fireblockers which are made from fibres of high inherent FR properties, such as Nomex, Panox, PBI and Inidex, are now being used increasingly in trains and coaches. These fibres are, of course, very expensive and could not be used in private cars. In fact the FR requirements of car seat fabric are not especially high, presumably because the risk of a fire inside a car is probably quite low and there are at least two doors for a rapid exit.

For car seat fabric the test used by OEMs is the USA standard FMVSS 302, a horizontal burn method, but the test performance standard required varies according to the OEM. Generally, this test is not especially difficult to pass, but sometimes an FR coating is needed on the fabric or an FR foam is necessary. In certain cases, the fabric alone could pass the FR test, but FR chemicals are required in the coating to reduce its flammability. Sometimes both FR coatings and FR foam are required to satisfy the test standard. Flame retardant chemicals both in the foam and the coating add to the cost, and for commercial reasons the concentration levels in both foam

Table 5.6 Selection of FR test standards and methods associated with coated and laminated fabrics

Product	Standard	Comments
Clothing	BS 6249 Part 1:1992	Performance standard based on Test 2 method in BS 5438:1976, three flammability indices
	BSEN 469:1995	Firefighters' clothing
	BSEN 470-1:1995	Welders' clothing
	BSEN 531:1995	Industrial workers' clothing
	BSEN 532:1995	Protection against heat and flame
	BSEN 533:1997	FR fabrics
Furnishings	BS 6807:1986	Ignitability of mattress and bedding materials
	BS 7176:1995	Spec for resistance to ignition of upholstery for non-domestic seating
	BS 7177:1996	Spec for resistance to ignition of mattresses, divans and bed bases
	BS 5852 Part 1:1979	Test for ignitability – different sources for different articles – fabrics coated for FR are subject to water soaking before testing
Upholstered furniture (domestic)	BS 5652:1979 BS 5652:1982 BS 5652:1990	Test methods for ignitability and performance standards. Domestic furniture must meet certain specifications in the Furniture and Furnishings (Fire) (Safety) Regulations 1988/89. Covers beds, sofa-beds, futons and other convertibles, nursery furniture, furniture in new caravans, etc, but not carpets and curtains
Curtains and drapes	BSEN 1101:1996	Burning behaviour of curtains and drapes
	BS 5867:Part 2:1980	Ignition of vertical specimens by small flame
Textile floor coverings	BS 4790:1972	Hot nut test
	BS6307:1982 ISO 6925:1982	Methenamine Tablet test
	BS 476 Part 7	Surface spread of flame
Tents and marquees	BS 6341:1983	Camping tents – 10 grades – flammability requirements
	BS 7157:1989	Ignitability of fabric used in large tent structures
	BS 7837:1996	Spec for FR properties for marquees and similar
Motor vehicle seat coverings	FMVSS 302	Each manufacturer has their own performance specification

Table 5.6 (Continued)

Product	Standard	Comments
Automotive carpets		ASTM D 4723-90 method and specs for heat and flammability ASTM D 2859-96 FR test method
Pile fabrics	BS 4569:1983 (1990)	Ignitability of fabrics with pile on surface (surface flash)
Ventilation ducts	BS 476-24:1987	Determination of fire resistance
Railways	BS 6853:1987	Code of practice for fire precautions for railway passenger and rolling stock
	BS 5852, crib 7	Railway seats in UK
	UIC 574-2 DR	International Union of Railways Specification
Marine	IMO Regulations e.g. IMO Resolution A471 (XII) for fire resistance	
Aircraft	FAR 25 853b	Ignition and propagation
	FAR 25 853c	Paraffin burn test and weight loss
	ATS 1000.001	Smoke opacity and toxic gases
Other tests		
	NBS Smoke Chamber	Smoke release
	NFX 70-100	Test for a series of toxic gases
	OSU 65/65	Ohio State University test for heat release

N.B. These tests are a selection and a summary. Certain tests are in the process of revision, and standardisation of test methods in Europe is being carried out.
For further information see References 69–77 especially the papers by Eaton (69), Johnson (71), and Bagnall (75) and the textbook by Troitzsch (76).

and coating are kept to a minimum. A fabric coating can cause stiffening as well as adding to the overall weight and cost. Flammability is assessed by the burn rate, which is the distance in centimetres burnt in one minute. The burn rate regarded as acceptable varies according to the OEM. Door panel fabric is usually assessed on the actual door panel itself. Variations in flammability test results are sometimes caused by uneven up-take of fabric finishes, or by variations in scour or stentering. Stentering may influence the flammability of stretchy knitted fabrics, because it could affect the manner in which the fabrics shrink away from the flame. Some OEMs require that pile fabrics be brushed in a particular way before testing.

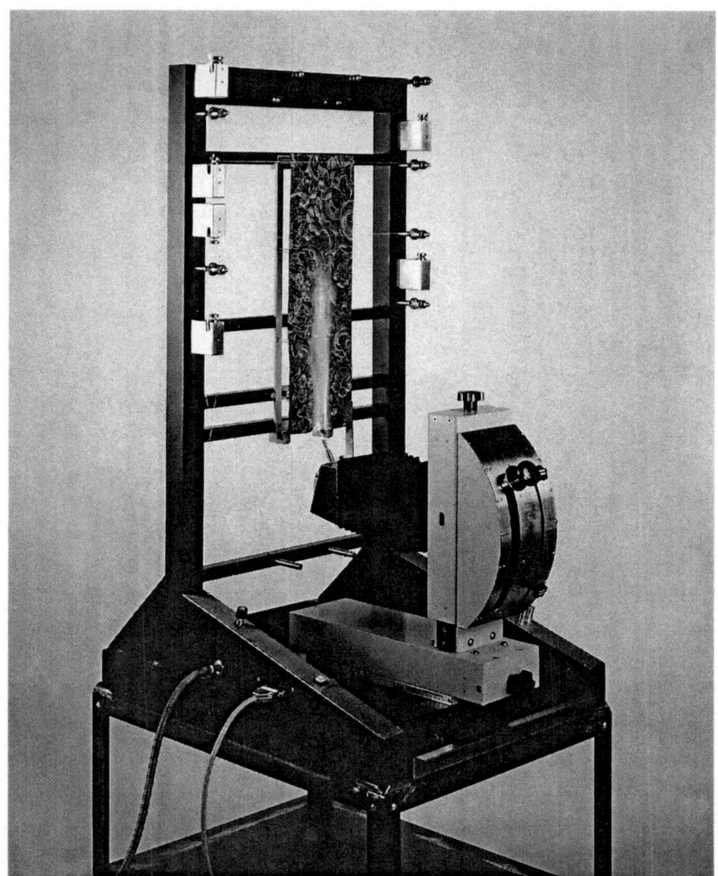

5.12 Rhoburn flammability tester, designed to measure the ease of ignition and flame spread of vertically orientated textile materials. Photograph supplied by James Heal and Company Ltd and reproduced with kind permission.

Coated apparel, apart from specialist protective clothing, does not generally have to pass FR tests, presumably because it is used mainly out of doors where fire risk is minimal. Certain medical products need to have FR properties, because they may be used in atmospheres rich in oxygen. Table 5.6 supplies details of a selection of FR tests applied to coated and laminated fabrics.

5.7.4 FR test methods

Flame retardancy tests now take the form of testing materials for ease of ignition, rate of propagation of flame, smoke generated and toxicity of

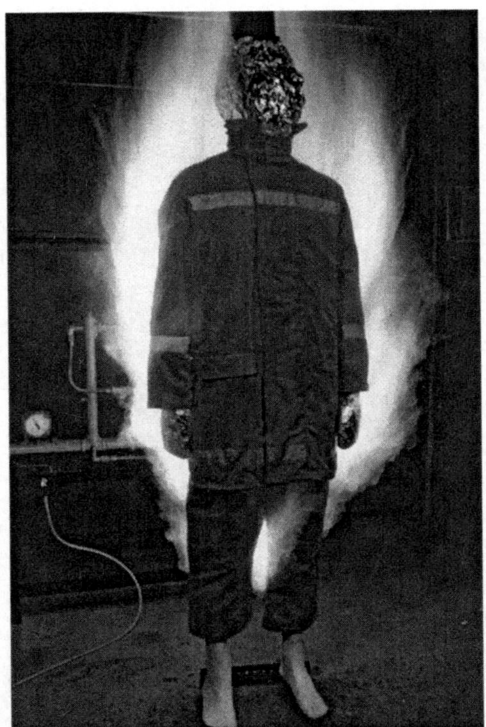

5.13 The BTTG Fire Technology Services RALPH manikin for full flame envelopment conditions. With each test, a burn injury prediction print-out is produced, providing information on first and second degree burn levels to allow users and garment manufacturers to evaluate performance under the most extreme of conditions. Fire-fighters' garments, FR coveralls for the off-shore oil industry and military clothing have been tested with this apparatus. Photograph supplied by BTTG and reproduced with kind permission.

fumes, opacity of the smoke and heat generated by the burning, see Fig. 5.12 and 5.13. The formation of burning debris and falling droplets is also important in some circumstances, such as in buildings. In many fire disasters, more people have died as a result of heat stress and suffocation by toxic fumes. Dense smoke can disorientate people and make escape routes difficult to see. In a real life situation, the FR tests should establish whether the persons concerned have sufficient time to evacuate the building, vehicle or vessel and reach places of safety. However, testing the individual article cannot always accurately establish the likely outcome in the event of a fire in a real life situation. Tests involving whole assembles of articles are carried out in research projects, and 'whole room burns' are sometimes conducted to observe what happens in actual domestic situations. The actual burning nature depends on the FR characteristics of the individual items, their position relative to each other, and on many other interrelating factors, including air supply and airflow and how these are affected by the positioning of furniture or other items such as doors or windows. No two fires are alike which makes the whole subject extremely complex.

For research and development, the LOI is a useful way of evaluating inherent FR properties of materials, because a quantitative figure is given. The amount of oxygen in the atmosphere is about 20% and this is a baseline. Unmodified cotton has an LOI of about 18%; treating with Pyrovatex or Proban can increase this figure to about 28–30%. Regular polyester has an LOI of about 21%, FR modified polyesters about 28%, whilst aramids have LOIs of up to 33%.

5.8 Specialist testing for chemical and biological hazards

The ASTM standard, 'Test Method for Resistance of Protective Clothing Materials to Permeation by Liquids and Gases' (ASTM-F739-91) and the 'Standard Guide for Test Chemicals to Evaluate Protective Clothing Materials' (ASTM-F1001-86) are used to evaluate material to be used in chemical protective clothing. ASTM-F1001-86 lists about 20 chemicals, including the potentially dangerous substances sodium hydroxide, concentrated sulphuric acid, nitrobenzene, and the gases chlorine and ammonia. Amongst the important factors to be considered in the test method is the 'breakthrough time', which is the time taken for the chemical agent to permeate through the test material.

In the 1980s a number of accidents occurred involving hazardous chemicals. This led to Federal involvement in the production of comprehensive standards to improve the performance of chemical protective clothing. The ASTM began to work on comprehensive specifications to document performance and manufacturer reporting requirements. The National Fire Protection Association (NFPA) set up a technical subcommittee and a draft ASTM standard, 'Standard Guide for Minimum Documentation and Test Methods of Chemical Protective Clothing', ASTM F23.50.03 (Z1416-Z), resulted. Three standards relating to vapour protective suits, liquid splash suits and suits for non-emergency and non-flammable hazardous chemical operations (NFPA 1991–3) were also issued and took effect in 1990. However, the quest for further improvements continues, and there are papers in the literature.[79–85]

Barrier textiles are also required in the medical and food industries against contamination and infection by micro-organisms, and the film and nonwoven manufacturers are putting much effort into both the development of new products and more effective means of testing. There is risk of infection of medical workers by contamination from patient's skin and blood, which is especially relevant with the presence of the AIDS virus. Test methods with a high standard of reliability are essential. However, protective clothing should be comfortable to wear and should not restrict movement. The EU is funding a project called BIOBAR for increased safety and health under its Standards, Measurement and Testing Programme, specifically to assess standard requirements and improve test methods for barrier materials to be used to protect people and goods against micro-organisms. There are four separate tasks which investigate the resistance of barrier materials to wet and dry contact with micro-organisms and also wet and dry airborne contact. BTTG is playing a leading role in the consortium of six research organisations.[84]

5.9 Recording of results – statistical process control

5.9.1 Process control charts

All results must be recorded and filed for a reasonable length of time so that if any problems occur in downstream processing or there is a customer complaint, the incident can be investigated. The results of individual tests are best plotted on a SPC chart where it is possible to see at a glance how consistently the product is being produced. Sudden changes in test results, trends and recurring cycles can sometimes give clues about what is happening if there is a problem, see Fig. 5.14. If the results are within acceptable limitations, it is known that the process is in control. If results begin to deviate significantly from the mean, this may be taken as a warning that something may have happened, and corrective action may be needed. Warning limits can be set, for example at the mean value plus or minus two standard deviations, while 'action' limits may be set at the mean plus or minus three standard deviations. This technique allows quality to be checked while the component is being made and not after completion, when it is usually too late. The philosophy is not one of detecting faults after production has been completed, but rather one of preventing their occurrence during production, as a means of continuous improvement with the objective of achieving zero defects.

5.9.2 Fabric examination

Every metre of coated or laminated fabric must be visually examined before it leaves the factory to go on to the customer or to the next stage of manufacture. Samples of each production batch, or whatever samples are

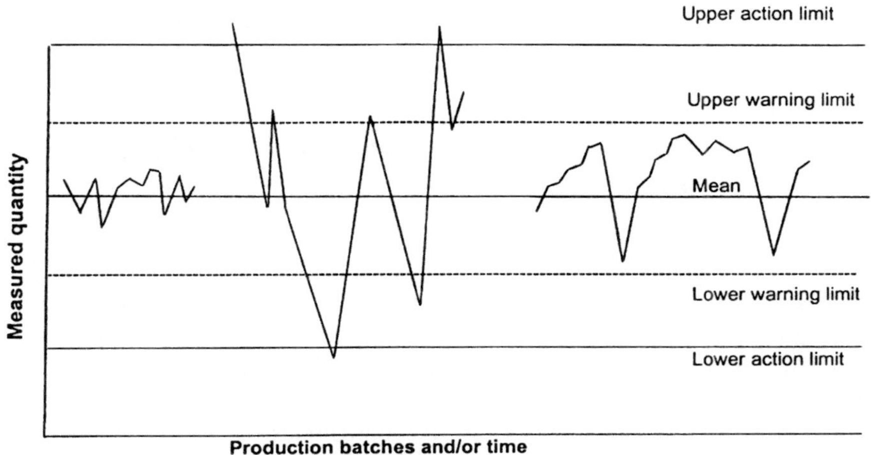

5.14 Statistical process control. This provides a method of detecting faults during production, as a means of maintaining continuous improvement with the objective of achieving zero defects.
N. B. Warning limits can be mean plus or minus 2 × standard deviations. Action limits can be mean plus or minus 3 × standard deviations.

decided with the customer, are sent to the laboratory for testing. Trained examiners inspect the material for width, shade/pattern regularity, visual coating or lamination faults, e.g. tackiness, delamination, oil marks or other soiling, weaving or knitting faults, handle/drape/stiffness and any other faults which could cause complaints from customers or problems in downstream processing.

Faults are marked with a coloured tag or label and a length allowance is given to the customer. Usually the faults are colour coded for rapid identification. However, even experienced examiners cannot see all faults at practical, commercial examination speeds, and ways of improving this situation are always under consideration, e.g. better quality examination frames, better illumination, etc. Some customers such as car companies or their suppliers specify the speed at which the material is to be examined – the slower the speed the better the chance of seeing all faults. The examiners also measure the length of fabric and put it on to rolls of the required size for despatch.

Examination of incoming fabric before coating and lamination may or may not be carried out; the responsibility for this is generally placed on the base fabric supplier. Automatic on-line examination systems can pick out basic construction faults, tears, creasing, resin add-on, etc., but more sophisticated apparatus is necessary to identify every possible fault, especially pattern distortions.

5.10 Investigation of faulty material and customer complaints

5.10.1 Introductory comments

This subject comprises three categories: investigation of a fabric fault arising from within the coating or laminating factory; investigation of material returned from the immediate customer such as the garment maker or auto seat maker; and last, investigation of a complaint from the general public – the ultimate customer. In the last two mentioned situations, the first step is actually to confirm the identity of the material – was it actually produced in *your factory*. Is the fabric construction, colour, fibre type and chemical type of coating within your range of products?

The competent technologist will confirm everything for himself or herself. The simplest – even bizarre or banal – reason may be responsible. Every factory manager has their own anecdote which may or may not be true, but can draw attention to what is actually possible. One incidence of mass contamination of coated material turned out to be due to the cleaner mistakenly putting his floor sweepings into a Banbury mixer, believing it to be a dustbin. In another incident, a short operative, not tall enough to empty the adhesive powder from the sack into the hopper of a screw extruder, put the plastic sack in together with the contents and extruded the plastic sack in addition to the hot melt adhesive. Frequently the technologist is handed a problem which has *just started*, the base fabric is the same and it has been processed under *exactly* the same conditions – absolutely nothing has changed! Unfortunately, costs are very tightly controlled in modern industry. If a supplier admits responsibility for a particular fault or evidence is available to show that they are responsible, debit notes or even litigation for damage costs may follow. Even different departments in the same factory, each with their own individual budget, may be reluctant to admit responsibility. These factors may make the investigating technologist or quality control manager's job a little more difficult. However, it is the responsibility of higher management to ensure that accurate information is available to technical staff, because the sooner a fault is actually cured, the better it will be for all concerned. If information is concealed, the problem is likely to recur, entailing more cost in lost material and staff time.

There is no substitute for actually seeing the process at first hand. However this should not just be a brief glance, but a thorough process, actually asking the operative what he or she actually does, irrespective of what is in the standard operating procedure. The first step is to confirm that the coating has actually been produced in the factory and that it is not a case of mistaken identity, the material in fact having been produced somewhere else. The technologist should confirm details and fabric description for him or herself, leaving nothing to chance. The starting point is fibre type; is it nylon 6, nylon 66 or polyester? These three fibres account for the base fabric of well over 90% of the more common coated fabric products. Melting point determination will identify nylon 6 which melts at around 215 °C – nylon 66 and polyester melt around 260 °C. Nylon 66 should be easily stained with an acid dye, but if the fabric is already dyed to a dark shade the outcome of this test may not be easy to establish, in which case the work should be given to a fibre identification laboratory.

5.10.2 Delamination

Probably the most common problem in coated and laminated fabric production is investigating the cause of delamination of the product or insufficient bond strength. The importance of clean fabric and the factors affecting adhesion have been discussed earlier, and efforts should be made to obtain a sample of the base uncoated fabric, ideally from the same piece as the coated material undergoing investigation. Usually this is difficult, and it is only possible to obtain a piece from a 'typical' production. This is better than nothing and it should be examined for excessive amounts of finish, although it must be said that it may be difficult to draw any firm conclusions from a 'one off' incident. A laboratory coating can be carried out on the material and tested to see if it will show the same fault, and to examine closely if the resin 'wets' the fabric without any difficulty. If the problem is major, affecting a large amount of fabric, or recurrent (as problems of this type usually are), a more detailed investigation is justified. In this case, the fabric finish or lubricant content of about ten base fabric samples from different production batches from the dyeing and finishing factory, should be determined to see if it varies significantly. This exercise at least establishes the 'typical' level and may be useful for further incidences of delamination. Sometimes it is possible to establish a relationship between fabric finish content and coating or lamination adhesion levels (peel bond). Estimation of lubricant level needs careful consideration, however, prolonged extraction using solvent in a soxhlet apparatus (similar to a reflux condenser) will extract everything on the fabric, including oligomer in the case of polyester, and also some dyestuff. There are milder methods of doing this, but it may be worthwhile developing a simple 'in house' method, which is specific to a particular set of circumstances and requirements. A very simple technique is to wash 5 g samples of fabric in water (made alkali with 2 g/l sodium carbonate) for three minutes at 100 °C, weighing before and after. Care must be taken not to lose fabric by fraying, and weighing must be done after careful drying at room temperature in a conditioned atmosphere. The difference in weight is the 'level' of lubricant. Drying should not be carried out in an oven in this case, because heating in an oven will drive off further finishes and lubricants, depending on the temperature and time in the oven. Weighing samples before and after oven treatment at specific temperatures can be the basis for alternative 'in house' tests.

Fabric, especially polyester, generally requires some lubricant or anti-static finish on it for it to be processed efficiently. Static is annoying and can be quite painful to operatives, but it can also influence quality and production efficiency. For example, fabric charged with static can cling to the sides of machines causing creasing (and can also attract dust). Excessive anti-static agent is quite likely to cause poor adhesion or delamination, and it has been known for a finishing factory to increase anti-static finish levels to reduce static, thus giving rise to problems of delamination – curing one problem whilst causing another! In most cases, only a fairly low level of anti-static agent is necessary to overcome static problems, and this low level should have minimal effect on peel bonds. The best solution to the problem, however, is installation of anti-static bars after every main roller in both the finishing works and the coating and lamination plant. This can be expensive, but the most efficient bars can be restricted to those points where they are most needed, i.e. just before lamination rollers.

Lubricant levels can sometimes be checked very quickly by measuring the surface resistivity with a static meter. Many fabric lubricants are hydrophilic or, being water based, contain some hydrophilic material in them. Anti-static agents are, of course, strongly hydrophilic. The result obtained can be anything between approximately 1×10^{13} to 1×10^{6} ohms. A high level, say 1×10^{11} and above probably indicates low lubricant content, while very low levels, say 1×10^{8} and below, show a high anti-static finish level. The simplest way of *comparing* fabric finishes is by dripping water on the fabric surface. The result of course depends on the finish itself – it may be hydrophilic or hydrophobic – but this very simple test is sometimes useful because it establishes a *difference* between two fabrics which should be the same or are claimed to be the same. Another, very simple qualitative test is to try to stick office adhesive tape on to the fabric being examined and compare the ease with which it sticks with a fabric known to have given excellent or satisfactory peel bonds. The laws of adhesion apply to all adhesives, and if there is a problem sticking office adhesive tape to a particular fabric, then there could well be a problem in coating or lamination production.

In the case of a coating made up from two or more layers, the first step is to examine whether the delamination is actual separation from the base fabric or if it is interlayer delamination, i.e. one layer of coating separating from another. In the latter case, the reason is probably insufficient heat in application of the final layer, i.e. not a high enough temperature, or running the machine too fast which reduces dwell time. This particular problem is usually associated with poorer abrasion resistance, because the resin has not had sufficient heat to crosslink properly. Have the correct machine settings, as specified by the technical department, been used? Was there a stoppage or a breakdown and was coating or lamination restarted (owing to production pressure) before the ovens reached the correct temperature? Are all of the oven bays working correctly? If these enquiries draw a blank, attention must now turn to the compound itself. Did it contain everything specified in the correct amounts – especially the crosslinking agent or catalyst (if they were specified)? Other factors which may cause poor adhesion are the substrates being disturbed or moved before the bond is formed – a roller or rollers not running true may contribute to this. Nip rollers may not be parallel or pressure may vary across the width.

5.10.3 Contaminants

Another frequent problem is contaminants or marks on the coated fabric. These problems are usually due to lapses in housekeeping – but the technologist is usually asked to trace the source. Sometimes it can be established that soiling – frequently oil marks – was already on the fabric before it arrived in the coating or lamination department. Coating or laminating over an oil mark is likely to lead to poor adhesion in that particular area, and examination of peel bond could be a way of providing evidence that soiling was already on the fabric beforehand. Solid contaminants are also a problem, but are generally more easy to investigate – is it solid resin or some other substance? Chemical analysis should provide the answer quite quickly, the problem is usually explaining how it got there. Different samples of oil taken from different parts of the factory, including the weaving or knitting department, can be characterised by IR spectroscopy, or perhaps a chromatography technique. Comparing an analysis of oil taken from a contaminated area with the reference samples should quickly identify the origin. This could be extended to sampling oils in supplier factories.

Solid resin could be the deposits which sometimes build up on the reverse side of a doctor blade or on the insides of tubing or piping. Water-based resins are easy to clean until they dry, and then the only effective way is by manual cleaning. Deposits can build up unnoticed in pipework until they become sufficiently large to breakaway of their own accord. 'Foreign objects' such as fabric 'fly' or dust can cause problems, especially if there is a draught in the area. Resin can sometimes get on to the stenter brushes which are used to push fabric edges on to the stenter pins. This resin will dry, build up on the brushes and eventually become detached as a powder or small deposits; these will fall on the surface of the coated fabric as the brushes rotate. During flame lamination, ash may build up on the burner and drop on to the fabric. Some qualities of foam are more prone than others to causing this and, if this is known, suitable precautions should be taken.

5.10.4 Related faults

Many fabric faults are related to each other. In coated water-resistant protective clothing fabric, poor hydrostatic head is usually associated with low resin add-on, delamination or poor abrasion resistance of the coated surface. The same factor – insufficient heating in the last layer of coating – could cause all three faults. In flame lamination of automotive fabrics, laminates which are thicker than normal are frequently associated with poor peel bond, because of low burn off or ineffective nipping together. Stiffer than normal laminates are sometimes associated with discoloration or, if it is a raised fabric, with pile crush. All of these faults could be related to excessive heat. Sometimes during wear, a garment is rejected because of leaking through a seam which has not been taped properly. This may be due to some contaminate on the coating surface, or perhaps to some additive included in the top layer as a detackifier but present in excessive amounts. As a matter of interest, it can sometimes be difficult to establish where the water is actually coming into a garment. Sometimes a tape may be leaky at the top of a garment, but the water appears near the bottom, having run down the seam. One garment manufacturer has referred to this as the 'drain pipe effect'.

5.11 References

1. Slater K, 'Comfort or protection; the clothing dilemma', *Fifth International Symposium on Performance of Protective Clothing*, San Francisco 25–27 January 1994. Papers published by ASTM as *Performance of Protective Clothing; Fifth Volume (STP 1237)*, West Conshohocken PA, USA 1996, 488.
2. Merkel RS, *Textile Product Serviceability*, MacMillan, New York 1991, 77–91.
3. Topf W, 'Quality management for the textile industry', *JCF*, Vol 25 Apr 1996, 285–300.
4. Taylor HM, 'Physical testing of textiles', *Textiles*, Vol 14 No 1 1985, 21–8.
5. Eichert U (AKZO), 'Weather resistance of coated fabrics for the automotive industry', *IMMFC*, Dornbirn 17–19 Sept 1997.
6. Norris CA, 'The testing of water repellent finishes on textiles', in *Water-proofing and Water-repellency* (Editor Moilliet JL), Elsevier, Amsterdam 1963, 278. Original paper, Wenzel RN, *American Dyestuff Reporter*, Vol 25, 1936, 598.

7. Kawabata S and Masako N, 'Fabric performance in clothing and clothing manufacture', *J Textile Institute*, Vol 80 No 1 1989, 19–50.
8. Kim CJ, 'The Kawabata System use in the fabric hand and evaluation of automotive textiles', IFAI Symposium, *2nd International Conference on Automotive Test Procedures*, Atlanta GA, USA 1992.
9. Harwood RJ, Weedall PJ and Carr C, 'The use of the Kawabata evaluation system for product development and quality control', *JSDC*, Vol 106 Feb 1990, 64–8.
10. Stearn AE, D'Arcy RL, Postle R and Mahar TJ, 'Statistical analysis of subjective and objective methods of evaluating fabric handle', *Journal of the Textile Machinery Society of Japan*, Vol 34 No 1 1988, 13–18.
11. Hearl JWS, 'Can fabric enter the dataspace?' Part 1, *Textile Horizons*, Apr 1993, 14–16 and Part 2, *Textile Horizons*, Jun 1993, 16–20.
12. Stewart A, 'Understanding hypothermia', *The Great Outdoors*, Dec 1985, 24–7.
13. Keighley JH, 'Breathable fabrics and comfort in clothing', *JCF*, Vol 15 Oct 1985, 89–104.
14. Holmes DA, 'Waterproof breathable fabrics', in *Handbook of Technical Textiles* (Editors Horrocks AR and Anand SC), Woodhead, Cambridge 2000, 461–89.
15. Van Roey M, 'Water-resistant breathable fabrics', *JCF*, Vol 21 Jul 1991, 20–31.
16. Umbach KH, *The Comfort of Clothing*, Hohenstein Institute Technical Report, 1985.
17. Gretton JC, 'MVT – a continuing controversy', *World Sports Activewear*, Autumn 1997, 34–9.
18. Overington Y-H and Croskell R, 'Standards for breathable fabrics', *International Dyer*, May 2001, 23–7. A similar paper by same authors also appears in *TTi*, Apr 2001, 9–13.
19. Congalton D, 'Heat and moisture transport through textiles and clothing ensembles – utilising the Hohenstein skin model', *JCF*, Vol 28 Jan 1999, 183–96.
20. Ruckman JE, 'Water vapour transfer in waterproof breathable fabrics' Part 1, *International Journal of Clothing Science and Technology*, Vol 9 No 1 1997. Part 2 detailing 'Windy conditions' in same journal 23–33.
21. Anon, 'Breathability testing – SATRA', *Textile Month*, May/Jun 2001, 13.
22. Milligen B, 'The degradation of automotive upholstery fabrics by light and heat, *Review Progress Coloration*, Vol 16 1986, 1–7.
23. Horsforth A, 'One big technical headache on wheels', *JSDC*, Vol 108, May/Jun 1992, 243–6.
24. Parsons MA (Rover), 'Fabric requirements for automotive use', *Autotech*, Seminar 9, NEC, Birmingham 1991.
25. Kowalski M (Guildford), 'Automotive fabric presention', *Autotech*, Seminar 9, NEC, Birmingham 1991.
26. McCallum JB (Ford), 'Engineering requirements for automotive textiles', in *Automotive Textiles* (Editor Ravnitzky M), PT-51 SAE, SAE Inc, Warrendale PA, USA 1995.
27. IWS Interior Technical Information Letter No 31, *Technical Specifications for Automobile Industry*, Aug 1986.
28. IWS Interior Technical Information Letter No 32, *The Styling and Performance of Wool Automotive Upholstery*, Sept 1986.

29. Fung W, 'Present and future technical requirements of automotive fabrics', *Industrial, Technical and High Performance Textiles,* World Textile Congress, Huddersfield University 15–16 Jul 1998.
30. Pickett D (General Motors), 'Challenges for 10 year automobile textile service durability', *IMMFC*, Dornbirn 17–19 Sept 1997.
31. Rink GS (Opel), 'Harmonisation of textile testing in the German automotive industry', *IMMFC*, Dornbirn 20–22 Sept 1995.
32. Weber E (Opel), 'Harmonisation of test procedures of automotive textiles; influence of globalisation', *IMMFC*, Dornbirn 15–17 Sept 1999.
33. Card AH, 'The road to harmonisation', *Financial Times World Automotive Manufacturing Monthly Analysis*, Issue 2 Jun 1998, 4.
34. Butts K (Datacolor), 'Colour tolerances for consistent pass/fail decisions', *AATCC Symposium, Yarn Dyeing '96, Meeting the Challenges*, Sunset Beach NC 18–19 Apr 1996, AATCC, Research Triangle Park NC, USA 1996.
35. Wootton A, 'Light fastness and weathering tests in the automotive industry', *JSDC*, Vol 108 May/Jun 1992, 239–42.
36. Hibbert M, 'Throwing light on automotive trim testing', *JSDC*, Vol 108 May/Jun 1992, 253–5.
37. Watanabe Y (Nissan), 'Estimation of the light durability of automotive interior materials with the value of equivalent total sunshine energy', in *Automotive Textiles* (Editor Ravnitsky M), PT-51 SAE, SAE Inc, Warrendale PA, USA 1995, 13–20.
38. Bird LA (Heraus), 'USA automobile testing yesterday, today and tomorrow', in *Automotive Textiles* (Editor Ravnitsky M), PT-51 SAE, SAE Inc, Warrendale PA, USA 1995, 31–8.
39. Smith TL, 'Taking the heat', *Automotive and Transportation Interiors*, June 1996, 32–5.
40. Park J, 'Assessment of fastness properties', *Review Progress Coloration*, Vol 10 1979, 20–24.
41. Midwest AATCC Section Committee, 'Accelerated lightfastness testing of disperse dyes on polyester automotive fabrics', *Textile Colorist & Chemist*, Vol 25 No 12 Dec 1993, 25–32.
42. Bachor M, Lampe T and Brinkmann A (VW), 'Foreign pilling of car upholstery caused by external influences such as clothing materials', *Textiles in Automobiles*, VDI Congress, Dusseldorf 14–15 Oct 1992.
43. Francke G and Henkel A, 'Friction wear on car upholstery materials, causes and possibilities of avoiding it', *Textiles in Automobiles*, VDI Congress, Dusseldorf 14–15 Oct 1992.
44. Hurten J, 'How pilling in polyester weaves can be controlled in the finishing process?', *Textil Praxis International*, Vol 33 1978, 823–36.
45. Bosch M, 'Pilling on textiles, fundamentals, extent of influence and test procedures', *Textiles in Automobiles*, VDI Congress, Dusseldorf 14–15 Oct 1992.
46. Baetens E and Albrecht E, 'Fogging characteristics of automotive textiles', *Techtextil*, Frankfurt 14–16 May 1991.
47. Baetens E and Albrecht E, 'Reducing the fogging effect in cars', *TuT*, Vol 4 1992, 42–4.
48. Hardt P, 'Estimation of the amounts of volatile substances as applicable to stenter frames for automotive fogging', *IMMF*, Dornbirn 22–24 Sept 1993.

49. Behrens W and Lampe T (VW), 'Fogging behaviour of textile materials', *Textiles in Automobiles*, VDI Congress, Dusseldorf 30–31 Oct 1991.
50. Ehler P, Schreiber H and Haller S, 'Emissions from textiles in vehicle interior trim – causes and assessment of short term and long term fogging', *IMMFC*, Dornbirn 22–24 Sept 1993.
51. McCallum JB, 'Ford Motor Co develops its own test method for predicting light-scattering window film', *Textile Colorist & Chemist*, Vol 21 No 12 Dec 1989, 13–15.
52. Behrens W, 'Fogging behaviour of car interiors', *Technische Textilen*, Vol 36 Mar 1993, E25–E27.
53. Anon, 'Kanebo car seat cloth', *JTN*, Jun 1986, 59.
54. Anon, 'Kanebo, Bellatron conductive fibre', *JTN*, May 1991, 86–7.
55. Lennox-Kerr P, 'Reducing static shock', *Inside Automobiles International*, Oct 1998, 20–21.
56. Hall P, 'Motoring', *Daily Telegraph*, 14 May 1994, 15.
57. Finke H, 'Antisoiling treatment of automotive fabric from the viewpoint of the user', *Textiles in Automobiles'*, VDI Congress, Mannheim 25–26 Mar 1998.
58. Hilden J, Msiyah M and Niederrhein FH, 'Antisoiling finish for polyester car upholstery, effects and problems', *IMMFC*, Dornbirn 15–17 Sept 1999.
59. Bollinger H and Duwel KR, 'New concepts in seating', *Textiles in Automotives*, VDI Congress, Dusseldorf 30–31 Oct 1991.
60. Knozinger GT, Theysohn H and Vogt H, 'Physiology of seat comfort', *Textiles in Automotives*, VDI Congress, Dusseldorf 30–31 Oct 1991.
61. Fung W and Parsons KC, 'Some investigations into the relationship between car seat cover materials and thermal comfort using human subjects', *JCF*, Vol 26 Oct 1996, 147–76.
62. Fung W, 'How to improve thermal comfort of the car seat', *JCF*, Vol 27 Oct 1997, 126–45.
63. Umbach KH, 'Parameters for the physiological comfort of car seats', *IMMFC*, Dornbirn 15–17 Sept 1999.
64. Bartels VT and Umbach KH, 'Laboratory tests of thermophysiological seat comfort', *Comfort in the Automotive Industry*, ATA Symposium, Bologna 6–7 Nov 1997, ATA/Bologna University.
65. Anon (input from GM, Ford and VW), 'Measuring seat comfort', *Automotive Engineering*, Jul 1993, 25–30.
66. Luessmann-Geiger H (Audi), 'Emissions from vehicle interior components containing textiles and how to reduce them', *IMMFC*, Dornbirn 15–17 Sept 1999.
67. Kennerley S, Jones D and Gagne W (Aromascan), 'Electronic nose helps quantify odours in PU foams and raw materials', *Urethanes Technology 1998*, Oct/Nov 1998, 40–42.
68. Sassmannshausen J, 'The electronic nose: more than just an instrument for objective olfactory', *Textiles and Surfacing Materials in Automotive Engineering*, VDI Congress, Mannheim 25–26 March 1998.
69. Eaton PM (BTTG), 'Flame-retardancy test methods for textiles', *Review Progress Coloration*, Vol 30 2000, 51–62.
70. Paul KT (RAPRA), 'Fire testing of upholstered furniture, current and possible future test methods', *Textile Flammability; Current and Future Issues Conference*, Textile Institute, Manchester 30–31 Mar 1999.

71. Johnson D, 'Textile flammability – where are we!!', *Textile Flammability; Current and Future Issues Conference*, Textile Institute, Manchester 30–31 Mar 1999.
72. Barrow CC, 'Standards for textiles used in commercial aircraft', *Textile Horizons*, Apr/May 1992, 30–34.
73. Benisek L, 'Innovation in flame resistant wool transportation furnishings', *Textile Asia*, Aug 1998, 36–42.
74. Benisek L, 'Burning issues'(TI Flammability Conference, Salford), *Textile Month*, Jul 1999, 19–23.
75. Bagnall J, 'Testing the reaction of textiles to fire', *Textiles Magazine*, Issue 4 1995, 12–17.
76. Troitzsch J, *International Plastics Flammability Handbook*, Second Edition, Hanser, New York 1989, 299–310.
77. Jones HR, 'Textiles in the railway passenger environment', *Flammability*, BTTG Conference, London 1–2 Dec 1993, BTTG, Manchester.
78. Horrocks AR, 'Flame retardant finishing of textiles', *Review Progress Coloration*, Vol 16 1986, 62–101.
79. Eichinger H (DuPont), 'Certification testing for chemical protective clothing', *IMMFC*, Dornbirn 13–19 Sept 1997.
80. Wadsworth LC and Allen HC, 'Development of highly breathable and effective blood/virial barrier laminates of microporous films, staple fibers and nonwovens', *JCF*, Vol 28 Jul 1998, 12–28.
81. Raheel M, 'Chemical protective clothing', in *Protective Clothing Systems* (Editor Raheel M), Marcel Dekker, New York 1994, 39–78.
82. Stull JO, 'Assessment of chemical barrier properties', in *Modern Textile Characterisation Methods* (Editor Raheel M), Marcel Dekker, New York 1996, 393–468.
83. Stull JO, 'A review of the ASTM F1001 battery of chemicals and its effect on the chemical protective clothing industry', in *Performance of Protective Clothing; Fifth Volume (STP 1237)*, ASTM Publications, West Conschohocken PA, USA 1996, 110–22.
84. Brown PL, 'Assessment of the protective properties of textiles against microorganisms', in *Modern Textile Characterisation Methods* (Editor Raheel M), Marcel Dekker, New York 1996, 469–550.
85. Anon, *BTTG Independent*, April 2000, 9.

5.12 Further reading

1. AATCC, *AATCC Technical Manual*, AATCC, Research Triangle Park/NC, USA 1995.
2. ASTM, *Annual Book of ASTM Standards* (Vol 7.01/7.02 which relate to textiles), ASTM, Philadelphia PA, USA.
3. ASTM, *Annual Book of ASTM Standards* (Vol 9.02 which relates to rubber products), ASTM, Philadelphia PA, USA 1999.
4. Booth JE, *Principles of Textile Testing*, Third Edition, Heywood, London 1968.
5. *Comfort in the Automotive Industry*, ATA Symposium, Bologna 6–7 Oct 1997, ATA/Bologna University.
6. Fourt L and Hollies NRS, *Clothing, Comfort and Function*, Marcel Dekker, New York 1970.

7. Fung W and Hardcastle M, *Textiles in Automotive Engineering*, Woodhead, Cambridge 2001.
8. ISO 9000 series. *Quality management systems*, 1994.
9. Johnson JS and Mansdorf SZ (Editors), *Performance of Protective Clothing; Fifth Volume (STP 1237)*, ASTM Publications, West Conshohocken PA, USA 1996.
10. Kothari VK, *Testing and Quality Management* (Progress in Textiles, Science and Technology), IAFL Publications, New Delhi 1999.
11. McDonald R (Editor), *Colour for Industry*, Second Edition, Society of Dyers and Colourists, Bradford 1997.
12. Mahall K, *Quality Assessment of Textiles; Damage Detection by Microscopy*, Springer-Verlag, Berlin 1993.
13. Merkel RS, *Textile Product Serviceability*, Macmillan, New York 1991.
14. *QS9000*, Third Edition available in North America from AIAG on 01-248-358-3003; outside North America from Carwin Continuous Ltd, West Thurrock, Grays, UK. +44 (0) 1708 861333.
15. Raheel M (Editor), *Modern Textile Characterisation Methods*, Marcel Dekker, New York 1996.
16. Ravnitsky M (Editor), *Automotive Textiles*, SAE PT-51, SAE Inc, Warrendale PA, USA 1995.
17. Ross JE, *Total Quality Management*, Kogan Page, London 1994.
18. *SAE Automotive Textiles and Trim Standards Manual*, SAE HS-2700, SAE Inc, Warrendale PA, USA 1996.
19. Saville BP, *Physical Testing of Textiles*, Woodhead, Cambridge 1999.
20. Scott JR, 'Testing procedures and standards', in *Rubber Technology and Manufacture* (Editor Blow CM) Newnes-Butterworths (Plastics and Rubber Institute) London 1971, 446–74.
21. Slaten LB and Hall DM, 'Textiles testing', in *Wellington Sears Handbook of Industrial Textiles* (Editor Adanur S), Technomic, Lancaster PA, USA 1999, 651–712.
22. Slater K, *Physical Testing and Quality Control*, Textile Progress, Vol 23 1–3, Textile Institute, Manchester 1993.
23. Smith LP, *The Language of Rubber*, Butterworth Heinemann (in association with DuPont), Oxford 1993.
24. Taylor MA *Technology of Textile Properties*, Third Edition, Forbes, London 1990, 163–244.
25. Wetherill GB and Brown DW, *Statistical Process Control*, Chapman and Hall, London 1991.

6
Coating and laminating effects on the environment

6.1 Introduction

All human activity has some effect on the environment and the coating and lamination industry is no exception. The main effects are effluent and emissions to atmosphere during manufacture and eventual disposal of the finished article. The problems of effluent and polluting emissions to atmosphere, including VOCs, are gradually being addressed, with more effective methods of abatement and more environmentally friendly materials and methods of processing, but there is still much to be done. Coating and laminating factories are generally service industries to others, such as the clothing, automotive, furniture and carpet industries, and as such may not be directly involved in recycling and disposal issues. However, we are all living on the same planet, and the environmental issue of global warming, together with the effects of climate change and rising sea levels, affects everybody and should be the concern of all. In addition, certain chemicals are known to be harmful to human health, and others are destroying the ozone layer in the upper atmosphere. This high level ozone filters out the harmful UV radiation which contributes to skin cancer. These too are survival issues, which must be addressed.

All industrial processes require energy, generated chiefly by the burning of coal, gas or oil. All are non-renewable resources and all produce carbon dioxide and, especially in the case of coal, other harmful emissions. Many raw materials used in the coating and lamination industry, both base fabrics and coating polymers, are derived from oil. Coating and laminating thus have an effect on the environment by the depletion of non-renewable resources and by the generation of carbon dioxide, the main ‘greenhouse’ gas causing global warming, see Section 6.2.1.

There was a sharp increase in public awareness of environmental issues during the 1970s and 1980s, and this has been reflected in increasingly stringent UK legislation. In addition, the EU, since the early 1970s when it first adopted a formal environmental policy, has passed well over 200 laws, and further impetus was applied by the 1987 Single European Act. This act demands that full account of the environment and of the principle of 'sustainable development' is taken whenever Community policy is put forward for legislation. The term 'sustainable development,' which can be defined as 'the objective of meeting the needs of the present without compromising the ability of future generations to meet their own needs', was first put forward in the 1987 Brundtland Report, *Our Common Future*. This document was produced by a UN Conference on Environment and Development (UNCEO) and was chaired by Gro Harlem Brundtland. A European Environmental Agency has been set up in Copenhagen to provide independent, reliable, objective and comparative information on the state of Europe's environment. This unit publishes periodic reports on the state of the environment, and is examining a range of issues including air quality and nature conservation. The agency is unusual for an EU institution because it includes some non-member countries such as Norway and Switzerland.

Both the EU and the UK government have published reports defining their interpretation of and their strategies for achieving sustainable development. Most large companies want to appear environmentally friendly and socially responsible and to show that they are in agreement with the principle of sustainable development. Indeed, some have declared ethical policies, and many now issue annual environmental reports and are either ISO 14001 certified or EMAS (Environmental Management Audit Scheme) accredited, or are working towards this. In addition, an increasing number are assuming 'product stewardship' responsibility for the impact of their products on the environment based on life cycle analysis (LCA). The general public worldwide is now aware of, and concerned about, the environment, and evidence of this is the existence of over 2600 environmental organisations in more than 200 countries of the world.

As living standards increase, the populations of the world have higher disposable incomes, more leisure time and more consumer goods will be produced. There is likely to be increased depletion of the world's resources and more waste generated. Environmental problems are going to intensify, because increased living standards produce more pollution as summarised by Meadows and his co-workers[1] in the equation:

$$\text{impact on environment} = \text{population} \times \text{affluence} \times \text{technology}$$

All these three dependent factors are increasing, and without drastic action an environmental disaster seems inevitable. The management of waste is a formidable problem, and legislation has been passed to control means of disposal. However, the solutions are by no means clear, and debate continues about the best overall answers, taking into consideration environmental impact, effect on human health and social and economic factors. The existing situation in each country of the world is different, and each has its own particular needs. However, the overall guiding principle, agreed by everyone, to protect the environment is to 'reduce, re-use, repair or recycle', and actual disposal of waste should be a last resort. The issues of recycling and methods of disposal involve controversial factors that need careful examination.

6.2 The effects of pollution

6.2.1 The greenhouse effect and global warming

Human life depends on the 'greenhouse effect', which is caused by the presence of certain gases and water vapour in the atmosphere. If they were not there, the surface of the Earth would be about 30 °C lower than it actually is. Small changes in their concentration are believed to cause changes in the surface temperature of the Earth. It is a very delicate balance, because an increase in surface temperature of only 1 °C is thought likely to change the world's ecology and cause very significant changes in climate and sea water levels.

The greenhouse effect happens because the Earth's atmosphere and surface retain heat. The atmosphere as a whole, including greenhouse gases, allows sunshine, made up of both short and longer wavelength radiation, to pass through to heat the Earth's surface. The warm Earth then radiates heat back, but this new heat radiation is at a longer wavelength, some of which cannot pass back out through the greenhouse gases to outer space. Instead, the greenhouse gases are heated up by this radiation energy, thus retaining the heat on the Earth and giving rise to global warming, see Fig. 6.1. The main greenhouse gases are carbon dioxide, methane, nitrous oxide, ozone and CFCs (fluorine compounds used in aerosol sprays and refrigeration but now banned by the Montreal Protocol). Particulate matter also contributes to global warming.

The amount of carbon dioxide in the atmosphere is increasing at the rate of about 0.5% every year, and motor vehicles are responsible for 15% of the world's total carbon dioxide emissions. Research shows that, since the middle of the nineteenth century, increased amounts of greenhouse gases have raised the average global temperature by about 0.5 °C. If greenhouse gas levels continue to increase at the rate expected, a further warming of about 1.5 °C within the next 40 years could be caused, with serious effects on global climate.[2] Global warming, in reality excess global warming, is therefore the concern of every nation on the Earth. Some researchers believe that the process has already started with extreme and unusual weather, such as the severe winter in Canada during 1997/98, the damaging

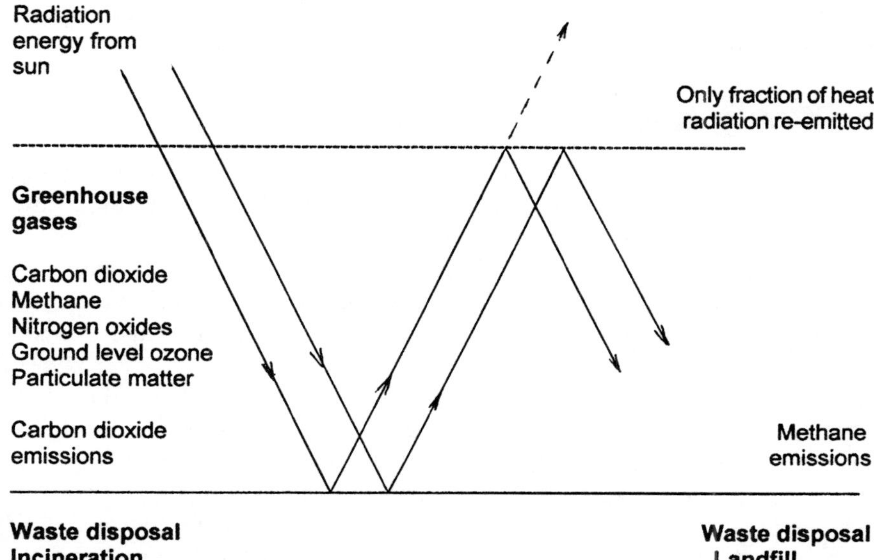

6.1 The 'greenhouse effect'. Global warming causes changing weather patterns and rising sea levels. Incineration of solvent fumes and emissions from coating and lamination operations must be carefully controlled and abatement of fumes is essential. All incineration processes, including municipal waste disposal, produce carbon dioxide, the main 'greenhouse' gas.

storms in Western Europe in 1999 and the heavy rain causing flooding in Africa and England in 2000 and 2001.

World leaders from 160 countries met in 1992 at Rio de Janeiro to discuss global warming. At a second meeting held at Kyoto in 1997, the general target was set for developed nations to reduce their greenhouse gas emissions by 5% below 1990 levels by the years 2008–2012.[3] Within the same time frame, the EU was committed to reducing emissions of six greenhouse gases by 8%, and the UK agreed to reduce its emissions by 12.5%. Although carbon dioxide is the chief greenhouse gas, responsible for approximately 75% of global warming, methane, nitrous oxide and ozone, in fact, have a far greater effect than would be expected from their relatively low concentration.[4] Methane gas is 30 times more effective at retaining heat in the atmosphere than carbon dioxide and accounts for 20% of global warming. Some CFCs are 17000 times more effective than carbon dioxide, and if their use is unchecked they could account for 33% of global warming by the year 2030. To add to this, VOCs under the action of sunlight can react with nitrogen oxides to produce photochemical oxidants which have greenhouse effects. The control and reduction of greenhouse gases is, therefore, one of the most important issues that has ever faced the human race.

6.2.2 Air pollution

Harmful gases or vapours and solid particulate matter cause air pollution. There are many air polluting gases which are harmful to human, animal and plant life, but the main ones are oxides of sulphur, carbon monoxide, hydrocarbons and oxides of nitrogen.[5–7] The last three are present in significant concentrations in car exhaust fumes, although sulphur is being phased out of petrol and diesel. Particulate matter is also present in car exhaust fumes, especially those from diesel engines. The catalytic converter has reduced pollution, but older cars not equipped with them are still on the road. Oxides of sulphur, produced mainly from industrial burning of fossil fuels, give rise to acid rain; this is harmful to human health and also damaging to agricultural crops, vegetation and materials including buildings. Oxides of nitrogen contribute to acid rain, and produce adverse physiological conditions including eye and throat irritation if present in high concentration. Carbon monoxide is toxic to human and animal life, because it combines with haemoglobin in the blood resulting in less oxygen being carried to body organs, heart patients being especially vulnerable. Hydrocarbons, VOCs and oxides of nitrogen are the main constituents of traffic smog and under the action of sunlight they produce ozone, which is an eye and throat irritant. Their ability to create ozone is measured by their photochemical ozone creation potential (POCD). This ozone should not be confused with the UV filtering ozone layer, which is several miles above the Earth's surface. In addition, all combustion processes involving organic compounds produce carbon dioxide, the chief greenhouse gas.

6.2.3 Particulate matter

Particulate matter is mainly soot (carbon) in a wide range of particle sizes, but the most harmful to health are those which are smaller than 10 micrometres in diameter.[8–11] These small particles do not settle rapidly as dust, but stay in the air for long periods and are, therefore, more available for breathing into the human body. The American Environmental Protection Agency classified these particles as PM10s (particle matter less than 10 μm), and PM10 pollution is the total weight in micrograms of all particle matter of 10 μm in diameter or less contained in one cubic metre of air. In the USA the legal limit is 50 μm of PM10 per cubic metre of air as an average in a year.[10] Many cities of the world have peak PM10 values of 100–200. Particle size is important because particles larger than 10 μm penetrate no further than the nose or throat, but those smaller than 5 μm can enter the bronchial tubes at the top of the lungs, and the smallest, 2.5 μm and less, can penetrate into the deepest (alveolar) lung tissue where gas exchange occurs between the air and blood. If soluble, the particle matter enters the blood stream; if not soluble, it may remain in place for months or years. Particle matter is believed to be the cause of bronchitis, asthma, other chest and breathing conditions and to contribute to lung cancer. In the UK motor vehicles are responsible for 26% of PM10s.[12]

6.2.4 Land and water

Pollution of waters by the textile industry is caused by manufacturing processes such as dyeing and finishing and the washing of pesticides from cotton. Toxic liquors will harm marine life and could be harmful to humans on beaches. Pollutants are broken down chemically and biologically in the water, but both mechanisms require action by the dissolved oxygen in the water. This reduces the amount of oxygen available for fish and other marine life. Measures of these effects are the oxygen demand indices, biological oxygen demand (BOD) and carbon oxygen demand (COD).

Disposal of industrial waste by landfill is one of the least attractive options for a number of reasons, not least the appearance of such sites. Decay of organic substances results in the formation of methane gas which contributes to global warming. There is also the risk of explosion and of uncontrollable fire. Leachate from landfill could carry toxic materials including heavy metals into underground waters, which may in turn contaminate rivers and even drinking waters. Heavy metals such as lead and mercury accumulate in the body until present in sufficient quantity to act as poisons. Lead is believed to cause damage to the brain and to the central nervous system, mercury also affects the central nervous system, while cadmium is believed to produce adverse bone conditions.[13,14] Heavy metals were once widely used in PVC stabilisers and, although being phased out, they are present in articles which are still in use and which will need to be disposed of at the end of their working life.

6.3 Environmental legislation

The earliest environmental laws in the UK were passed during the thirteenth century to control the burning of coal. However, times change and accepted standards become more demanding with the increase in the quality of life. Environmental regulations have become stricter to meet new circumstances and, in addition to UK law, there is now EU legislation, which takes the form of directives and regulations. Directives are guidelines to be used by individual countries to form their own laws, while regulations are themselves legislative acts which apply across the EU without further action by individual countries. The EU in 1973 (then called the EEC) started the mechanism for programmes of action to reduce pollution and nuisances, to tackle environmental problems caused by depletion of natural resources, to promote education and awareness of environmental problems, and also to improve the natural and urban environment. Since this date the EU has been one of the leaders in the world community in protecting the environment.

There are also UN environmental initiatives, through their Environmental Programme (UNEP – founded 1973), which include the Montreal Protocol in 1987 to protect the ozone layer, and the decisions taken at the 'earth summits' at Rio de Janeiro and Kyoto. The UN was also instrumental in setting up the Convention on long-range transboundary air pollution, adopted in Geneva during 1979 and which came into force in 1983, to reduce transboundary air pollution over Europe and North America. The main concern at the time was acid rain over Scandinavia, but the Convention now also targets other pollutants including heavy metals and VOCs which have transboundary effects. Some scientists believed that emissions from UK power stations contributed to the Scandinavian acid rain.

In addition to EU and UN initiatives, there have been since 1984 meetings of countries which border the North Sea to discuss steps to reduce the build up of toxic chemicals and other pollutants in this area.[15] Measures agreed at the Third North Sea Conference in 1990, to cut levels of a number of specific chemicals and to treat sewage more effectively before discharge, became legally binding. Switzerland, which has industries along the Rhine which flows into the North Sea, attended this conference. A similar arrangement exists for the Mediterranean Sea with the issuing of the Barcelona Protocol in 1980; this banned the release of toxic chemicals such as mercury and cadmium and restricted other potentially harmful chemicals.

6.3.1 United Kingdom laws

In recent times, new laws have been implemented which update, but do not always completely replace, earlier ones. Thus a process may be governed by a series of laws applying to different aspects. The 1974 Control of Pollution Act was one of the most important environmental laws passed, although key aspects covering effluent discharges only came into force during 1985. The National Rivers Authority (NRA) created by the 1989 Water Act took over the regulatory duties of the individual (soon to be privatised) water companies. This new body tightened up on regulations, added further restrictions and set up mechanisms to monitor and control effluent discharges into inland surface waters, estuaries, coastal waters and underground waters. Clearly defined and absolute limits were set for pollutants, together with other actions to implement the new requirements which included a named manager responsible for effluent control in each organisation.[16] Much of the 1989 Act was re-enacted by the Water Resources Act and the Water Industry Act which, together with three further Water Acts in 1991, consolidated all the previous legislation involving water. The Water Acts formalised the principle of 'the polluter pays' which was a condition of the Treaty of Rome (a forerunner of the EU), but was first mentioned at a UN conference on the Human Environment in Stockholm in 1972. The NRA has also implemented European laws and North Sea conference decisions.

The Environmental Protection Act (EPA), which was first placed on the UK Statute Book during November 1990, set up an industrial pollution control system which included 'Integrated Pollution Control' (IPC) for the 5000 potentially most polluting industrial factories. The act applied to all major solid, liquid and gaseous emissions to land, waters and air. A timetable for gradual implementation was drawn up which began on 1st April 1992 with the fuel and power industries, and progressed through to November 1995 when it was extended to all other industries, including textiles. Her Majesty's Inspectorate of Pollution (HMIP) ran IPC, and all existing operators had to register and were given between three and eight years to bring their plant up to the new standards or face closure. After January 1991 all new processes and significantly modified processes prescribed for IPC had to register immediately for authorisation.

Air pollution control (APC) also became law as part of the 1990 EPA to control smaller scale polluters, which were to be regulated by local authorities. Industries such as textile finishing, coating and lamination and solvent joining operations generally came under this category. Thus the EPA was implemented by two mechanisms: IPC controlled by HMIP for large scale polluters which had national implications (so called Part A) and APC for smaller polluters regulated by local authorities so that any relevant local factors could be taken into consideration (so called Part B). Both systems applied the same regulations and used the same guidance notes. Applicants for authorisation for a process had to show that attempts had been made to prevent or minimise emissions or to render them harmless using the 'best available techniques which do not entail excessive cost' (BATNEEC). These regulations applied to both Part A and Part B processes, and to all operations including design, staff training and qualifications, operating procedures and emission levels. The interpretation of BATNEEC, however, was open to qualification and definition depending on individual circumstances.[17] If the cost of the best available technique is considered to outweigh the environmental benefit, this will be taken into account. The pollution inspectorate has drawn up technical notes for the Part A processes and the Department of the Environment and Local Authority concerned drew up the notes for the Part B processes. These notes include abatement technologies, monitoring techniques, storage, handling aspects and emission controls.

The 1995 Environmental Act established the Environmental Agency in England and Wales (EA) and the Scottish Environmental Protection Agency in Scotland (SEPA). From April 1996, the new agencies were made responsible for waste regulation and control of water pollution by combining the activities of HMIP, the NRA, some government agencies and the local authorities' waste regulation authorities (WRAs), which were responsible for regulating the handling and disposal of waste. The objective of this reorganisation into a single agency was to simplify matters for industry and to produce more effective control. The EA is responsible for air and water monitoring, and advises the government on environmental standards and the means necessary to ensure that they are met. However, some responsibility for air pollution control remains with local authorities and they have a duty to manage air quality in their own areas, which includes monitoring, modelling and establishment of emissions data bases.[18] The EA in England is now part of the Department of the Environment, Food and Rural Affairs (DEFRA).

6.3.2 European legislation

The European Integrated Pollution Prevention and Control Directive (IPPC), based on Part 1 of the UK EPA 1990, concerns major industrial operators who are to be licensed in an integrated way to control emissions to the air and water and to manage waste, in order to protect the environment as a whole. In addition IPPC is also concerned with the use and nature of raw materials, including water, energy efficiency, prevention of accidents, low waste technology, noise, and also restoration of a factory site to a satisfactory condition when production ceases. The overall concept is to make industry aware of the process as a whole, in order to ensure adoption of clean technology, careful management of resources (especially energy) and acceptance of the principle that the 'polluter pays'. This EU Directive of 1996 applied to all new installations within the EU from October 1999, and will apply to all existing installations by October 2007. It is broader in scope than IPC, which it replaces, and it is being phased in gradually. There are expected to be about 5000 major sites which will be affected – some of which fell outside the scope of the EPA.[19] The conditions of licensing are an adherence to EU environmental quality standards or laws, and a commitment to high standards of environmental protection to protect the environment as a whole using the best available techniques (BAT). However, cost and the advantages gained are taken into consideration with the EU BAT and it is, in effect, similar to BATNEEC. The monitoring will need to be done by an independent test laboratory to CEN Standard 45001.

The EU in 1997 put forward a Directive to apply a carbon dioxide/energy tax on energy in order to encourage a change to fuels emitting less or no carbon dioxide. This has appeared in the UK in the form of the Climate Control Levy on non-domestic fuel. Further detail is given in Section 6.5.

6.3.2.1 Air quality

Deteriorating urban air quality is one of Europe's two major environmental problems (the other is agriculture). The EU publish guidelines for air quality in Europe based on World Health Organisation information. The European parliament issued a Directive during March 1999 by which member states will be required to have national reduction plans for VOCs. The objective is to reduce VOC emissions by 50% when compared to 1990 levels by the year 2010. Estimates show that, on average, 24% of Europe's VOCs are from the use of solvents, which can in certain weather conditions produce 'summertime smog'. Power generation, transport, combustion processes and natural vegetation account for the rest of VOCs.

This so called 'solvents directive' is affecting 400000 European companies, 90% of which are small or medium sized enterprises. There have been very significant contributions to clean air by voluntary actions, which have also produced cost savings. An example is the reduction of VOCs from paint spraying by energy management and use of new technology by Vauxhall at Ellesmere Port. They cut their electricity bill by 6%. Ford at Halewood reduced VOC emission by 40% and saved £60000 on chemicals and waste disposal in the process.[20]

In the UK, air quality in the general public environment is monitored for the main pollutant gases, lead and particulate matter, PM 10, at various urban sites. This is co-ordinated by the National Environmental Technology Centre under EU regulations, and reports are available on UK Freephone 0800 556677 (or on the Internet), on levels of nitrogen dioxide, sulphur dioxide, ozone, benzene, 1,3-butadiene, carbon monoxide and also particulate matter. Under the Environmental Act 1995, air quality standards were set in the National Air Quality Strategy, and local authorities will be responsible for ensuring that the levels set are not exceeded in their area after the year 2005.

Solvents are still used in textile coating processes despite the development of water-based resins, because in some cases it is difficult to achieve the same performance at the same price or with the same add-on of compound. Factories with solvent emissions above certain thresholds have been obliged to install some form of abatement such as incineration. The actual design and type of abatement depends on the chemical nature of the emissions. Many common organic solvents produce only carbon dioxide on combustion, but if nitrogen is present in the solvent, oxides of nitrogen are produced. If the emissions produced contain anything other than carbon dioxide and water, the incinerator fumes may require treatment. The automotive interior trim industry still uses flame lamination which produces potentially toxic fumes; these are best controlled by carbon adsorption. Further details are given in Section 6.4.3.

6.3.2.2 *Disposal of waste*

The EU is in the process of promoting the management of waste in a way that does not jeopardise public health or the environment. The problem was identified by study of health reports on nausea, headaches, skin and eye irritation, and also by public complaints about landfill sites – 60 incidents since 1989. In March 1998, the EU ministers agreed on legislation to reduce the adverse effects of existing and new landfill sites and to encourage the development of alternative ways of disposal such as composting and incineration. Landfill safety standards would be increased and waste categorised and separated. All waste would have to be treated beforehand; other waste would be sent to specific sites for hazardous materials, and liquid and more hazardous waste would be banned altogether from landfill. Member states would have to reduce the amount of biodegradable municipal waste going to landfill to 75% of the volume of 1995 by 2006, 50% of the 1995 volume by 2009 and 35% by 2016. To meet these targets the EU has indicated that composting be increased as a way of treating biodegradable waste. The UK, which does not have the facilities at present nor the tradition of composting, is being allowed a four year extension to 2020 to meet the 35% requirement, i.e. 65% of biodegradable must be composted by this date.[21] This issue is discussed further in Section 6.5.6. Further EU Directives on waste control can be expected, and already directives on composting and the incineration of waste are being discussed and formulated.

Landfill charges are increasing in many countries of the EU and have tripled in Germany since 1992 to about 400 DM per tonne. Landfill charges were introduced for the first time in the UK in 1996. Waste classed as 'inactive', e.g. bricks, concrete or glass, was charged at £2 per tonne, while all other waste that decays (active) was charged at £7 per tonne. The active waste charge has now risen to £14 per tonne (March 2001) as part of a five-year programme of increases, and £25 has been recommended as a minimum to promote better waste management.[22] The Customs and Excise (with DEFRA) must now license all landfill operators in the UK.

Waste Directives have been issued since 1975 to provide a framework to reduce waste generation, encourage recycling and ensure the safe disposal of waste. The EU's Fifth Environmental Action Programme was aimed at stabilising waste production by the year 2000 at 1985 levels. There has been much activity on reducing and managing packaging waste and scrap cars – 'end of life vehicles' now referred to as ELVs – see Section 6.5.2.

6.3.3 United States legislation

The situation in the USA has followed a generally similar pattern to that in the UK, with much environmental legislation activity and public concern from the about the 1970s onward.[23,24] Major acts and amendments have included the Environmental Protection Acts of 1970, 1980 and 1995, the Clean Air Acts of 1977 and 1990, the Pollution Prevention Act of 1990 and the Clean Water Act of 1987. Control details vary from state to state, but the main framework is broadly similar to Europe. Under the Clean Air Act Amendments of 1990, the US EPA developed maximum achievable control technology (MACT) standards to control hazardous air pollutants (HAP) from textile processing operations. A total of 189 HAPs were identified, and the MACT regulations apply to factories with an individual HAP emission of 10 tons a year or more, or a total of 25 tons a year of any mixture of HAPs. The MACT standard is determined by the best performing 12% of a source category. The US EPA was scheduled to issue the standards by November 2000 and factories must comply by 2003.[25]

The US also has a Climate Change Tax, and provision was made in the 2000 budget for the Climate Change Technology Initiative (CCTI) to fund research and development and tax incentives for energy efficiency improvements, renewable technologies, more efficient energy generation and carbon sequestration research. In the USA there has also been special emphasis on control of insecticides, fungicides and rodenticides with an act passed in 1988. Public concern has also been influenced by international scale disasters such as Bhopal, Chernobyl and the Exxon Valdez oil spillage; the Greens are now the third largest party in the USA.

6.4 Manufacturing concerns

In common with most major industries, textile producers take environmental and health and safety issues very seriously. All major manufacturers have established environmental policies and environmental management systems. BS 7750 issued in 1992 was designed to enable industry to establish effective environmental management systems for sound environmental performance, for participation in environmental auditing procedures and for management reviews. This standard had links with the quality management standard systems BS 5750, and it was also consistent with the then draft EC regulation to set up voluntary schemes for environmental management, which became known as the EMAS. Four years later, the first of the ISO 14000 series was published by the Swiss based International Organisation for Standardisation, which draws from 111 countries. ISO 14000 is a group of voluntary international standards, which have the objective of providing consistent and effective environmental management systems for all operational procedures. ISO 14001 is the first of some 20 separate standards, which cover all issues from environmental auditing to life cycle analysis of products. ISO 14001 requires identification of environmental aspects, policy and objectives and a commitment to comply with all relevant legislation and regulations.[26,27] Companies qualifying for ISO 14001 are audited regularly to ensure compliance, and the standard also requires continuous improvement. BSI have also issued guidelines for environmental performance evaluation (EPE), in the form of ISO 14031. EPE helps to 'describe the environmental performances of an organisation', compared with LCA which deal with products and services. Environmental performance evaluation can be used independently or to support ISO 14001.

6.4.1 Wet processing

Textile processes attempt to use the minimum amount of water for economic as well as for environmental reasons. The textile manufacturing industry is potentially environmentally polluting in two ways: discharge of effluent from wet processing and air emissions from stentering, finishing, fabric coating and lamination. Both are now very carefully controlled by the environmental pollution laws which have become progressively stricter in recent years; informative articles have appeared in trade journals.[16,28–34] Manufacturers may not discharge trade effluent into a public sewer without authorisation from the local water authority. The application form for permission should include details of the effluent, chemicals, suspended solid matter, biological content, etc. and information on the amount to be discharged in a day, including the peak rate of discharge. In April 1989 a list of 23 chemicals which were to be subject to stricter control was announced by the government, the so called 'red list'. These chemicals were already on the EC 'black list' and are now subject to control under BATNEEC to minimise input into the environment. The 'red list' includes mercury and its compounds and cadmium and its compounds. Discharge levels are also subject to agreements reached in North Sea Conferences, especially the Third in 1990 and the Fourth in 1995 at which there were further commitments to reductions.

The environmental effects of textile chemicals have come under scrutiny, and certain chemical types have had to be replaced with more environmentally friendly ones. Certain anionic sulphonates used as surfactants and scouring agents have branched molecular chains which are not biodegradable; they are, thus, being replaced by straight chain types which can be broken down by bacteria in the effluent plant. Also certain phosphate chemicals which support algae growth in rivers and waters are being phased out. The algae 'blooms' block out sunlight and reduce the oxygen content of the water – which is harmful to both plant and marine life. The chemical manufacturers have been very active and can supply much useful information. Any damage to the environment caused by harmful discharges can be made good and the bill sent to the offender under the 'polluter pays' principle. There are also restrictions on pH and temperature of effluent discharges. The rate of discharge and peak disposal rates must be notified to the local sewage undertaker. The discharge of prescribed substances, such as mercury and its compounds, cadmium and its compounds, and certain organic compounds, are very severely restricted. Consent levels are agreed with the local water company under conditions set by the EA who also set monitoring levels procedures. Process Guidance notes (PG series) are available from the DEFRA through HMSO.

Lubricant content of loomstate or unfinished fabrics can sometimes be up to 2% by weight, and much of this 'oil' is removed during scouring and stentering. Most fabric, however, is now scoured, and spent liquors should be treated before discharge to drains. pH, temperature and solids content are the usual factors requiring attention. All stenter fumes should be abated before discharge to atmosphere, and the emissions should be monitored regularly in accordance with standards arranged with the EA. Coatings applied to automotive fabrics are now invariably water based, and the material driven off during processing is mainly water, but there may be organic chemicals, which have to be removed before the emissions are released to atmosphere. Guidance Notes relating to air quality (AQ series) have been prepared by the DETR and are available from HMSO.

6.4.2 Coating and lamination

Joining methods using solvent-based adhesive spray application methods and flame lamination are potentially highly polluting, and many solvent-based adhesives have been replaced and continue to be replaced with hot melt, high solids content and water-based varieties. However, fabric coating and joining using solvent-based resins and adhesives is still widespread. In the fabric coating industry, water-based resins are used whenever possible, but in some cases it is proving difficult at present to obtain the high standards of performance and durability normally achieved with solvent-based types. However, solvent use is subject to closer restrictions, and targets have been set by the EC to reduce VOC emissions by 66% compared to 1990 levels with a compliance date of 2007.[34–37]

Textile coating factories with VOC emission thresholds above a certain level must have abatement facilities to reduce VOCs emitted to the atmosphere to within a concentration agreed by the local authority. Typically this may be 150 mg/m^3 of air. PVC factories need abatement for the fumes emitted during the gelling process. The fumes are likely to contain plasticiser and traces of the stabilisers and possibly other additives.

Flame lamination fumes are monitored and must be treated by very effective methods, if necessary by carbon adsorption techniques. Under the EPA 1990 and the Environment Protection (Prescribed Processes and Substances) Regulations 1991 Di-isocyanate Processes, operators of potentially polluting industrial plant must be registered with the EA before beginning production and must regularly monitor levels of potentially harmful substances.[38–40] Guidance notes for operators are available from government bookshops; the relevant pamphlet for flame laminators is IPR6/5, entitled *Toluene di-isocyanate use and flame bonding of polyurethanes*.

6.4.3 Methods of emission abatement

The methods of abatement fall into two categories: destructive techniques, where the undesirable fumes are destroyed by incineration into non-toxic gases and capture techniques, where the undesirable material is removed from the airstream by various techniques.[41–49] Capture techniques include electrostatic precipitation, activated carbon adsorption, filtration and 'scrubbing'. Scrubbing involves passing the emissions through water solutions, such as caustic soda to neutralise acidic material or, in the case of flame lamination, isocyanates. However, scrubbing techniques may give rise to stack 'plumes' which may consist almost entirely of harmless water vapour, but they look alarming and might, therefore, give rise to complaints from the neighbours.

PVC coatings are heated at around 200 °C to gel the PVC and plasticiser, and fumes of vaporised plasticiser and traces of other materials in the plastisol recipe are produced. The amount varies depending on the volatility of the plasticiser being used, but abatement is essential whatever the recipe. Techniques used include aerosol or electrostatic filtration and incineration, but effective incineration depends on the temperature, turbulence to ensure good thermal mixing and contact, and the length of time the fumes are actually in the incinerator. Incineration has to be very carefully controlled to ensure complete oxidation, because potentially toxic gases may be produced and released to atmosphere. This is especially relevant when chlorine containing material is incinerated, because dioxins, a group of extremely toxic materials, may be produced if the process is not at a sufficiently high temperature. Incineration temperatures must therefore be maintained precisely at the high temperature necessary for complete oxidation and to ensure public safety. This requires careful process control, especially if the waste fume stream is not consistent, and extra fuel gas is needed in these circumstances which adds to the running costs. The incineration process can sometimes be carried out at lower temperatures by the use of catalysts made from precious metals such as platinum, but they need to be protected from species which can render them inactive.

Volatile organic compounds used in direct or transfer coating of polyurethane, such as MEK and toluene, are fully oxidised to water and carbon dioxide, but DMF, however, also produces oxides of nitrogen. Incineration of sulphur compounds can produce oxides of sulphur, which like oxides of nitrogen, are acidic; these could be harmful to buildings and vegetation in addition to being harmful to human health. Incineration of chlorine compounds, as already mentioned, requires special care and control to prevent the formation of dioxins, but hydrogen chloride gas or even chlorine may be produced. It may, therefore, be necessary to wet scrub incinerator fumes for acidic species using sodium hydroxide or calcium hydroxide, although oxides of nitrogen have low solubility in water and cannot be effectively removed by a wet process.

Incinerators can be designed to produce useful heat by the incorporation of heat exchangers which will contribute significantly to the running costs, but this depends on a consistently high volume of combustible organic material in the waste stream. Initial capital outlay is higher for an incinerator with primary heating, in which heat generated from the incinerated gases is used to preheat gases entering the incinerating chamber. If volumes of VOCs to be incinerated are sufficiently high, so called secondary heating may be possible. Secondary heating means that the heat from burning of the waste gases is used for useful purposes such as heating and raising steam. The DEFRA is actively promoting combined heating and power systems (CHP) because of their overall energy efficiency and contribution to reduced carbon dioxide emissions. Operators of qualifying CHP systems may be exempt from the climate change fuel levy. Some factories apparently operate very successful CHP systems.[50]

Carbon absorption is considered to be one the most effective means of abatement of fumes from flame lamination, which include iso-cyanates and hydrogen cyanide. Capital installation costs can be high, but running costs are generally moderate to low – so much depends on the production volume. Oily and particulate matter must first be removed from the laminator production fumes, because they would soon deactivate the carbon, and it is necessary to install pre-filters as part of the abatement unit. Potentially toxic gases are adsorbed on to the activated carbon, and if the unit has been properly designed and is working effectively the emissions to atmosphere should be within levels permitted by the EA. The filters need servicing and replacement periodically, and eventually the carbon will become saturated and will also need to be replaced. The spent carbon can be reprocessed under controlled conditions by specialist operators, but is believed to have been disposed of by landfill – which must now be regarded as an unsatisfactory practice.

6.4.4 Health and safety aspects

6.4.4.1 Control of hazardous substances

All chemicals and materials used in the textile and automotive industries are subject to the Control of Substances Hazardous to Health regulations (COSHH) of 1994 (which replaced the original 1988 regulations) and March 1999, which cover all aspects of purchase, handling, transportation, storage, use and disposal. The most important issue is that an employer cannot carry out any activity which exposes employees to any hazardous substance unless a 'suitable and sufficient' assessment has first been made. The assessment must be reviewed regularly and whenever any significant modification to the process has been made. Hazardous material may include gases, vapours, liquids, fumes, dusts, solids or micro-organisms. When stored or transported they are also subject to the Classification, Packaging and Labelling of Dangerous Substances Regulations of 1984. Guidance notes on safe handling, maximum exposure limits, occupational exposure limits and monitoring strategies are regularly updated, published annually under the EH40 series, and are available from the Health and Safety Executive.

Manufacturers and suppliers must supply documented information on chemicals, adhesives and other raw materials in material safety data sheets. These documents contain information on chemical composition, physical and chemical properties, toxicological and ecological aspects, potential hazards together with first aid measures, storage and transport, protective clothing and other controls necessary for safe handling and disposal. They should conform in format and content to EC Directive 91/155/EEC. EU regulations have generally tightened control of potentially harmful chemicals.

6.4.4.2 Hazardous material content

Some environmental and customer pressure groups are now concerned about the presence of potentially toxic chemicals in consumer items, and an increasing number now require information on this. Hazardous material assessment forms are issued which list chemicals such as lead, nickel, cadmium, mercury, certain organic chemicals including bromine, etc., and their usage in the manufacturing process or concentration, in parts per million, in the finished article must be declared. There are halogen and phosphorus FR chemicals in some FR qualities of both fabric and polyurethane foam. While the risk to health from this source is likely to be quite low compared to similar chemicals in some apparel or even domestic furniture, consumer concern must be taken seriously. Considerable effort has been made by the chemical industry to replace these materials, but so far with limited success. Higher concentrations of alternative chemicals are required to produce the same performance. However, the situation is not yet clearly resolved, especially because bromine compounds have been found in the marine environment and in mammals.[51,52] Some researchers, however, believe that the benefits of FRs outweigh the risks to human health and the environment.[53–55]

6.4.4.3 Eco-labelling

In a strict interpretation of the term, eco-labels are applied to consumer goods as an indication that the product has been manufactured under environmentally friendly conditions, from environmentally friendly materials, and that will not pose any threat to human health or to the environment during its useful life nor at disposal. This process is a life-cycle analysis, also called a 'cradle to grave' approach (ISO 14040 group), but in practice is almost open-ended for many products and, at best, is extremely costly and time consuming to carry out. However, this has been the basis of the EU eco-label, which started with only a limited number of products including T-shirts and bed linen. There have been recommendations for simpler procedures, which concentrate on the more important characteristics of products. There are several eco-labels in Europe such as the 'Blue Angel' in Germany and the 'White Swan' in Scandinavian countries, while in the USA there is the 'Green Seal'. Some of these labels are based on environmental audits of the manufacturing process alone, such as energy efficiency. Others are based on assessments of the content of hazardous material in the textile as an assurance that it will not harm human health.[56–59] One such label is the Oeko-Tex Label run by the International Association for Research and Testing in the Field of Textile Ecology, which includes the Hohenstein Institute in Germany and BTTG in England.[59] To obtain an Oeko-Tex label, textile products are analysed for content of a range of potentially harmful chemicals such as formaldehyde, cadmium and mercury. A label is granted if they are below maximum permitted levels. Well over 18000 Oeko-tex certificates have been issued, including some in the USA, the Far East, Japan and South Africa, see Fig. 6.2. It is probably the most widely recognised textile environmental safety standard.[60,61] Another example is the Toxproof mark offered by TUV Rheinland, Cologne; this is issued to textile goods only after they have been tested to TUV criteria which include heavy metals, chlorinated phenols, etc. The EU eco-labelling scheme encourages companies to design and manufacture environmentally friendly products,

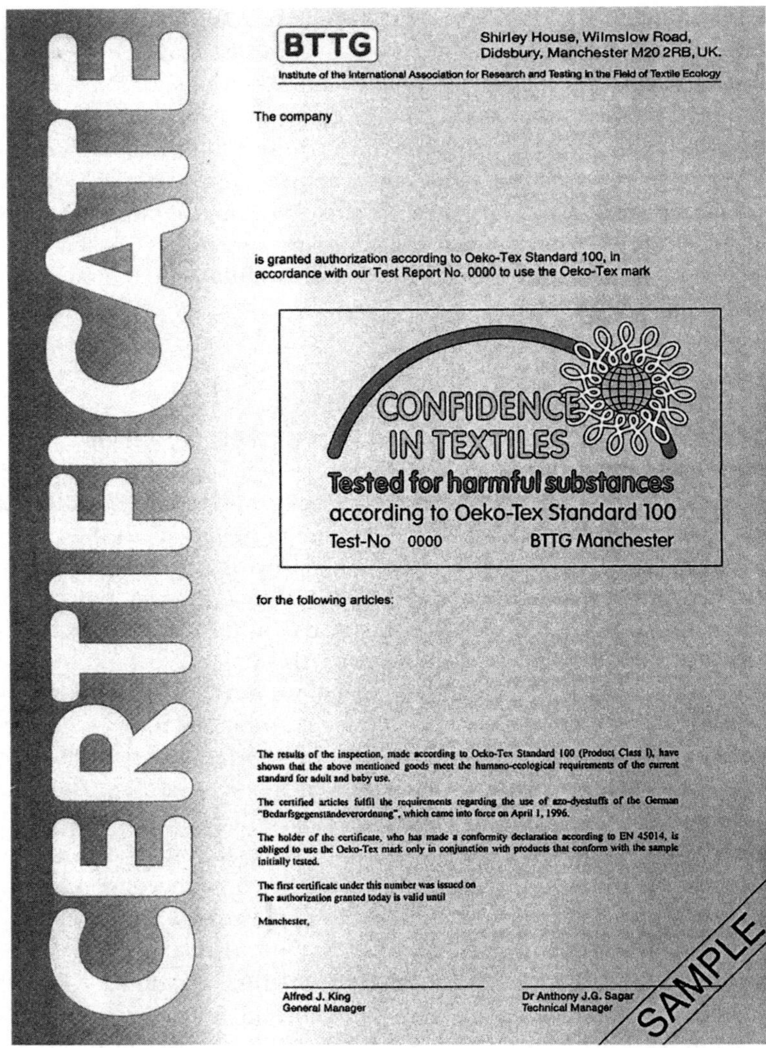

CERTIFICATE

BTTG

Shirley House, Wilmslow Road,
Didsbury, Manchester M20 2RB, UK.

Institute of the International Association for Research and Testing in the Field of Textile Ecology

The company

is granted authorization according to Oeko-Tex Standard 100, in accordance with our Test Report No. 0000 to use the Oeko-Tex mark

for the following articles:

The results of the inspection, made according to Oeko-Tex Standard 100 (Product Class I), have shown that the above mentioned goods meet the humano-ecological requirements of the current standard for adult and baby use.

The certified articles fulfil the requirements regarding the use of azo-dyestuffs of the German "Bedarfsgegenständeverordnung", which came into force on April 1, 1996.

The holder of the certificate, who has made a conformity declaration according to EN 45014, is obliged to use the Oeko-Tex mark only in conjunction with products that conform with the sample initially tested.

The first certificate under this number was issued on
The authorization granted today is valid until

Manchester,

Alfred J. King
General Manager

Dr Anthony J.G. Sagar
Technical Manager

SAMPLE

6.2 The Oeko-Tex Standard 100 is a product safety standard and prohibits or regulates the use of harmful substances such as formaldehyde, phthalates (in baby products) and emissions of volatile components. It has become the world's leading eco-label for textiles tested for harmful substances. Specimen Certificate supplied by BTTG and reproduced with kind permission.

and provides customers with information with which to choose between 'green' products and others. More notable than this could be the suggestion in an EC Green Paper on integrated product policy that there should be lower VAT on eco-labelled products.[62]

6.5 Sustainable development

6.5.1 Introductory notes

The UK and most of the other major countries of the world gave a commitment at Kyoto in 1997 to reduce carbon dioxide emissions in efforts to reduce global warming. Much is being done by the EU and the UN and by governments of the world to foster environmental responsibility and the concept of sustainable development. In recent years the world has witnessed severe weather causing loss of life and damage to property that is believed to have been caused by global warming. Environmental pressure groups call for even firmer and wide-ranging action, and much still needs to be done and done quickly.

The UK government in 1990 set non-statutory targets for local authorities to recycle 25% of municipal waste by 2000, but the vast majority fell far short of this target and mandatory targets may be necessary. Recycling has been a major cause for concern in the packaging industry since before 1990, and much progress has been made both in the reduction of packaging and in the collection and actual recycling of used packaging. Some of the lessons learnt are being applied to the automotive industry, which is now facing similar problems together with some which are unique to it. At the time of writing there does not seem to be significant pressure on the recycling of coated and laminated products – with the possible exception of car tyres, if they are regarded as being coated materials. A summary of the situation in the automotive industry is included here because the automotive industry is the largest single user of coated and laminated textiles which are used extensively in cars. The automotive industry was a pioneer of the mass production industry, and there may be lessons to be learnt in the disposal and recycling of automobiles.

The least expensive and least adverse effect on the environment is when a component can be recycled into its original product, i.e. so called 'closed loop' recycling. The second best is when it can be used in another article which usually requires less demanding properties, for example face car seat fabric being recycled into backing material. The third option is to incinerate the material to generate useful heat, but this generates the inevitable carbon dioxide and other gases and emissions, which generally have to be treated before release to atmosphere. Plastics, in fact, generally have high calorific values and are efficient fuel. The fourth and least satisfactory disposal method is landfill, which as has already been mentioned may produce environmental problems of its own. Some components consume so much energy and other resources to recycle them that it is judged more environmentally friendly overall to use the two latter options. This is why life-cycle analysis ('cradle-to-grave') is essential. Each case has to be thoroughly assessed on its own individual circumstances and merits. In some countries with a shortage of fossil fuels, such as Japan, incineration is a preferred option.

On 1st April 2001 the UK government, in line with the EU Directive of 1997, introduced the 'Climate Change Levy' on non-domestic fuel to encourage industry to use less fuel and become more energy efficient, thus reducing the amount of carbon dioxide produced. The levy could increase electricity bills by 8% and gas bills by 15%, but to avoid adding extra cost to industry, the employer's contribution to National Insurance was cut by 0.3%. Companies using approved CHP schemes are exempt, and there are also 80% discounts available for some energy intensive industries which enter into legally binding energy efficiency agreements with the government.[63,64] In addition, companies which use energy efficiently may also qualify for discounts under IPPC schemes.[65]

6.5.2 Automotive recycling

6.5.2.1 ELVs – 'end of life vehicles'

Proposed EU legislation will require a system of collection – at no cost to the last owner – and disassembly to be created for re-use, recovery and recycling at the end of the vehicle's life. An amendment states that 'producers must meet all, or a significant part of the costs'. The cost, of course, will ultimately be passed on to the consumer. Needless to say, the European car industry considers these measures unreasonable because the car is already 75% recycled by weight, and actual waste from cars represents only 0.2% of all European industrial waste.[66] However, the EU considers ELVs a priority, and these measures are likely to affect the textile industry eventually, because fabric and fabric laminates are major interior components of motor vehicles. Car industry spokespersons have commented that the new law, in effect, encourages the production of heavy cars using more metal – if the ELV directive had already been passed, the 3-litre car could not have been developed. The 3-litre car uses three litres of petrol to cover 100 km, i.e. 92 mpg, and it contains a high proportion of weight saving plastic material. Recycling analyses of automotive plastic and textiles have been carried out.[67–71]

In the UK at the time of writing about 25% by weight of the car goes to landfill with about 75%, mainly metal, being recycled. The ELV Directive requires 80% recycling and 85% recovery (5% by energy recovery) by 2006, with no more than 15% to go to landfill. These levels will be raised to 85% recycling and 95% recovery (10% energy recovery) with no more than 5% to landfill by the year 2015.[72,73] The UK government is in the process of actually detailing how the Directive will be applied. Throughout Europe there are about 12 million ELVs disposed of every year, and this is increasing at the rate of about 3% per year. The metal parts are recycled, but the remainder, about 25% by weight consisting of tyres, glass and an assortment of plastics, textiles and other materials called automobile crusher or shredder waste (ACR or ASR) mostly goes to landfill. Car seats and other interior items making up ASR pose serious challenges for the industry, and this issue will become more pressing as the deadlines for reduced landfill approach. The OEMs are supporting schemes to identify and sort scrap material and are also trying to increase the recycled material content of their cars. Some are making recycling a key design consideration and specifying that parts supplied must be made from a certain percentage of post-consumer recycled material, thus putting pressure on their suppliers.[74–77] Eventually Ford want to attain 90% vehicle recyclability (by weight), while others such as DaimlerChrysler want to do even better – 95% by the year 2005.[75] The UK government is contributing to the funding of certain projects concerned with car recycling.[78]

6.5.2.2 *Automotive fabric recycling*

The car seat laminate is generally made up from polyester face fabric, polyurethane foam and a scrim fabric which is either nylon or polyester. These chemically dissimilar materials are not easily separated and, therefore, cannot be easily recycled. The use of polyester scrim reduces the number of chemical types to two, but even this presents a problem. Chemical hydrolysis can be used to break down the three polymers into simpler chemicals which can be used as fresh raw materials, but at present this is not commercially feasible.[68] Several different types of nonwoven and knitted fabric (for example 'spacer fabric', Kunit, Multiknit (by Karl Mayer) and wool/polyester blends) have been evaluated as substitutes for laminate polyurethane foam. Karl Mayer manufacture the machines on which they are made. Some materials have been made from garment waste and are being used commercially in German made cars.[79–87] These foam substitutes do not need a scrim backing and using one made of polyester together with a polyester face fabric produces a seat cover laminate all in one polymer type.

These materials provide a soft touch to the laminate and compression properties initially comparable to foam, but they all lose significant thickness in compression and heat ageing tests.[88,89] The thickness loss would be easily noticeable with thick layers, but may be less noticeable in thinner layers. Fibre producers have demonstrated the possibility of using recycled polyester face fabric into nonwoven material. Shredded face fabric is mixed with 30% of virgin polyester polymer, melted and re-extruded into a nonwoven fibre which, although discolored, can be used as the foam substitute in a new seat cover. When this seat cover comes to the end of its life, it can in turn be shredded, melted and extruded again, but this time with a higher proportion of virgin polymer to compensate for the used polyester being recycled a second time. Alternatively, it can be used in a less demanding end-use. Thus, the same polymer is reused, but each time in a progressively lower specification application. Nonwoven polyester spun from recycled polyester bottles by Wellman is currently being used in some production models. Bottle manufacturers continue to develop and improve polyester bottles for many other end-uses, and there could soon be a surplus of polyester bottles available for recycling.

6.5.2.3 *Recycling of polyurethane foam*

The manufacturers of polyurethane throughout the world have responded to the challenge by showing how waste foam can be reused, recycled or disposed of with minimum environmental effect.[90–93] The methods involve shredding into crumbs and smaller particles and reprocessing by compression moulding, and adhesive or thermoplastic rebonding into useful articles, including backings for carpets, rugs and other items which make use of the acoustic and shock absorbing properties. Foam has been ground into powder and added into new compound mixes as filler. Studies have been carried out on the feasibility of chemolysis, breaking the material down chemically into the original raw materials for use in existing or new products. Composting and incineration as a useful fuel to recover the energy has also been considered. The Polyurethanes Recycle and Recovery Council (PURRC) of the (American) Society of the Plastics Industry was set up as early as 1990, and has tackled the problem of both process and post-consumer waste. In the USA there is a market for foam for recycling into new products such as carpet backing. The European Isocyanate Producers' Association (ISOPA) provides a focus for environmental responsibilities in Europe, and has issued comprehensive technical information sheets on recycling possibilities.[93]

Automotive seat makers have explored the use of ground up foam from car seats as filler for virgin foam in new seats. Physical properties appear to be reduced in quality, but may be satisfactory in actual use. The problems of interior trim foam being joined to other materials and the cost of dismantling and handling, however, remain. Foam processing waste in fabric lamination plants is taken back by certain manufacturers.

6.5.2.4 *Logistics of automobile recycling*

The first step in the recycling process is collection of the ELV. A single scrap car has little value to its last user, and occasionally abandoned cars are seen littering the landscape. These are not only an eyesore, but also constitute a health and safety hazard. About eight million ELVs are scrapped each year in the EU. With a certain amount of pressure from governments, voluntary accords have been set up since the early 1990s to address the problem, amongst them the Automotive Consortium on Recycling and Disposal (ACORD) in the UK, Project Altfahrzeugverwertung Deutscher Automobilhersteller (PRAVDA) in Germany, and the Environmental Car Recycling in Scandinavia (ECRIS).[94–96] The Association des Constructeurs Europeens d'Automobiles (ACEA) in Brussels promotes research to improve the design of vehicles with a view to ease of disassembly and also to the selection of materials which can be recycled and minimal use of those which cannot. The Consortium for Recycling (CARE), established in the UK in 1996,[97] works with government bodies and other organisations to produce specific results from practical work. Recytex, previously a subsidiary of the car upholstery manufacturer Viktor Achter has processed textile waste and has been co-operating with Mercedes-Benz and others on the use of recycled nonwovens for sound insulation.[98,99] Car manufacturers are co-operating with each other, for example Volvo will scrap Mercedes cars in Sweden and Mercedes will scrap Volvo cars in Germany and Toyota are co-operating with VW and General Motors.

The second step in the recycling process is disassembly, which must be accomplished quickly if it is to be economic. Plastics must be identified and sorted and different types separated from each other; this is not easy, and it may not always be economically possible. The actual impetus for recycling has arisen only since the 1990s, and many of the cars currently being scrapped were not actually designed with recycling and disassembly in mind. In addition, some components made maybe between 1985 and 1990 contain substances which are now considered to be toxic and are therefore prohibited from being used in a new car. After many years of use, seat covers are likely to be heavily soiled, which means there will be high levels of contaminants with which to contend.

The non-metallic parts of the car total about 25% of the weight and comprise glass, rubber, plastics and textiles. These materials, about 300 kg in an average vehicle, represent the biggest challenge, with adhesives, paints, coatings and fasteners further complicating matters. At present, the cost of dismantling, sorting and transporting components is not generally commercially viable, but the pressure is on to change this or to find alternative solutions. Vehicles being built at the present time benefit from these lessons, and future ELVs should gradually become more and more easily dismantled. Cars being built now will not become ELVs until the year 2010 onwards, and it has been suggested that when a polymer type is chosen for a car part, a second future use for it, at the end of the car's life, should be decided in advance. To assist with dismantling, the EU has requested OEMs to code car parts and to produce dismantling manuals. One notable exception to the trend of easier disassembly is the move towards directly joining car seat covers to the foam cushion and squab in seat making. DaimlerChrysler, Ford and BMW are believed to be planning a pilot identification and sorting plant for automotive plastic waste in Germany. Car carpets are generally one polymer, but with high levels of binders and a coating of bitumen or other material to assist with vibration and sound absorption. They can also contain very large amounts of dirt – one report records up to 1 kg of dirt per square metre of carpet.

6.5.2.5 *Reduced emissions by reduced weight*

Efforts are being made to reduce pollution by making the car lighter in weight. Some OEMs have requested reductions in car seat cover fabric weight of 30%. Generally this is not possible without compromising the fabric performance, especially abrasion resistance. However, the add-on of coating to the back of car seat fabric has been significantly reduced since around 1995 to reduce both cost and weight. The effect on fabric abrasion has been minimal. The car industry is responsible for 12% of all carbon dioxide emissions in Europe, and its co-operation was requested so that the EU can comply with its commitment to reduce carbon dioxide emissions. In response the ACEA proposed very significant reductions in fuel consumption. Cars are now available emitting carbon dioxide corresponding to an average fuel consumption of 4.91 per 100 km (equivalent to 48.2 mpg). Further improvements up to the year 2008 will result in car fuel consumption at levels 25% less than 1995 levels.

Textile fibre in the form of composites should contribute substantially to reduced traffic fumes by allowing considerable weight savings. However, there are technical problems to be overcome before large metal replacement composites can be made commercially in the quantities required by the high volume automotive industry. Advanced tyre cords and textile fibres/rubber combinations are, in fact, contributing to reducing rolling resistance in tyres, leading to economies in fuel consumption and prolonged tyre life. Replacement of relatively heavy fibreglass in headliners or other parts of the car by polyester, or polypropylene which is even lighter, can reduce weight further. In addition, fibreglass is unpopular because skin dermatitis is caused during manufacture.

6.5.2.6 *Disposal of tyres*

Coated rubber products, and tyres in particular, are difficult to recycle or dispose of on a large scale. The vulcanisation process cannot be reversed easily or economically. The tyre companies and others have researched means of safe disposal by incineration with energy reclamation. The Blue Circle Cement Company and Michelin are co-operating, under the name Sapphire Energy Recovery, to use tyres as fuel in part replacement for coal and coke in a number of Blue Circle's cement works. The plan could use half of the annual 40 million scrap tyres for useful energy by the time the EU Directive banning landfilling of whole tyres comes into effect in 2003. The Landfill Directive will also prohibit disposal of shredded tyres in 2006,[100,101] but the UK hopes to postpone this until 2009 in the case of existing landfill sites. Every year the UK disposes of 450000 tonnes of scrap tyres, 30% by landfill, and the last few years have seen significant steps taken to solve this problem, with cement works taking tyres for incineration to produce useful energy. Two sites are in operation and trials are being conducted at three more. However, public consultative procedures, which are part of the application process, still need to be formulated by the EA.[102] The manufacture of cement requires high temperatures of about 1200 °C, and at these temperatures the risk of toxic emissions is reduced, but the public needs to be reassured.

The EA issues licences for the tyre burning process only when it is satisfied that the environmental controls are effective. Sapphire believe that their process uses the latest technology in abatement and hope eventually to have five plants operating in the UK and Northern Ireland. The UK is a leader in reclaiming energy from waste tyres. The problem of tyre disposal is also being helped by the prolonging of the service life of tyres by retreading and by more advanced design. All of these measures also conserve natural resources, and there is possibly an opportunity for parallel developments in the belting and associated industries which also rely on rubber/textile composite technology.

6.5.3 PVC recycling

The use of PVC has been questioned by some environmentalists, especially Greenpeace, because of the possible risk to health of phthalate plasticisers, and also because of the possible formation of dioxin chemicals during PVC manufacture and eventual disposal.[103–105] BMW and General Motors are reducing or have discontinued the use of PVC in cars. General Motors state that PVC is being phased out mainly on technical rather than environmental grounds. Several other car makers also intend reducing the amount of PVC, but this is likely to take some time because up to the late 1990s, PVC was used in 70–80% of car dashboards and door interior trim. Amendments to draft proposals to an ELV EU Directive which would have reduced or phased out the use of PVC in EU built cars were defeated in the European Parliament. PVC requires a whole range of stabilisers and processing aids as detailed in Chapter 2, and these add to the problems of disposal. Potentially toxic additives have been or are being phased out in current production. However, tarpaulins and other coated PVC products made maybe 10 or 15 years ago, which may still be in use or which are being disposed of at the present time, could contain additives prohibited today. Marks and Spencer, the UK chain store, is phasing out the use of PVC packaging and products by about February 2003. In response to consumer concern, the PVC industry is making efforts to remove the problems associated with PVC, and holds the view that, for many applications, there are no alternatives which pose less of a hazard. PVC is difficult to ignite, is an extremely versatile material and it is probably the most researched plastic in existence.

Disposal of PVC is difficult because it cannot be landfilled or incinerated without certain risks. The high chlorine content produces hydrochloric acid gas (HCl), and if it is not controlled, the combustion process may also produce dioxins. Many incinerators are not built to cope with large amounts of HCl gas which could not be released to atmosphere without treatment. Landfilling PVC poses the risk of additives being leached out and eventually entering water courses and even drinking water. In addition, the covering properties of discarded tarpaulins could change soil morphology, and could be harmful to earthworms and other soil dwelling creatures by reducing the level of oxygen or by restricting the flow of water. PVC coated fabric biodegrades slowly, and its disposal has become a problem which has contributed to the search for ways of recycling the material and also for alternative covering materials.

The recyclability of PVC coated polyester tarpaulins has been researched both in the USA and in Europe.[106–111] In Europe, the Solvay company has examined grinding the tarpaulin material into a powder at low temperature (−196 °C) and using the powder as a filler in new PVC plastisols. The work was apparently successful, but it was judged not to be economically feasible. Separating the PVC from the polyester has been tried using heat and using different solvents. PVC melts at 150–200 °C, which is a lower temperature than the melting point of polyester (250 °C), but it begins to decompose before this temperature is reached. Thermal methods are, therefore, not possible. In addition, the polyester becomes embedded into the PVC like a filler. Solvay examined the use of tetrahydrofuran (THF), a solvent which dissolves PVC but not polyester, but concluded that, because of the price of the solvent and its flammability and toxicity, and owing to the presence of additives and soiling material in the extracted PVC and its quality in general, the solvent route did not justify further examination. The company then explored incineration in a rotary furnace with both energy recovery and chlorine recovery in the form of hydrogen chloride. They concluded that this was economically feasible and worthy of a pilot plant. The intention is that the hydrogen chloride gas is fed to supply a vinyl chloride production plant for conversion back into PVC. Other work has examined freeze grinding, at –160 °C, and reusing the powder as filler in new PVC coatings. The Ferrari company of France, with the assistance of Solvay, have developed the VINYLOOP^R. This is a closed loop recycling process for PVC coated fabrics which first grinds the material and then separates the components by selective dissolution in different solvents. The recovered polyester fibre is claimed to be pure and reusable, while the PVC is in the form of granules which are homogeneous and also suitable for reuse. Ferrari is to build a plant at Tour du Pin, Isere in France during 2001.[109]

The University of Auburn in the USA investigated the use of separation solvents and chose MEK for more detailed work. The MEK swells the PVC and allows it to be detached from the polyester fabric by mechanical agitation. The recovered polyester fabric still had adhesive residues on it, but it retained its physical properties and it was possible to use it in epoxy composites. If the glue residues were removed with DMF solvent, the polyester fabric was suitable for the preparation of needlepunched nonwoven fabric. The recovered components were reported to be suitable for reuse in PVC coatings. A process of automated separation, the recovery of the wash liquors and the DMF/glue residues, and the design of pilot equipment are the subjects of further research at the Auburn University.[108,109] The recycling figure for the PVC industry as a whole has been quoted (at the time of writing) as between 3 and 4%, and so the amount of PVC coated fabric actually being recycled at present must be even lower than this figure. The PVC industry is committed to voluntary phasing out of potentially harmful chemicals, and it maintains that in environmental assessments, LCA methods should be used to examine items made from PVC and also to compare them with items made from alternative materials.[105]

6.5.4 Recycling of carpets

Some industries, such as the automotive industry, are high profile in environmental terms, but the problem of carpet disposal, about four million tons every year, which at one time made up approximately 2% of all landfill in the USA, has also been recognised as one that required an urgent solution. In Western Europe about 1.6 million tonnes of carpet waste are disposed of each year. The figure for Germany is half a million tonnes, but after 2005, landfill of waste carpet will not be allowed in Germany. Additional pressure was put on to producers in Germany in 1996 with the passing of the Recycling and Waste Management Act which focused the responsibility for waste on to the producer. However, industry has been working on the problem for a number of years, and as will be seen, solutions are available. As far back as 1993, the European carpet industry declared (Declaration of Vaals) that it recognised that the recycling of carpets was basic to its producer responsibility. In 1995 a three-year project was initiated to develop and introduce a closed loop production system for carpets. The necessary technology was developed by the RECAM Project (Recycling of Carpet Materials), which was supported by the EU under their BRITE EURAM (Basic Research for Industrial Technologies in Europe, European Research in Advanced Materials) Programme. Amongst the project partners were the European Carpet Industry, the German Institute for Carpet Research, DSM and EniChem.[112,113] The objectives included research into sustainable waste management, methods of treating post-consumer and industrial carpet waste to replace solid waste incineration and landfill, recovery and re-use of high quality raw materials from used carpets. Life cycle analysis of the carpet material revealed that recycling of polyamide is economically feasible because of the high energy consumption in the production of virgin polyamide. However, in the case of polypropylene the calorific value of waste is quite close to the energy needed to produce virgin polypropylene.

In 1998 five European carpet production associations formed Carpet Recycling Europe GmbH (CRE) to collect, identify, sort and market recyclable fractions. An automatic sorting plant, the world's first, has been built at Ginsheim-Gustavsburg, and processes 25 000 tonnes of carpet waste per year using fibre identification technology developed by DSM and the Dutch research institute TNO working in the RECAM Project. Waste carpet is made up of 30% polypropylene, 25% each of nylon 6 and 66, together with wool and other materials. Polypropylene is a cheap material which cannot at present be economically recycled either mechanically or chemically and is prepared for incineration. Nylon 6 can be chemically recycled back into caprolactam from which it was produced, and this is being carried out at a number of plants both in Europe and the USA.[114,115] Nylon 66, which is produced from two starting chemicals, is more difficult to depolymerise and is being mechanically recycled into a lower value material. Wool and polypropylene carpet residues are being converted into insulation material. CRE has plans to build ten sorting plants in Europe over the next five years.

Carpet Recycling Europe has a recycling contracts with the Evergreen Nylon Recycling (ENR) plant at Augusta in Georgia, USA. This plant, set up by AlliedSignal, now Honeywell International, and DSM Chemicals North America (DCNA) in November 1999, depolymerises 100 000 tonnes of nylon 6 carpet a year and produces from it 45 000 tonnes of high quality caprolactam. This is a truly 'closed loop' recycling operation and is believed to be capable of reclaiming 20% of all scrap nylon 6 carpet in the USA.[116] Evergreen Recycling is collecting old carpet in 75 US metropolitan areas, and carpet retailers are saving many thousands of dollars in landfill fees and disposal costs by participating in the collection network. DCNA and

Honeywell have developed a portable instrument called CarPID™ which quickly identifies carpet type, and which is based on near IR spectroscopy. ENR continues to improve techniques for and reduce the costs of recycling.

During 2001 Polyamid 2000 AG plans to open a 120000 tonne nylon carpet recycling plant near Berlin, where caprolactam will be produced from nylon 6 and nylon 66 will be mechanically processed. This will actually be the first carpet recycling plant in Europe.[114] In the USA, carpet recycling has been conducted in a limited way since the mid-1990s. In Europe Rhodia Performance Fibres are recycling waste nylon 6 into caprolactam and mechanically processing nylon 66 at three plants. However, carpets are not yet processed and the scrap material is waste fabric and fishing nets, but CRE are believed to be in communication with Rhodia.

Since 1994, BASF in North America has been accepting used nylon 6 carpet back under their 'Six Again-program'. Although the Canadian recycling plant capacity is only about 1000 tonnes per annum, new carpets have been produced from the recycled material for a number of years. BASF are reported to have guaranteed not to incinerate or landfill collected carpet. DuPont has been operating a nylon carpet processing plant at Chattanooga since 1995 mainly for nylon 66 using mechanical recycling. DuPont also maintains a network of franchised carpet retailers to offer services including recycling. Further development work on chemical recycling by DuPont may lead to a pilot plant at Maitland, Ontario in Canada.[112]

The problem of carpet waste disposal has been researched at a number of universities, including the Georgia Institute of Technology in the USA.[117] Polypropylene carpet waste is being examined for possible use as the thermoplastic matrix in glass mat reinforced, thermoplastic composites for use, for example, in the automotive industry. Carpet waste fibres are also being examined for use in injection and press moulding, for concrete reinforcement and for soil reinforcement in road construction. In the UK, some carpet is disposed of by incineration, for example in cement works. Despite the energy release this is now considered to be a last option because of the carbon dioxide produced and the possibility of toxic emissions. Other methods of disposal and recyling have been reviewed.[118,119] Eduard Kusters Maschinenfabrik GmbH in Germany are believed to have developed a method to process unwashed used carpet, together with wood chips, into laminated board.[120]

Amoco, which is part of BP, has developed the AdBac process which replaces latex as the backing for tufted carpets.[121] The process involves binding in the tufts with thermoplastic olefins, and is in commercial production in the USA. Several advantages over latex are offered, including energy savings in production because there is no water to dry off, weight reductions of about 500 g/m^2 with savings in transport costs, and significantly less emissions because polyolefins do not contain VOCs. In addition, the finished carpet has better water-resistant properties, and polyolefin is easier to recycle than latex and also contains no fillers to complicate issues. Latex is very effective as a tuft binder because it penetrates between individual filaments and fuzz is prevented. This penetration is more difficult to achieve with thermoplastic binders, which means powder coating as well as extrusion technology, because thermoplastic binders tend to encapsulate filament bundles. The problem can be minimised by a number of process adjustments.

Interface Inc. of Atlanta are developing biodegradable carpets using polylactic acid derived from cornstarch. This company has had a policy of sustainable development for some years now and aspires to zero emissions in manufacture. Interface is also one of the participants in a new company called WM Fibres Ltd which is setting up the first facility in the UK to recycle 12000 tonnes of carpet waste. The project, located at Knowsley, Merseyside is costing £1.5 million, and a significant feature is the ability to treat latex backed carpets. Eventually other textile waste will be processed.[122]

6.5.5 Recycling of textiles

At present, 'recycling' of textiles generally means cotton rags into paper, recovered wool (e.g. shoddy, mungo), charity shops selling second-hand clothing – a £15 million industry – and sending garments to third world countries – heavy clothing is sent to Eastern Europe, lighter garments to Africa. Clothing which is too degraded for use is made into rags for wipes.[123–126] However, these 'rags' will eventually have to be disposed of – probably with heavy soiling on them – and in addition disposable paper and non-woven textiles are reducing the need for rags as wipes. Other textile waste is shredded and used for filler and insulation in automotive components, and some is used in the flocking industry. Political problems have arisen over the export of used clothing because the EU classify ungraded mixed textiles as waste, and certain countries can impose additional import tax as waste control taxes.[127] In addition, some pressure groups believe used clothing should not be dumped on third world countries. A quality grading system for used clothing is being considered by the Textiles Recycling Association (TRA), Recyclatex, the Association of Charity Shops and other charities. There do seem to be a growing number of commercial organisations and associations concerned with textile recycling. The Bureau of International Recycling (Office in Brussels), affiliated to 600 companies in over 50 countries, has a Textiles Division. A 'Forum on Recycled Textiles' (FORT) has recently been formed by members of the TRA to discuss issues and co-ordinate effort. The Secondary Materials and Recycled Textiles Association (SMART) was actually founded in 1932, and has members from all over the world which represent a wide range of textile industries.[128,129]

Recycling of wool has been carried out for centuries, and for many years cotton rags have been made into paper. The problem of recycling of technical textiles as such does not seem to be a prominent issue at present, apart from the recycling of PVC coated fabrics. Despite work done in this area, there are as yet no actual PVC coated fabric recycling plants in commercial operation, and the recycling rate of PVC as an industry is believed to be only 3%. Light weight tarpaulins made from all olefinic material, i.e. polyethylene or polypropylene coated polyethylene or polypropylene yarns are being sold against lighter weight PVC coated material. The all olefinic material is easier to recycle than the PVC coated polyester. In another approach, research work is being conducted to develop water-based polyesters as coating resins for polyester fabric.[130] If this work is successful, the coated fabric will be made from a single plastic type and will be more suitable for recycling. In addition quicker, more economical and more convenient methods of identifying polymer type are being developed.

Recycling of coated and laminated fabrics is generally not easy, because by definition they are frequently made from at least two different polymers or materials and separation is likely to be a difficult if not an impossible commercial process for certain articles. Even standard articles of clothing such as shirts are likely to be made from polyester and cotton – two very different fibres. Much clothing and even old shoes are 'reused' by sending them to third world countries, and there may be some scope for this as a means to dispose of anoraks and cagoules which have become unfashionable or are only partly worn and which are still serviceable. However, the volume of coated jackets is quite small compared to the total amount of general garment waste. Some garment waste is shredded and used as sound and vibration insulation by the automotive industry and others. Coatings sometimes contain FR agents such as bromine compounds and antimony, and metallic additives and stabilisers such as those found in PVC and some rubbers. Waste material being disposed of now may have been made as early as 1990 or before when these chemicals were not considered harmful, and these materials may require special consideration. This factor could cause complications in recycling because these chemicals may now be considered harmful and their use prohibited. In addition, azo dyes were first banned in Germany in 1994 and are now also prohibited in France and the Netherlands. Clothing, especially industrial protective, may contain small amounts of potentially toxic chemicals; even smoking residues and traffic fume particulate matter may be harmful. They may have to be removed if the fibre is to be recycled into apparel, household textiles or automotive applications.

As has been seen in the case of automotive recycling, the recycling process begins with the logistics of collection, disassembly, if appropriate, identification and separation of the various fibre types, and then cleaning and disinfecting in the case of clothing. Material of the same fibre and quality has to be gathered together to make any recycling process economically feasible. Because of central government pressure on local authorities (see below), far more determined attempts are now being made to reduce the amount of waste which goes to landfill and to increase the proportion recycled and composted. The increased landfill charges, possibly £24 per tonne or more within the next four years, will change the economic feasibility of recycling and waste disposal processes. The Climate Change Levy on energy is also likely eventually to have an effect on the economics of recycling, reusing and certainly reducing. Everyone agrees that recycling, reusing and reducing are the best procedures, but they are not applicable to everything at present. As more and more articles are recycled, the issue of coated and laminated fabrics may well come under scrutiny and manufacturers of these articles could be affected under producer responsibility considerations. WRAP, a non-profit-making company, has recently been set up to create stable and efficient markets for recycled material. The government is to injected £40 million during the next three years – there may be opportunities for textiles?

6.5.5.1 Ecolog ™ Recycling GmbH

In Germany VAUDE Outdoor Sport GmbH, a sportswear manufacturer, together with a number of suppliers to the garment industry, have formed Ecolog Recycling GmbH with the objective of producing an ecological and economical recycling system.[131] The system is based on the concept of producing garments made entirely out of polyester, including the buttons and cord stoppers. The polyester is thus pure and is conveniently recycled into new buttons, fleece fibre, threads and fibres. The consortium includes Sympatex Technologies, the makers of the waterproof and breathable polyester film. The Ecolog system was first formed in 1994, and a year later a subsidiary was formed in Japan which includes Toray Industries, Itochu and the Japanese Ministry of Commerce. This concept of single-fibre clothing is gradually gaining strength in the struggle to conserve non-renewable Earth resources.

6.5.6 Recycling and waste disposal

In the UK, local authorities were set a non-statutory target in 1990 to recycle 25% of municipal waste by the year 2000. No local authority actually attained this figure, and the average for England and Wales was just

9.5%. Statutory targets have now been set, and in addition to this the EC landfill directive is looming. UK landfill tax that was first set at £7 in 1996, had reached £12 by 2001 and will rise every year until 2004. Some analysts have forecast an eventual figure of £34. This policy is an attempt to reduce the amount of waste landfilled. British landfill levels are the highest in Europe; 1998/99 figures show 83% landfill, 8% incinerated with energy recovery, 1% composted and only 8% recycled. In comparison, during 1996 the figures for Germany were 34% landfill, 18% incinerated with energy recovery, 10% composted and 38% recycled.[132–134]

The EU propose a Directive on Waste designed to reduce emissions to air, water and land from the incineration of non-hazardous waste. This will tighten up and extend the scope of existing Waste Directives. An EU Composting Directive is being proposed which may include mandatory source separation of biodegradable waste, limits on contamination in composted material and possible targets for composting. An EU discussion document proposes biodegradable landfill waste to be reduced by 25%, taking 1995 levels as the baseline and 2006 as the target date. Countries such as the UK with existing high landfill levels would be allowed four extra years. A working document used by the EU suggests a hierarchy for managing waste which can be summarised as prevent or reduce, reuse, recover (when environmentally justified), compost, mechanically or biologically stabilise unsorted biodegradable waste – energy recovery is the final option. The paper suggests separate collection of biodegradable waste to maximise the scope for composting. Even composting needs careful control and siting away from residential areas because of unpleasant odours. Composting in the UK has grown by 23% per year in the last few years, and stood at about one million tonnes at the end of 2000, but of course this is still very small compared some other European countries.[135–137]

Textiles on the whole do not appear to be a major problem – from figures available, textiles constitute only about 1.5 to 2% of municipal waste. A proportion of this is likely to be carpet for which, as has been detailed above, there is a technical means of recycling. Coated domestic and apparel textiles must account for only a small percentage of this small figure, and no reference or comment has yet been seen which is specific to coated or laminated textiles. Coated PVC appears to be the only coated fabric receiving attention at the moment, and this is probably more associated with the environmental pressure groups' reservations in connection with PVC polymer material than the coated article.

The debate over whether to dispose of unuseable materials by landfill or incineration continues, see Fig. 6.3. The problems associated with incineration are greenhouse gases, such as carbon dioxide, and the possible formation of dioxins. However, if managed effectively, incineration can provide useful energy, and the process of incineration with energy release is referred

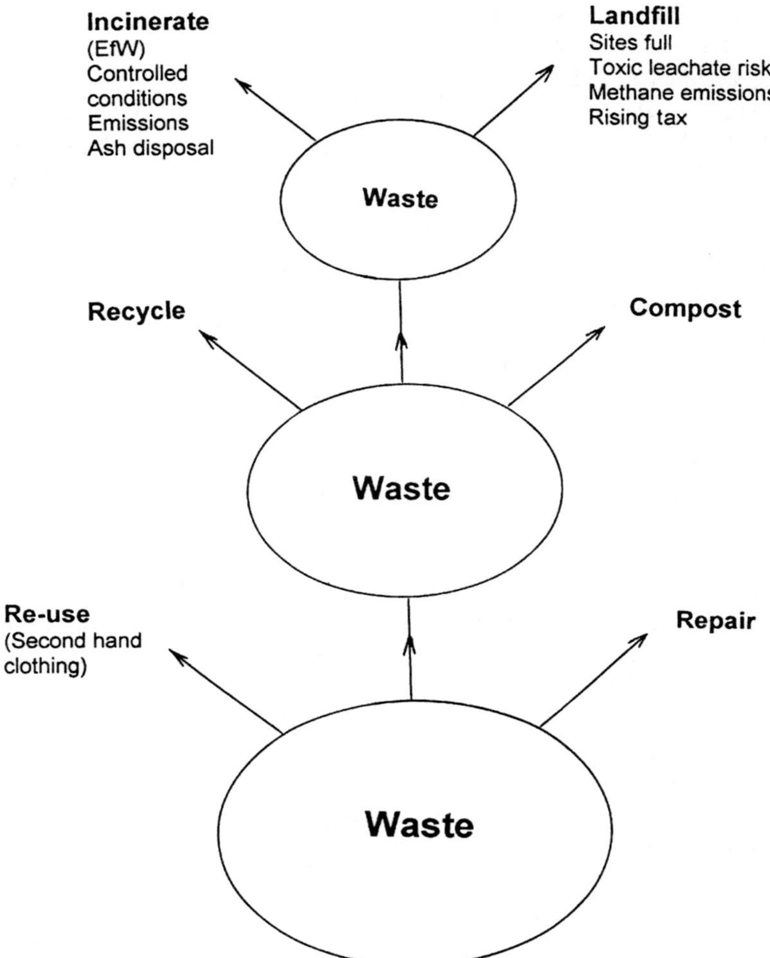

6.3 The problem of waste disposal. Material remaining after reusing (second hand clothing), recycling and composting must be disposed of either by incineration or by landfill. Incineration with energy recovery (EfW — energy from waste) requires careful control to prevent the formation of dioxins; emissions must be rigorously treated according to legislation, and preferably with the approval of local residents. Incineration residues, which may contain a high concentration of toxic substances, must be disposed of in some way. Landfill sites are filling up and landfill poses the risk of toxic leachate to underground waters, which may eventually reach rivers and drinking water. Landfill taxes are due to rise significantly. In addition, methane, a 'greenhouse' gas, is produced in landfills. The logistical problems of waste collection, dismantling, sorting and identification of different chemical types, and the transport costs are formidable. Routes may be possible, but they need to be commercially viable.

to as energy from waste (EfW). Landfill has the problems of rodents, insects, emissions of methane, the risk of explosion and possible leaching of liquors to watercourses. Leached liquors can contain high concentrations of toxic material, including ammonia and heavy metals, and can have high BOD, all of which are harmful to aquatic life and also pose risks to humans. However, methane emissions from landfill can also be used to generate electricity, and there are in fact 150 sites involved in power generation. On the other hand, the potential is likely to be reduced as less biodegradable waste is sent to landfill, and the generation of landfill gas needs 10 years for material to decay and produce useable levels of gas.

What cannot be recycled or composted needs to be disposed of in some way. If it has calorific value, incineration with energy recovery seems to be the logical means of disposal. To meet EU targets for reduced landfill, the increased use of EfW has been proposed, and this will require the construction of a number of EfW incinerators. The environmental pressure groups and others who fear toxic emissions and the effect on human health and the environment oppose this. The EU Directive on waste incineration contains a requirement that information on emissions from plants above a certain capacity, should be made public. The chief concerns of incineration are that waste not burnt at a high enough temperature could release dioxins and other toxic species into the atmosphere, and also that there is a problem disposing of residual ash which is likely to contain a concentration of toxic heavy metals. Any waste containing sulphur or nitrogen if incinerated is likely to produce the acidic emissions of sulphur dioxide and oxides of nitrogen which, as has already been noted, will require fume abatement. European Union regulations have also been applied to the technical design of municipal incinerators, and limitations have been set for toxic emissions. Incinerators are reputed to reduce waste volume by 90%, but some reports claim that the reality is nearer 50–70%, and that incinerator residual ash, which can only be disposed of by landfill, may contain toxic material in a more concentrated state than before. The emissions of municipal incinerators have been known to include dioxins, furans, oxides of nitrogen and even heavy metals, and this is one of the main reasons for the environmental pressure groups' concern. It is likely that there will be long lead times before new incineration capacity becomes available, because of the need for public enquiries and because of strong local opposition.

The UK government has set out its strategy to maximise the amount of value from the 106 million tonnes of commercial, industrial and municipal waste by increased recycling, composting and energy recovery. The amount sent to landfill must be reduced substantially, producers must increasingly expect to arrange for recovery of their material and new and stronger markets need to be developed for recycled materials.[138,139] Technical textile and coating polymer manufacturers, chemical and adhesive producers and marketing organisations way well be asked at some time in the future to contribute to disposal of coated or laminated fabrics.

6.5.7 Use of natural fibres and materials

In recent years there has been a revival of the use of natural materials in the automotive industry. The reasoning is that use of raw materials from renewable sources is more environmentally responsible than the use of synthetic fibres and plastics produced from non-renewable oil reserves.[140,141] In addition, natural materials are more biodegradable, and there are also technical advantages. Bast fibres, flax, sisal, kenaf and hemp are being explored as replacements for the more expensive glass as long fibre reinforcement in polyurethane injection mouldings for door panels and in other applications. Some products are already on the market and are being used, notably by German OEMs. Indeed the use of natural fibres in German cars quadrupled between 1996 and 2000 to 20000 tonnes and this figure is likely to double by 2005. Bast fibres are claimed to need less chemicals in growing and are therefore more environmentally friendly. Hemp is reported to grow over most weeds and can reach a height of 4m in 70 days. It is essential to use LCA techniques to determine impact on the environment; natural fibres are not necessarily the most environmentally friendly. Cotton, for example, while it is renewable and biodegradable, is believed to be sprayed with 12% of the world's insecticides and 26% of the world's pesticides, and in addition requires large amounts of water to grow.

The Crea Tech Process developed by Alpha Plastics and Haas Kunstoff uses natural thermoplastic material extracted from certain plants which is filled with derivatives from wheat, oats and soya. The resultant compound resembles and has similar properties to ABS, but will biodegrade under high humidity after 10 years. A further novel feature is that the bond between the covering decorative material and the rigid part can be controlled to be satisfactory in use, but can be separated at the end of the life of the car to facilitate recycling.[142,143]

DuPont and other major chemical companies are researching the manufacture of biodegradable polymers using plants as the starting material.[144,145] Work of this type is likely to increase, and it is possible that we will eventually see significantly less non-renewable oil being made into textiles. Already the cellulosic based Lyocell fibres, e.g. Tencel (Acordis), are being made in increasing amounts from wood grown in sustainable forests. Perhaps we should be researching biodegradable polymer materials as coatings.

6.6 References

1. Alloway BJ and Ayres DC, *Chemical Principles of Environmental Pollution*, Blackie-Academic, London 1997, 10 (after Meadows DH and DL and Randers J, *Beyond the Limits*, Kogan Page Earthscan, London 1992).
2. Watkins LH, *Air Pollution from Road Vehicles*, HMSO, London 1991, 83.
3. Anon, *The unfinished business after Kyoto, ENDS Report 275*, Dec 1997, 16–20.
4. Alloway BJ and Ayres DC, *Chemical Principles of Environmental Pollution*, Blackie-Academic, London 1997, 168.
5. Hodges L, *Environmental Pollution*, Holt, Reinhart and Winston, New York 1977, 70–1.
6. Arnold AE, 'Air regulations affecting the textile industry in New England', *AATCC Symposium, Coated and Laminated Fabrics; New Processes and Products*, Danver MA, USA 3–4 April 1995, AATCC, Research Triangle Park NC, USA.
7. Devine TW, 'Air permits; effect on manufacturing', *AATCC Symposium, Coated and Laminated Fabrics; New Processes and Products*, Danver MA, USA 3–4 April 1995, AATCC, Research Triangle Park NC, USA.
8. Watkins LH, *Air Pollution from Road Vehicles*, HMSO, London 1991, 66–82.
9. NSCA information leaflet, *Air Pollution and Human Health*, NSCA, Brighton Apr 1998.
10. Watson A, 'Generation problems, power generation and health', *Health and the Environment*, SERA, London 1996, 16–17.
11. Chanlett ET, *Environmental Pollution*, McGraw-Hill, New York 1973, 204–7.
12. NSCA information leaflet, *Motor Vehicle Pollution*, NSCA, Brighton Apr 1998.
13. Hodges L, *Environmental Pollution*, Holt, Reinhart and Winston, New York 1977, 429–30.
14. Vesilind PA, Peirce JJ and Weiner RF, *Environmental Pollution and Control*, Third Edition. Butterworth-Heinemann, Boston MA, USA 1990, 169–77.
15. Murley L (Editor), *1998 NSCA Pollution Handbook*, NSCA, Brighton 1998, 374.
16. Cooper P, 'Overview of the effect of environmental legislation in the UK textile wet processing industry', *JSDC*, Vol 108 April 1992, 176–82.
17. Murley L (Editor), *1998 NSCA Pollution Handbook*, NSCA, Brighton 1998, 8.
18. Gould R, 'Turning the screw; tightening up air quality standards', *EBM*, Sept 1998, 24–5.
19. Farrow L (interview), 'IPPC project manager', *EBM*, Apr 1999, 12–13.
20. Gould R, 'Voluntary schemes cut emissions', *EBM*, Jun 1998, 37.
21. Anon, 'Commission airs ideas for strong push on composting', *ENDS Report 309*, Oct 2000, 49.
22. Rayner J, 'News analysis', *Material Recycling Week*, 6 Mar 2001, 6.
23. Wagner SD (Ciba), 'Regulatory issues impacting the textile industry', *AATCC Symposium, Yarn Dyeing '96: Meeting the Challenges*, Sunset Beach NC, USA, AATCC, Research Triangle Park NC, USA.
24. Martin RL, 'Do the right thing with hazardous waste', *ATI*, Mar 1992, 60–2.
25. Fleming JS, 'Development of MACT standards for fabric dyeing, printing and coating operations', *AATCC Symposium, Coated and Laminated Fabrics*, Braintree MA, USA 7 Apr 1998, AATCC, Research Triangle Park NC, USA.

26. Jackson SL, 'ISO 14000: What you need to know', *ATI*, Mar 1997, 118–24.
27. Steadman L, 'Setting the standard; development in the ISO 14000 series', *EBM*, Apr 1999, 22–3.
28. Shaver JA, 'Applying ISO 14001 to environmental management', *Textile Chemist & Colorist*, May 1999, 27–30.
29. Shaw T, 'European Union Integrated Pollution Prevention and Control Directive and its impact on the wool textile industry', *JSDC*, Vol 114 Sept 1998, 241–6.
30. Laing IG (Ciba) 'The impact of effluent regulations on the dyeing industry', *Review Progress Coloration*, Vol 23 1991, 56–71.
31. Jackson K, 'Water pollution – environment and civil liability', *JSDC*, Vol 110 Apr 1994, 134–5.
32. Holme I, 'Flammability – the environmental and the Green movement', *JSDC*, Vol 110 Dec 1994, 362–6.
33. Grund N, 'Environmental considerations for textile printing products', *JSDC*, Vol 111 Jan/Feb 1995, 7–10.
34. Moran C, 'Reducing the toxicity of textile effluent', *JSDC*, Vol 114 Apr 1998, 117–8.
35. Gould R, 'Applying a little give and take (impact of Solvents Directive)', *EBM*, May 1999, 18–19.
36. Dobson I, 'Solvent abuse days are numbered', *Environmental Times*, Winter 1999, 46–51.
37. Anon, 'Over six thousand UK processes caught by Solvents Directive', *ENDS Report 283*, August 1998, 41–2.
38. Webb J, 'Polyurethane plant emissions control', *Urethanes Technology*, Oct/Nov 1995, 20–2.
39. Webb J, 'Flame laminators watch out – you are on the list', *Urethanes Technology*, Oct/Nov 1995, 23.
40. Anstey M, 'MACT for flame lamination', *AATCC Symposium, Coated and Laminated Fabrics*, Braintree MA, USA 7 Apr 1998, AATCC, Research Triangle Park NC, USA.
41. Brodhead Jr DJ, 'Practical environmental solutions to solvent coating processes', *JCF*, Vol **21** Oct 1991, 112–22.
42. Gleaves RJ and Dean JF, 'Control of air emissions', *JCF*, Vol **24** Jan 1995, 230–43.
43. Meyers D, 'Advantages of precipitators', *Textile Maintenance Engineering*, May 2000, 39–41.
44. Dertinger M, 'VOC abatement systems in the coating and laminating industry', *6th International Conference on Textile Coating and Laminating*, Dusseldorf 4–5 Nov 1996, Technomic Lancaster PA, USA.
45. Becht E, 'Specifying and selecting fume oxidizers for your process exhaust', *5th International Conference on Textile Coating and Laminating*, Williamsburg VA, USA 13–14 Nov 1995, Technomic, Lancaster PA, USA.
46. Berglund E, 'Exhaust control, industrial', *Kirk-Othmer, Encyclopaedia of Chemical Technology*, Fourth Edition, Vol 9, John Wiley, New York 1994, 1022–60.
47. Crocker BB, 'Air pollution control methods', *Kirk-Othmer Encyclopaedia of Chemical Technology*, Fouth Edition, Vol 1, John Wiley, New York 1994, 749–825.

48. Freiberg H, 'Clean exhaust air for textile machines-what are the options?', *Melliand English*, 9/1993, E323-E325.
49. Gottschalk K-H, 'Possibilities for cleaning exhaust air – an idea for economical methods', *Melliand English*, 9/1993, E321–E322.
50. Anon, 'CHP improves efficiency', *TTi*, May 2000, 4.
51. Anon, 'Fire research points finger', *PRW*, 24 Jul 1998, 1.
52. Anon, 'Oestrogens research fingers flame-retardant chemical', *ENDS Report 267*, Apr 1997, 9–10.
53. Anon, 'Bromine industry strengthens its defences', *ENDS Report 273*, Oct 1997, 9–10.
54. Anon, 'Life cycle benefits of flame retardants', *PRW*, 10 Dec 1998, 2.
55. Anon, 'DTI backs FR cause', *PRW*, 12 Feb 1999, 3.
56. Fuad-Luke A, 'The green grossers', *The Guardian*, 11 Mar 1999, 14.
57. Neitzel H (Federal Environmental Agency), '20 years of experiences of the German Environmental Labelling Scheme, Blue Angel', *Consumers Council Conference*, Washington DC, USA 24–5 Apr 1998.
58. McCarthy BJ and Burdett BC, 'Eco-labelling and textile ecology', *Review Progress Coloration*, Vol 28 1998, 61–70.
59. Zippel E, 'Oeko-Tex Labelling', *Eco-Textile'98 – Sustainable Development Symposium*, Bolton Institute 7–8 April 1998, papers edited by Horrocks, AR, Woodhead, Cambridge 1999.
60. Sager A, 'Heimatextil 2000; a 30 year milestone', *Textile Asia*, Mar 2000, 14.
61. Bide M, 'AATCC's Environmental Symposium', *Textile Chemist & Colorist and American Dyestuff Reporter*, Jan 2000, 21–3.
62. Anon, 'Draft paper moots lower VAT for eco-labelled products', *ENDS Report 312*, 44–5.
63. Anon, 'Energy examples (Climate Change Levy), *Environmental Times*, Spring 2000, 30–1.
64. Green D, 'Combined heat and power uptake set to soar', *Environmental Times*, Winter 2000, 34–8.
65. Anon, 'Companies clamour to come under IPPC to win climate levy discount', *ENDS Report 312*, Jan 2001, 33–4.
66. Anon, 'End of life vehicles (ELVs) recovery; European situation', *Automotive Textiles Newsletter*, Apr 1997 Marvel/Rhone-Poulenc Setila, 4.
67. Ehler P and Schreiber H, 'Textile waste again in automotive applications', *R'97 International Congress*, Palexpo 4–7 Feb 1997, EMPA, St Gall, Switzerland.
68. Weber A (BASF), 'Potential for recycling plastics from scrap cars', *PRW*, Apr 14 1990, 13.
69. Rebboah S and Smith GF, 'Recycling implications in the motor industry' (C524/142/97), *Autotech '97*, 4–6 Nov 1997, Mechanical Engineer Publications, Bury St Edmunds, UK 1997.
70. Anon, 'Scrap car challenge for recycling regulators', *ENDS Report 316*, May 2001, 23–6.
71. Anon, 'Identifying opportunities of the end-of-life-vehicles legislation', *MRW*, Apr 2001, 15.
72. Thyer R, 'Options for ELVs released', *MRW*, 17 Aug 2001, 6.
73. Rayner C, 'Time running out on ELV Directive', *MRW*, 17 Aug 2001, 9.
74. Pryweller J, 'Ford sets tough new guidelines for recycled plastics', *Automotive News Europe*, 5 Jul 1999, 22.

75. Anon, 'Compounders respond to recycling demands', *MPI*, Apr 1999, 12.
76. Anon, 'Ford is targeting 50% use of recycle-content resin by 2002', *MPI*, Jul 1999, 26.
77. Anon, 'Daimler-Chrysler ups recycling stakes', *ISATA Magazine*, Jun 1999, 12.
78. Anon, 'Research projects for car recycling', *ENDS Report 318*, Jul 2001, 16.
79. Kmitta S (Fehrer), 'Polyester nonwovens – an alternative for car seats', *IMMFC, Dornbirn* 20–22 Sept 1995.
80. Wilkens C, 'Raschel knitted spacer fabrics', *Melliand English*, Vol 10 1993, E348–E349.
81. Karl Mayer/Malimo technical information leaflet, '*Manufacture of Fabrics for Automotive Interiors Using Warp Knitting and Stitch Bonding*', We 75/1/93.
82. Fuchs H and Bottcher P, 'Textile waste materials in motor cars – potential and limitations', *Textil Praxis International*, Vol 4 Apr 1994 4, II–IV.
83. Hirschek H, 'Recycling of automotive textiles', *IMMFC, Dornbirn* 22–24 Sept 1993.
84. Costes M (Rhone-Poulenc), 'Use of textiles in vehicles and recycling; state of the art and outlook', *IMMFC, Dornbirn* 22–24 Sept 1993.
85. Kiefer B, Bornhoff A, Ehrlern P, Kingenberger H and Schreuber H, 'Assessing second hand automotive textiles for use in new vehicles', *IMMFC, Dornbirn*, 20–22 Sept 1995.
86. Coll-Tortosa L, 'Recyclable upholstery textiles for the automotive industry' (BRITE-EURAM project), *IMMFC, Dornbirn* 16–18 Sept 1998.
87. Schmidt G, 'Replacing foam by using 'Kalitherm' technology and flameless laminating', *IMMFC*, Dornbirn 15–17 Sept 1999.
88. Schmidt G and Bottcher P, 'Laminating nonwoven fabrics made from or containing secondary or recycled fibres for use in automotive manufacture', *Index '93*, EDANA, Brussels.
89. Fung W, 'Properties required for nonwovens for use in motor cars as substitutes for other materials', *Index '99*, Geneva 28–29 May 1999, EDANA, Brussels.
90. O'Toole K, 'Waste issue takes centre stage', *European Plastic News*, Nov 1991, 60–66.
91. Hillier K, 'The recycling of flexible polyurethane foam', in *Chemical Aspects of Plastics Recycling* (Editors Hoyle W and Karsa DR), Royal Society of Chemistry, Cambridge 1997, 127–36.
92. Caligen Foam Fact Sheets, *The Environmental Choice – Polyurethane* and *A Clearer Focus on Recycled Foam* – with ARCO.
93. ISOPA – European Isocyanates Producers' Association – seven fact sheets on recycling options for polyurethane foam, Oct 1993.
94. Anon, 'ACORD deal agreed', *MRW*, 18 Jul 1997, 3.
95. Anon, 'PRAVDA – the moment of truth?', *EPN*, Sept 1991, 58–9.
96. Eminto S, 'Vehicle recycling – UK to take voluntary route', *MRW*, Jan 1997, 12–14.
97. James B, 'Rover takes CARE to meet European goals', *PRW*, 7 Feb 1997, 7.
98. Anon, 'Automotive textiles', *Kettenwirk-Praxis*, Vol 3, 1995, E30–E32.
99. Trautmann J, 'Recycling of automotive textiles', *Textile Asia*, Mar 1998, 45–6.
100. Reed J, 'Directive moves to ban tyres from landfill', *EBM*, Sept 1998, 21.
101. Reed J, 'Draft landfill directive; main provisions as they affect UK', *EBM*, Jun 1998, 21.

102. Anon, 'Tyre burning decisions to speed up under new consultation procedure', *ENDS Report 316*, May 2001, 41.
103. Anon, 'Parliament targets PVC in vote on incineration Directive', *ENDS Report 302*, Mar 2000, 44.
104. Anon, 'Green paper on PVC includes option of substitution', *ENDS Report 306*, Jul 2000, 40.
105. Anon, 'UK opposes EC moves on PVC' *ENDS Report 310*, Nov 2000, 45.
106. Saffert R, 'PVC plastisols for fabric coating – environmental aspects', *JCF*, Vol 21 Apr 1992, 281–300.
107. Saffert R, 'Recycling of PVC coated fabrics', *JCF*, Vol 23 Apr 1994, 274–9.
108. Krummheuer WR and Scobel M, 'Recycling and disposal of industrial fabrics', *JCF* Vol 23 Oct 1993, 105–23.
109. Perillon J-L and Bourbon E, 'Recycling of PVC coated textiles; a new source of technical materials', *10th International Conference on Textile Coating and Laminating*, Lyon, France 9–10 Oct 2000, Technomic, Lancaster PA, USA.
110. Adanur S, Zhenwei H and Broughton RM, 'Recovery and reuse of waste PVC coated fabrics, Part 1; experimental procedures and separation of fabric components', *JCF*, Vol 28 July 1998, 37–55.
111. Adanur S, Zhenwei H and Broughton RM, 'Recovery and reuse of waste PVC coated fabrics; 'Part 2, Analysis of the components separated from PVC coated PE fabrics', *JCF*, Vol 28 Oct 1998, 145–68.
112. Bohnhoff A and Petershans J, 'Product responsibility in the carpet industry; from recycling capability to recycling management', *IMMFC*, Dornbirn 15–17 Sept 1999.
113. Anon, 'Fibres for carpets', *Textile Asia*, Mar 2000, 22–5.
114. Anon, 'Carpet recycling into PA monomers', *Chemical Fibers Inernational*, Vol 49 May 2000, 204.
115. Anon, 'Carpet recycling offers marketing advantage to nylon', *ENDS Report 302*, Mar 2000, 18.
116. Whaley P, 'Evergreen makes nylon live forever', *ATI*, Oct 2000, 32–40.
117. Wang Y, 'An overview of activities on recycling of fibrous textile and carpet waste at the Georgia Institute of Technology', *Ecotextile '98 – Sustainable Development Symposium, Bolton Institute 7–8 April 1998*, papers edited by Horrocks AR, Woodhead, Cambridge 1999.
118. Anon, 'Carpet recycling gets off the ground', *MRW*, 21 Feb 1997, 23–4, 39.
119. Lennox-Kerr P, 'A practical process for the recycling of unwashed carpets,' *TTi*, Nov 1998, 20–1.
120. Mohsen M, Horrocks R and Woods C, 'Carpet waste, an expensive luxury we must do without!', *Ecotextile '98 – Sustainable Development Symposium, Bolton Institute 7–8 April 1998*, papers edited by Horrocks AR, Woodhead, Cambridge 1999.
121. Wolff R, 'Adbac^{R}, a new concept for latex-free carpet', *IMMFC*, Dornbirn 15–17 Sept 1999.
122. Anon, 'WRACE against waste', *Textile Horizons*, Jul 2001, 43.
123. Lowe S, 'Ahead of the rag trade', *MRW*, 21 Feb 1997, 18–21.
124. Patel S, 'Getting sorted with OHS', *MRW*, 21 Feb 1997, 27, 39.
125. Walker DG, 'Taking stock of textiles', *MRW*, 21 Feb 1997, 28–31.
126. DETR, *Waste Strategy 2000; Sustainable Development*, The Stationary Office, London, May 2000, 141–3.

127. Anon, 'Trade restriction uncertainty for used clothing recyclers', *MRW*, Feb 23 2001, 18–19.
128. Bureau of International Recycling; http://www.bir.org/biruk/info/textiles.htm
129. http://www.nrf.org.uk
130. Jansen I, Adler H-J, Hardtke G and Fuchs H, 'Water-dilutable coatings derived from polyester for textile use', *JCF*, Vol 29 Oct 1999, 139–50.
131. auf der Brucken B, 'A system for returns', *Melliand English*, Vol 82 May 2001, E110.
132. Anon, 'UK remains a leading laggard in Europe on recycling', *ENDS Report 309*, 17.
133. Rayner J, 'Mixed response to waste report', *MRW*, 23 Mar 2001, 6.
134. Thurgood M, 'Identifying trends in recycling and residual waste management', *MRW*, 20 Oct 2000, 8–11.
135. Rayner J, 'Survey shows concerns and problems facing compost', *MRW*, 15 Dec 2000, 10–12.
136. Anon, 'Commission airs strong push on composting', *Ends Report 309*, Oct 2000, 49–50.
137. Cooper J, 'Changing the way Europe treats biodegradable waste', *MRW*, 24 Nov 2000, 8–9.
138. DETR *Waste Strategy 2000*' Part 2, The Stationary Office, London May 2000, 7–8.
139. Anon, 'Weighing up the new recycling challenge for councils', *ENDS Report 309*, Oct 2000, 25.
140. Anon, 'Car manufacturers embrace natural fibres', *ENDS Report 312*, January 2001, 24–6.
141. Mapleston P, 'Automakers see strong promise in natural fibre reinforcements', *MPN*, Apr 1999, 63–4.
142. Roth T (Alpha Plastics), 'Launch of new technology', Lecture, University of Sheffield 23 Feb 1993.
143. Anon, 'New German technology may promise easier recycling', *PRW*, 20 Mar 1993, 9.
144. Lunt J (Cargill Dow Polymers), 'Polyactic acid fibres – their promise for industrial textiles', *TTi*, Dec 2000, 11–13.
145. Anon,'Carpets from corn', *Textile Asia*, Jul 2000, 114–15.

6.7 Further reading

1. Adanur S (Editor), *Wellington Sears Handbook of Industrial Textiles*, Technomic, Lancaster PA, USA 1995, especially, Chapter 21, 'Waste management', 713–28.
2. Allaby M, *Green Facts; the Greenhouse Effect and Other Key Issues*, Hamlyn, London 1986.
3. Alloway BJ and Ayres DC, *Chemical Principles of Environmental Pollution*, Second Edition, Blackie-Academic, London 1997.
4. *Automobile Material Technology – Proceedings of Autotech'97*, 4–6 Nov 1997, Mechanical Engineers Publication, Bury St Edmunds, UK 1997.
5. BS EN ISO 14001: 1996, British Standards Institution, London.
6. Chanlett ET, *Environmental Protection*, McGraw Hill, New York 1973.

7. Connell DL, 'The environmental impact of the textile industry', in *Chemistry of the Textile Industry* (Editor Carr CM), Blackie-Academic, London 1995, 333–54.
8. Cooper P, *Colour in Dyehouse Effluent Control*, SDC, Bradford 1995.
9. Degobert P, *Automobiles and Pollution*, SAE and Editions Technip, Paris 1995.
10. DETR brochures, *The Environmental Impact of Road Vehicles in Use* and *Driving the Agenda*, The first report of the cleaner vehicle task force, both Jul 1999.
11. DETR, *A Better Quality of Life, A Strategy for Sustainable Development*, The Stationary Office, London 1999.
12. DETR, *Indicators of Sustainable Development for the United Kingdom*, The Stationary Office, London 1996.
13. DETR, *Solvent Capture and Recovery in Practice, Industry Examples*, Environment Technology Best Practice Guide GG100, Nov 1997.
14. Engineers' Employers Federation (EEF), *The EEF Register of Environmental Registration*, EEF, London, 1999.
15. Hester RE and Harrison RM, *Waste Incineration and the Environment*, Royal Society of Chemistry, Cambridge 1994.
16. Hodges L, *Environmental Pollution*, Holt, Reinhart and Winston, New York 1977.
17. Holme I, 'FR & the environment', *International Dyer*, Jan 2001, 34–7.
18. Horrocks AR (Editor), *Ecotextile '98 – Sustainable Development*, Proceedings of conference held at Bolton Institute 7–8 April 1998, Woodhead, Cambridge 1999.
19. Horrocks AR (Editor), *Recycling Textile and Plastic Waste*, Papers from 1995 Ecotextile Conference 'Wealth from waste in textiles', Woodhead, Cambridge 1996.
20. Kachadourian G, '"Green v Green", Automotive Recycling', *AI Automotive Industries*, Oct 1999, 41–4. Also www.ai-online.com (*Automotive Recycling*, USA).
21. Mackie G, 'Natural selection', *Textile Horizons*, Nov 1999, 8–11.
22. McKenzie J, 'Towards eco fashion', *Textile Horizons*, Mar 1999, 16–17.
23. McKenzie J, 'Chicken and eggs of growth', *Textile Horizons*, May/Jun 2000, 10–12.
24. Moran T, 'To recycle, or not to recycle? Is it truly viable?', *Automotive & Transportation Interiors*, Nov 1999, 26–31.
25. Murley L, *1998 NSCA Pollution Handbook*, NSCA Publications, Brighton 1998.
26. *R'97 Industrial Congress*, Palexpo 4–7 Feb 1997, EMPA, St Gall, Switzerland.
27. Roberts DL, Hall ME and Horrocks AR, 'Environmental aspects of flame-retardant textiles – an overview', *Review Progress Coloration*, Vol 22 1992, 48–57.
28. Royal Society of Chemistry, *Simple Guide to the Management and Control of Waste*, ECSC-EEC-EAE, Brussels 1996.
29. Sheldon C and Yoxon M, *Installing Environmental Management Systems – a Step by Step Guide*, Kogan Page, Earthscan, London 1998.
30. Shishoo RL, 'Environmental issues facing the technical textiles industry in Europe', *JCF*, Vol 24 Oct 1994, 117–28.
31. *Sustainability Report*, issued by Interface Inc., Atlanta GA, USA 1997.
32. *Time Magazine* special edition, 'Our Precious Planet' (includes feature 'The Green Car' – Toyota Hybrid), Nov 1997.

33. UNEP (United Nations Environmental Programme), *Global Environmental Outlook 2000*, Kogan Page, Earthscan, London 1999.
34. Vesilind PA, Peirce JJ and Weiner RF, *Environment Pollution and Control*, Third Edition, Butterworth-Heineman, Boston MA, USA, 1990.
35. Watkins LH, *Air Pollution from Road Vehicles*, HMSO, London 1991.
36. Weinberg JH, Kiezulas MP and Pacquette MT, 'The coated fabrics industry and hazardous chemical regulations', *JCF*, Vol 16 Jan 1987, 159–70.
37. Wilson A, 'The power and the glory' (CHP), *Textile Month*, Nov 1999, 6–9.
38. Wragg PJ, 'Where to now with FR?', *Eco-textile '98*, Bolton Institute, Conference proceedings, papers edited by 7–8 Apr 1998, Horrocks AR, Woodhead, Cambridge 1999.

7 Future developments and outlook

7.1 General survey

In common with every other industry, textile coating and lamination is undergoing change and 'rationalisation' – unfortunately from the employees' point of view, all too frequently a euphemism for company mergers and job losses. On the positive side, a fresh approach is frequently stimulated which results in solutions to old problems and the bringing of new benefits and opportunities. Larger units, and being part of a larger group, can produce advantages of economies of scale and a wider economic and geographical reach. Some sectors of the textile industry, such as garments, are labour intensive, and the larger American and Western European companies have moved production to those areas of the world which have lower labour costs, such as Eastern Europe, North Africa and Asia. In many cases, they have retained the research, development and design functions in the home country. European designers of both garments and other items such as car upholstery are widely regarded as world leaders. A significant proportion of the UK polyurethane coating production has been lost to competition in the Far East, where many of the base fabrics of nylon and polyester are spun and woven. This process has, however, led to new overseas markets for British and European machinery producers and the chemical manufacturers who make the coating and lamination materials. The European machinery makers have an excellent reputation, but must stay ahead by investing in research and development. Some developing countries already have a good insight into and understanding of the overall global textile industry; they have plans for intensive development with a view eventually to becoming industry leaders.[1,2]

The coating and lamination industry is generally regarded as being closely associated with the technical textiles sector; this is generally true both because of the specialist technical requirements of the products, but also because of the technical nature of coating and lamination processes. However, there are still significant parts of the coating and lamination industry which have a considerable fashion element, i.e. household textiles and especially sports protective clothing, some of which is now sold in chain stores with designer and seasonal associations. The boundaries between fashion sports clothing and high performance protective clothing for the keen sportsman appears to have become quite diffuse. A well-designed anorak containing the same fabric content as a plainer design can retail at perhaps three times the price. There is now more choice of clothes in different designs than at any other time, and 'mass customisation' is influencing each sector of the clothing market with a performance level and a price level for each article. Possibly there is scope for more levels of performance and design content in the sportswear protective clothing market, and this may be becoming true of any consumer item, e.g. curtains, upholstery, carpets, etc. The increase in the quality of life in the developed countries is raising standards all round with ever-increasing comfort and convenience. Many of the stiff, nylon coated, non-breathable anoraks and cagoules of the 1980s have been replaced by softer breathable coatings or laminates, and this trend for softer, lighter and more comfortable garments will continue, not just for the fashion sector but also for industrial garments. Mass customisation requirements are gradually spreading to all sectors of consumer industry, and the time will come when both consumers and industrial workers will require protective clothing in their own individual style and design. These garments will be tailored to fit their body size and shape exactly (via a 'body scan' taking perhaps 130 body measurements), will fulfil all the technical specifications required for protection and durability, and will be available within a day of ordering.

In the automotive industry, more responsibility for design and development is being passed on to the suppliers with whom the car makers have close relationships or partnerships. The immediate suppliers are known as Tier 1 suppliers, and indirect suppliers as Tier 2. Some of the largest Tier 1 suppliers are now so large and influential that they are termed 'Tier ½.' The automotive industry has become a truly global industry, where materials, including fabric, are out-sourced anywhere in the world. The car companies have become assemblers of components and modules produced by their suppliers. Modules are becoming more sophisticated, and the time will eventually come when a whole car interior will be supplied as a complete ready to install module. Competition is intense and suppliers are faced with annual 'cost downs', i.e. being required to lower their selling price to the OEMs. Delivery JIT is essential, and the time factor for new developments and time to market for new products is now, in many cases, considerably less than one year, while at one time the norm was considered to be at least three years. Of course time is critical in the fashion garment industry, and one of the distinguishing features of technical textiles is their reduced dependence on changing fashion. Having said this, however, we have seen that there is now an element of up to date fashion requirement in virtually everything which the general public sees or comes into contact with.

7.2 Processing

To be profitable, it is now necessary to produce first quality material virtually all of the time – as industry in general aspires to the ideal of 'zero defects', i.e. giving the customer exactly what is agreed in the sales contracts every time and delivery on time every time. However, much intermediate and final inspection of fabric is still done manually by trained operatives, the best of whom still miss up to 20% of faults even at relatively slow speeds. Some customers actually specify the speed at which examination of their particular product must take place. Faults in pattern and fabric construction are still found at final inspection in laminated automotive fabric when the base fabric's value has been increased by a factor of maybe 500%. Research continues to inspect fabric on-line,[3] during production, and some advanced military technology is being adapted for this purpose. In an age where a car number plate can be read from a satellite in orbit, the technology fully to automate all aspects of coated and laminated fabric inspection is probably available. The problems are the initial cost of development and how to make it affordable for the small and average size fabric producer.

As much as possible, the 'human factor' is being removed from the production process and replaced by automatic instruments with computer control to ensure that the product is manufactured to a consistently high standard. The 'human factor' has become the least controllable and the least reliable in the manufacturing process. More automatic controls are required on processing machines to sense fabric tension variations and then automatically make the appropriate adjustments.[4] This will reduce variations in resin add-on in coating operations, variations fabric dimensional stability and pattern distortions, and also help to lessen the occurrence of creases. Resin add-on is already measured automatically on-line, for example by beta gauges. Again, the technology probably exists to ascertain the reason for the variation and to make automatic adjustments in, for example, blade angle or gap size, resin viscosity or whatever is necessary to bring the process back within specification. Material, however, must be produced in sufficient volume to justify the expense of such sophisticated instrumentation, and only very large operations will be able to afford it.

Waste reduction of chemicals and resins, especially those with a limited pot-life, is another area where better customer communications, production planning and organisation, perhaps computer assisted, could allow improvements and cost savings. Optimisation of resin coating add-on and the amount of adhesive necessary to pass specifications should lead the way to more precise control of materials. If, say, the required peel bond can be achieved with a margin of safety with an adhesive add-on of 16 g/m^2 any thing over this is excessive cost. If the average add-on is even, say, 18 g/m^2, reducing it by just 2.0 g/m^2 would reduce the cost of adhesive by 11%. Precise automatic process control would be essential to enable this cost reduction.

Advances are being made gradually, and sometimes in quite ingenious ways. For example, advances in engineering are allowing the design and production of rollers which run truer without bowing and remaining completely parallel during the production process. This has been achieved on a thermobonding calender using electromagnetic forces.[5] Sensor Products have just introduced a novel roller pressure testing sensoring device in the form of a thin sheet, which changes colour depending on the pressure applied between two surfaces. These developments could have applications in pad mangles, calenders and nip rollers in coating and lamination machines, and improve the across-the-width uniformity of take-up and add-on of chemicals in padding operations. Uniform pressure across the width in nip rollers in lamination should help improve reproducibility and reduce variability of peel bond results.

Costs are being reduced by producing laminated material in the same production line as fabric. One manufacturer is film extruding polyolefin film on to freshly produced nonwoven material in a single manufacturing line. This of course eliminates a whole series of processes, packaging, loading, transportation, unloading, etc. The trend will also be for longer, wider production runs with less down time for changing machine settings and materials to produce different qualities, and for machine maintenance. More effective production planning should result from the communications revolution, perhaps eventually via the internet, and from combining related processes and by production under commission, possibly for external customers – even competitors! Future production costs will be reduced by more effective buying through better communications via the inernet, wider width processing, longer continuous running with much less down time, better process efficiency, less waste and making more first quality material by more precise automatic control of machine parameters. In addition, new management techniques, team working and the culture of continuous improvement will become more widespread.

7.3 New and novel materials

New materials, new yarns and yarn variants, speciality polymers, chemicals and films are introduced periodically to satisfy the ever-increasing requirements of a higher standard of living or to exploit some novel property, e.g. phase change material. Changing circumstances may also provide opportunities for chemical specialities; for example upper atmosphere ozone depletion has led to the use of UV absorbing agents in clothing to reduce the risk of skin cancer. Since the late 1980s there has been a boom in anti-microbial yarns and chemicals as people become more conscious of health and hygiene. Yarns with exceptional strength, such as the ultra high modulus polyethylene Spectra, have been developed. Variants of polyester have appeared which are reputed to have many of its advantages together with better moisture management and softer handle and drape. WL Gore have developed a new yarn based on PTFE with 50% higher tensile strength than existing yarns. This can be used in structural and architectural applications, and when coated with PTFE, it becomes a coated material made from the same polymer, which is likely to facilitate recycling or ultimate disposal. There could be opportunities to develop coated products with these new yarns. An attempt has been made to produce a matrix for innovative product development in Tables 1.8 and 1.9.

New 'foam-in-place' and 'one-shot' manufacturing techniques in the automotive industry have produced new applications for foams or films which are laminated to fabric. In the former technique a car head rest, for example, is made by sewing the head rest cover into a 'bag' into which liquid chemicals are poured. These chemicals react to form the polyurethane foam inside the headrest cover. The foam, which is usually a high density, low porosity type, prevents the liquid components seeping through the fabric before the reaction is complete. In the latter case the fabric, laminated to a film, is put into a mould and hot liquid polymer is injected on to it to form a rigid component such as a door casing. The film prevents the hot polymer seeping through to the face of the fabric. These techniques combine several steps into a single process. Variants of this technique could be used in the manufacture of other articles and components in other industries. However, it is notable that, even in these high tech operations, the trimming of edges is still done by hand. Engineers still strive to eliminate manual handling by the use of robots.

Anti-wicking yarns have been developed for tarpaulins and covering materials. There is much activity in re-examining the use of natural fibres for use in composite structures and for the automotive industry. Resins made from natural products are being developed; possible coating resins may be also be produced from naturally occurring renewable sources. Some of the major chemical companies are focusing on biotechnology for polymer production.[6] It has been suggested that eventually we will be able to tailor make a polymer of a fibre which with have virtually any property required. Biosynthesis of molecules is believed to be more versatile and more precise than chemical methods. In addition, the manufacturing processes could require less capital and also be environmentally cleaner with less pollution, and the products themselves perhaps more biodegradable. However, LCA should be applied to both existing and any alternative technology or product.

The weak points in any garment are the seams. This is especially so for those garments intended for protection against water or maybe some other more harmful liquid. Seam design is a subject for garment technologists, but at the end of the day, when the garment is sewn, holes are made. Seams are generally taped in up market garments, but as wearers know, some seams are better than others. Welded seams are better, but not all materials are suitable, and welding, if not done carefully, can result in creasing and puckering. The Welding Institute believes they have a solution for this problem in a new laser-based process which could replace sewing in garment production, including water-resistant garments.

Surface science research continues to lead to a better understanding of the factors which determine good adhesion, and newer techniques are being developed to pretreat surfaces including those of fibres.[7] Plasma treatment has already led to significantly improved adhesion in polyurethane fabric coating, and could also produce benefits in other areas of coating and lamination. These techniques may possibly lead, not only to new products, as the result of being able to join materials which could not previously be joined together satisfactorily, but also to coating processes taking place at reduced temperatures and faster speeds. In the case of lamination, reduced levels of adhesive may be possible, and these steps would produce significant economic as well as environmental benefits.

Small independent companies may not have the resources for research, but there may be opportunities to join a consortium for collaborative research, possibly with government or even EU funding. The textile departments of universities, and especially BTTG, are usually well informed on these matters. The UK government is currently funding the TechniTex Faraday Partnership to promote improved interaction between the UK science, engineering and technology base and the technical textile industry. A prime objective is to turn the scientific and technological knowledge available in UK universities and research organisations into commercial success. The involvement of small companies is emphasised in the network, which is being built up to produce a proactive interface between industry and academia. The intention is that this will lead to academic excellence, industrial innovation and best practice in technical textiles. The exercise is being co-ordinated by BTTG and involves Heriot-Watt, UMIST and Leeds University.

7.4 Environmental aspects

As discussed in Chapter 6, the environment is likely to have a higher profile from now onwards. This is going to affect both chemical manufacturers and coating and lamination factories, as potentially toxic chemicals are brought under more control and emissions to atmosphere are restricted more stringently. Technology already exists to control virtually all manner of emissions; applying it to commercial operations is a matter of cost. New EC legislation has further tightened up regulations which in the 1980s and 1990s were themselves regarded as exacting, and this trend is likely to continue. The use of solvents is likely to continue to fall and perhaps disappear completely. The use of even water-based products may become unacceptable at some stage because of their relatively high energy demand in the evaporation of water – 539 calories/g. Both solvent and water-based adhesives are likely to be replaced more and more by 100% solids or hot melt materials, which require less energy for processing and hence produce less overall emissions. The appearance of reactive, moisture curing polyurethane adhesives has allowed high peel bonds to be achieved at low levels of adhesive add-on and gives softer handle in laminated garments. These adhesives have formed a new benchmark for fabric adhesives. Ways of saving energy are being explored on a number of fronts. Rohm and Haas and other resin manufacturers have researched low temperature crosslinking acrylic resins since before the 1980s, but recently a consortium including Oil States Industries (Aberdeen), Freudenburg (Wallsend), BP and Robinson Brothers (West Bromwich) have initiated a project to develop ultra fast cure accelerators for rubber.[8] The objective is to enable rubber to be cured at much lower temperatures than normal, thus saving energy. The UK government is contributing to the cost of the research through its Energy Efficient Best Practice Programme.

Environmental protection can be combined with significant economic benefit, as has been shown by certain companies, e.g. Heathcoats. The government is encouraging combined energy saving schemes, and factories where these are already in place report significant savings in cost and energy. Allowances in the new 'Climate Change Levy' on fuel (the so called carbon tax) should result in more energy efficient processes. Increased landfill tax will make landfill the last option with more effort put into recycling and composting. In the long term, these measures may well have harmful effects on the coating and lamination industry, since it basically joins two or more materials together which, at present, are generally dissimilar chemically. The disposal of coated fabrics does not appear to be a serious issue at the time of writing, but it does need serious consideration, especially if the coating contains heavy metal stabilisers, other additives and FR chemicals. Incineration, if not conducted efficiently, could lead to harmful emissions, especially dioxins, which could arise from chlorine containing materials. The residual ash from incineration, which may contain potentially harmful materials such as heavy metals in a now more concentrated form, needs to be disposed of, and some type of landfill seems the only option, even with all the possible risks. The debate continues as to whether incineration with energy recovery, now referred to as EfW, is the way forward. Most environmentalists agree, however, that recycling must increase, and there may well come a time when coated and laminated materials will come under scrutiny.

More LCAs are now being conducted, and indeed this is a feature of the ISO 14000 group of standards. Natural products are not necessarily the most environmentally friendly materials. Cotton is being recognised as quite environmentally unfriendly when the pesticides required for the growth of the cotton plant and the considerable wet processing required to produce a cotton garment are taken into consideration. In addition, the energy and detergent necessary to launder cotton goods compared to synthetic fibres should not be forgotten. We may see more use of bast fibres and of hemp, regarded by some environmentalists as the most environmentally friendly fibre. Is there any information on the coating adhesion of fabric coating polymers on specialist natural fibres such as pineapple fibre or hemp? Biodegradable polymers or polymers derived from renewable resources continue to be developed – can any of these be used in fabric coating, either as a resin or filler?

Disposable garments, such as surgeons' gowns and other items used in the medical sector, are presenting a disposal problem, and the search is on for materials with the necessary performance standards which are also washable, sterilisable and reusable. Laminated fabrics are being considered as possibilities; the results of 'cradle to grave' analyses will be interesting. The expected growth in incontinence devices and garments is likely to lead to disposal problems, and there may be an opportunity for comfortable, washable, reusable items which could be produced from coated fabrics. There may be parallels with and lessons to be learnt from other areas, presenting further possibilities for coated and laminated fabrics. Before the appearance of disposable nappies and diapers, there was a market for coated baby pants which could be machine washed and reused many times. Products of this type may well reappear and the disposable culture may well recede.

The joining of two or more dissimilar materials may eventually be discouraged as not being environmentally friendly. At the present time, consideration is being given to reduced VAT on products bearing eco-labels, and this may well put certain coated and laminated materials at a disadvantage.[9] The Oeko-tex label is being used more and more, and it is gaining wider recognition not only in Europe but also in Japan and the USA. Probably the long term future lies in producing laminates from materials of a similar chemical type, an example being the replacement of polyurethane foam laminated to polyester fabric for car seat fabric by polyester nonwoven material. By doing this, a laminate with similar, but by no means identical properties is obtained from a laminate comprising two chemically similar materials which can be conveniently recycled. Research is being conducted to produce water-based polyester resins, which could be coated on to polyester fabric to give barrier properties. One material used quite widely in laminated fabrics is PTFE, which at present cannot be easily recycled. However, the actual volume must be quite small in comparison to, say, PVC, which has been the subject of recycling projects for a number of years. Possibly some laminated goods could be produced with a 'measured' amount of adhesion, so the two chemically dissimilar materials could be separated at the end of the articles' life to facilitate recycling. This concept has already been mentioned in the Crea Tech process, see Section 6.5.7. Government funding is available through WRAP, a non-profit-making company, which has been set up to increase recycling and to develop stable and efficient markets for recycled materials.

Another possibility is to extend the use of mechanical means of temporarily fastening two materials together, such as hook and loop fasteners which are much stronger than most people appreciate. As with the automotive industry, the time may well come when, before anything is actually made, the method of disposal or recycling possibilities will be considered. The public is now much more educated in environmental affairs, and some environmentally unfriendly processes or materials may have to be discontinued or replaced, not only due to government legislation or to pressure group action, but also because of pressure from major customers who themselves wish to appear as 'green' as possible.

7.5 New opportunities

Many problems remain unsolved in certain areas – for example American firemens' turnout uniforms weigh 50–70 pounds which is far from ideal – and answers are still sought for problems in ecologically friendly comfortable medical protective clothing. Airbags, probably the largest growth area in technical textiles, require coating, usually on both sides, with an expensive resin, and opportunities for further innovations probably exist in fabric, polymer and film development. New materials still appear, such as phase change material and memory shape materials, for which applications are still being developed. From papers delivered in the Aventex Symposium in Frankfurt, it seems that clothing sales in the developed world have gradually declined since the 1970s. The growth areas are now travel, computers, cellular phones, leisure activities and medical and health products. If cosmetics and body care can be combined with some textile article, it could produce new areas and outlets for technical textiles. People are more aware of health factors, such as dust mites and possible toxic materials in everyday items such as carpets. Perhaps coated or laminated products could be devised to control, prevent, screen or help remove these agents. The coating and laminating technologist should keep up to date with the world around him or her, because opportunities for new applications may arise at any time. Examples are the fuel tank linings made from coated Kevlar for Concorde and the use of PVC coated linen in place of sandbags for flood control. The fear of terrorist activity is boosting sales of gas masks and NBC type clothing, not only for the military but also for the general public.

Coating presents opportunities for the exploitation of new materials, which may be either made into a coating resin or a film for lamination, or added as a filler to a resin. There is scope for synergies with the film and packaging industries, which are larger than the textile coating industry and have larger development budgets. Whatever can be done with a film can be done with a textile via a coating process or by laminating the pre-prepared film on to a fabric. More lateral thinking is necessary to exploit inventions or novel properties in other industries and technologies, or even those occurring in nature, and apply them to textile coating and laminating. Some world changing inventions have been made using relatively simple established technology in a novel manner. An example is 'post it' note paper which exploits the novel properties offered by microencapsulation spheres and pressure sensitive adhesives (see Tables 1.8 and 1.9). Researchers need to be constantly alert to new possibilities and requirements which may be entirely unconnected with a job in hand. The search is on for new angles or something that will make a particular product stand out from the rest, e.g. the kettle which changes colour when it has boiled. The new science of nanocomposites is producing materials with quite extraordinary properties – could they be exploited in some way if combined with the flexibility and covering power of a coated or laminated fabric? An example of an ingenious application requiring considerable technical input is the mobile phone cover made from coated waterproof fabric, which incorporates electronic components and is also actually the keyboard for the mobile phone.

We have entered the exotic age of 'smart fabrics' and 'intelligent clothes', and with a tremendous variety of base fabrics and resins and novel substances to put into coatings as additives, there has never been a better time to innovate.[10–13] With NanoSphere (Schoeller) technology[14] plus a little more innovation, the 'Man in the White Suit', which never needs washing and lasts for ever, may well become a reality! Chameleon effects (i.e. changing colour or pattern according to the environment) could be exploited in novel fashion clothing as well as being of practical and strategic value for the military. The automotive industry and others are developing a culture of continuous improvement and innovation which, if sustained, will certainly take us into new areas and product development – in addition to allowing annual 'cost downs'! A single garment suitable for all weathers, hot or cold, dry or wet, and which never needs cleaning may well be possible in the fullness of time! Failure to solve a particular problem in the past should not deter another fresh attempt, because times change; new materials and technology appear all the time, and these may make a solution possible. As in any science the horizons are limited only by the imagination of the research and development scientist, but they must have the support of management and equally imaginative sales and marketing departments. There are going to be technical problems to solve, but to the true scientist and technologist these are merely challenges to be overcome to open the door to worthwhile benefits or at least to further development opportunities.

7.6 References

1. Yang J, 'Textile powerhouse coming up', *Textile Asia*, Dec 2000, 43–7.
2. Chen R, Wu W and Xiang X, 'Development strategy in China's textile machinery industry', *Textile Asia*, Dec 2000, 40–3.
3. Page D, 'On-line inspection closer to reality', *ATI*, May 1997, 50–2.
4. Whiteman R, 'On-line gauging, control and benefits for the coating/laminating process', *JCF*, Vol 23 Oct 1993, 87–105.
5. Bader H, 'Preventing roll deflection totally in the first place through electromagnetic forces', *Index 99*, Geneva 27–30 Apr 1999, EDANA, Brussels.
6. Anon, 'DuPont focuses on biotechnology for polymer production', *BPR* Jun 1998, 45.
7. Anon, '*Plasma Case studies – Technical Textiles*', Plasma Ireland Ltd Briefing note. And also web-site www.plasma-ireland.com
8. Anon, 'Government adds funding to fast cure research', *BPR*, Feb 2001, 20.
9. Anon, 'Draft paper on IPP moots lower VAT for eco-labelled products', *ENDS Report 312*, Jan 2001, 44–5.
10. Sung V, 'The future on show', *Textile Asia*, Dec 2000, 19–22.
11. Fischer G, 'Avantex shows that intelligent garments are the future', *TTi*, Jan/Feb 2001, 11–16.
12. Wagner P and Tilbury N, 'Clothes go smart', *Textile Asia*, Aug 2000, 66–8.
13. Lafee S, 'Geek Chic' (Smart clothes), *New Scientist*, Vol 169 No 2279 24 Feb 2001, 30–3.
14. Anon, 'Nanosphere technology', *Nonwovens International*, Jun 2001, 21 and also in *Textile Month*, May 2001, 50.

7.7 Further reading

1. Aneja AP (DuPont), 'New fibres for the Millennium', *Industrial, Technical and High Performance Textiles*, World Textile Congress, Huddersfield University 15–16 July 1998.
2. Anon, 'Smart future', *Textile Horizons*, Apr 1999, 16–18.
3. Behnke B, 'Formula for success; providing total solutions to textile coaters', *JCF*, Vol 27 Apr 1998, 289–93.
4. Benisek L, 'Fibres fighting the elements', *Textile Month*, Nov 2000, 46–8.

5. Braddock SE and O'Mahoney M, *Techno Textiles*, Thames and Hudson, London 1999.
6. Colchester C, *The New Textile*, Thames and Hudson, London 1993.
7. Conway R, 'Global trends in coating and lamination technology and application', *7th International Conference on Coating and Lamination*, Charlotte NC, USA 17–18 Nov 1997, Technomic, Lancaster PA, USA.
8. Elkington J and Hailes J, *Manual 2000*, Hodder & Stoughton, London 1998.
9. Fishburn D (Editor) *The World in 2001*, The Economist Publications, London 2000.
10. Hearle JWS, 'Genetic engineering and fibre production and properties', *Index '99*, Geneva 27–30 April 1999, EDANA, Brussels.
11. James B, 'Finding a future in a field full of dreams' (plant based polymers), *PRW*, 27 Oct 2000, 19.
12. O'Brien JP and Aneja AP, 'Fibres for the next millennium', *Review Progress Coloration*, Vol 29 1999, 1–8.
13. Page D, 'High-tech fabrics aim even higher', *America's Textiles*, Feb 1997, K/A 8–10.
14. Payne M, *World Textiles – Into the New Millennium*, World Textiles, Bradford 2001.
15. Rouette HK, 'Trends in coating and laminating', *JCF*, Vol 26 Jan 1997, 241–51.
16. Smith WC, 'Coating and laminating fabrics – putting the industry in perspective', *JCF*, Vol 28 Apr 1999, 292–9.
17. Woodruff FA, 'A comparative look at coating, laminating and flocking technology practice from the viewpoint of an international machinery supplier', *7th International Conference on Coating and Lamination*, Charlotte NC, USA 17–18 Nov 1997, Technomic, Lancaster PA, USA.

8
Sources of further information

8.1 General comments and suggestions

The coating and lamination sectors of the technical textiles industry are constantly changing and advancing, although not as fast as some other areas. However, new information, materials and applications continue to be produced and to present new opportunities. Keeping up to date is time consuming but need not be difficult, provided staff know where to look, i.e. which relevant journals to read and which conferences and exhibitions to attend. This section lists some sources of information likely to be useful to all involved with coating and lamination. Attending conferences and exhibitions is an excellent way of keeping up to date and making personal contacts. The coating and lamination technologist must be aware of current world affairs to be able to use his or her imagination and ingenuity to exploit new materials to solve a particular problem, or to offer an improvement on an existing product. Recent examples are the development of fabrics using the latest ultra high modulus polyethylene fibre to contain terrorist bombs on aircraft, the use of a flexible Kevlar composite to reinforce the fuel tanks of Concord against puncture, and the use of coated fabrics to protect property during the recent floods in England. Regular reading of general science magazines, such as *New Scientist* and *Scientific American*, and quality newspapers can stimulate innovative development work to improve existing products or to develop new ones.

Sportswear and equipment consume significant quantities of coated fabric, and regular reading of magazines devoted to pursuits such as walking, sailing and golf is recommended. The quarterly trade publication *World Active Sportswear* is written with significant scientific content and is essential reading for researchers and sales staff. The journal also issues an annual directory and buyers' guide. Regular conferences, for example the *Survival* conferences convened by Leeds University, and exhibitions devoted to sportswear and equipment provide an excellent forum for interchange of news and views. There are also specialist journals on industrial protective clothing, medical textiles, civil engineering, the building industry, the furniture industries and fire research and prevention. These journals could stimulate novel ideas and new applications. The paper, film and foil converting industries, whose production volumes are much greater than the coating and laminated textiles sector, could also provide inspiration, especially on production and handling aspects. These industries face some of the same problems, such as static build up, creasing and rollers not running true, cutting and general conversion and environmental challenges, etc.

The prime publication for coated fabrics was undoubtedly the *Journal of Coated Fabrics*, which was published for 28 years by Technomic Publishing of the USA. This journal first appeared in July 1971 with the objectives of disseminating information on the coating and lamination industries and associated areas and providing a forum for discussion amongst those working in them. During these 28 years of publication, high quality articles covering virtually every aspect of fabric coating and lamination appeared in the journal, and the back issues are a tremendous mine of information. Unfortunately the October 1999 edition, Vol 29 No 2, was the last issue to be published, but material on coating and lamination will continue in Technomic's *New Journal of Industrial Textiles*. The publisher will continue to sponsor the annual International Conference on Textile Coating and Lamination, which it has done since the first took place on 6–7 November 1991. A coated fabrics symposium is also organised regularly in the USA by the Association of American Textile Chemists and Colorists (AATCC), and BTTG (Shirley Institute) have hosted conferences in England. In September 1999, Unitex and Centexbel organised the 1st European Symposium on Textile Coating and Laminating. These regular forums signify the importance and potential of the coating and lamination of textiles.

Publications of the Textile Institute and World Textile Publications Ltd are invaluable in keeping up to date with both the academic and the more industrial aspects of textiles in general. World Textile Publications also publish a range of commercial directories on chemicals, textile companies and textile markets, e.g. Eastern Europe and China. For general wet processing, the *Journal of The Society of Dyers and Colourists* and its annual *Review of Progress in Coloration* are excellent sources of information with extensive references. The monthly *International Dyer* features topical news and views and well presented articles of a more practical nature. World Textile Publications also produce invaluable directories such as *Index to Textile Auxiliaries*, and directories of textile agents and dyers and finishers in the UK and abroad.

Fabric coating and lamination are closely linked to plastics, and journals of the plastics industry regularly feature news and information. *Plastics and Rubber Weekly* reports the latest news and developments and regularly features technical items and specialist articles. *British Plastics and Rubber*, *European Plastics News* and the American publication *Modern Plastics*

International all regularly feature news items and technical papers of relevance to the textile coating and lamination industry. News on the related area of fabric composites also appears frequently. The Institute of Materials, now incorporating the Plastics and Rubber Institute, is a professional body, which keeps its members informed of the latest developments through publications and conferences.

In addition to the conferences and symposia listed below, many additional ones are organised regularly by the Textile Institute, the SDC and the Institute of Materials. Local and regional branches of these bodies meet regularly. Universities and research organisations such as BTTG (British Technology Textile Group), SATRA (Shoe and Allied Trades Research Association), and RAPRA Technology also organise conferences and run short courses. The University of North London offer workshops on rubber compounding and processing, which have the support of the Institute of Materials. The Center for Professional Advancement regularly runs courses on technical subjects, including polymer compounding, plastics additives and coating processes in the USA and Europe.

The IFAI, whose activities are at present mainly in the USA, has a growing number of members in Europe and Japan, and eventually regular meetings of the local branches will be held. The IFAI have specialist divisions, the relevant ones being Marine Fabricators Association, Awning Division, Truck Cover and Tarpaulins Association, Safety and Protective Products Division and the Transportation Division. Composites symposia are organised by universities and by various professional bodies including the Institution of Mechanical Engineers. Nottingham University has a 'Composites Club'.

Mention must be made of the wealth of information contained in the ASTM Volumes on test methods, Vols 7.0 and 7.1 for textiles and Vol 9.02 for rubber products, including rubber coated fabrics. There are of course British Standards that are equally useful. Some researchers consider the patent literature to be the most extensive single source of information, both historical and up-to-date. Last but by no means least is the vast 'ocean' of information freely (specialist information may have to be paid for) available on the internet, which is being added to every day. The revolution in communications, information and ease of international travel is fostering 'global synergy' in research, development and innovation.

8.2 Conferences and exhibitions

8.2.1 General textile

1. *Fibre to Finished Fabrics*, organised annually by the Textile Institute usually in December in Manchester.
2. *International Man Made Fibres Congress*, held annually at Dornbirn,

Austria. In recent years automotive textiles have been featured every two years. Organised by Osterreichisches Chemiefaser-Institut Tagungsburo Dornbirn, Rathausplatz 1, A-6850 Dornbirn, or Kolingasse 1, A-1090 Vienna, Austria. The conference is sponsored and arranged in association with the CIRFS, Avenue E Van Nieuwenhuyse 4, B-1160 Brussels and the City of Dornbirn.

3. *Techtextil*, held annually in Frankfurt, Germany where about 100 papers on all aspects of technical textiles are presented. Organised by Messe Frankfurt GmbH, Postfach 15 02 10, D-60062, Frankfurt am Main, Germany. Techtextil Conferences are now also held in other regions of the world, e.g. Asia and South America.
4. *Textiles in Automobiles* (sometimes run in conjunction with the *Plastics in Automobile Engineering Conference*). Held on average every two to three years. Organised by Verein Deutscher Ingenieure (VDI), Graf-Recke-Strasse 84, D-40239 Dusseldorf, Germany.
5. *World Textile Congress*, Huddersfield, England held annually featuring different aspects. Organited by Department of Textiles, University of Huddersfield, Queensgate, Huddersfield HD1 3DH. The 2001 Conference is being hosted by Bolton Institute.

8.2.2 Other relevant conferences

1. *Camping and Outdoor Leisure Association* (COLA) Exhibition, usually held in October at Harrogate.
2. *Intabond*, conference on hot melt adhesive technology, organised every two to three years by Dermil Research Ltd, 24 Buckingham Square, Wickford Business Park, Wickford, Essex SS11 8YQ.
3. *Interplas* and *Kunstoff* are international chemical exhibitions, organised every two to three years in Birmingham (NEC) and Dusseldorf (Dusseldorfer Messegesellschaft mbH-NOWEA, D-4000 Dusseldorf, Germany).
4. *Rubber Bonding* and other conferences are organised periodically by RAPRA Technology Ltd (see Section 8.4.3), in conjunction with partners such as *European Rubber Journal.*
5. *Survival*, occasional conference for sportswear and protective clothing, organised in recent years by the Department of Textiles of the University of Leeds.
6. *Textile Coating and Lamination International Conferences*, organised annually by Technomic Publications Inc., 851 New Holland Avenue, Box 3535, Lancaster PA 17604, USA.
7. *Urethanes Technology*, organised by Crains Communications Ltd, New Garden House, 78 Hatton Garden, London ECIN 8JQ – in both Europe and Asia.

8.2.3 Materials and machinery

1. Conferences and exhibitions setup by the organisers of *Techtextil*, the Messe Frankfurt, GmbH, Postfach 15 02 10, D-60062 Frankfurt am Main, Germany, are held in Europe, but recently additional events have been held in other parts of the world including Asia and South America. They include: Interstoff – yarns for fashion and performance, *InterYarn* and *Heimatex* (household and hospitality textiles).
2. *Expofil* – European yarn exhibition, held twice annually, Rue de Neuilly, BP 121, F92113 Clichy Cedex, Paris, France.
3. *Index* – conference for non-woven fabrics, organised in Europe, usually in Geneva by EDANA – see Section 8.4.1.
4. *ITMA*, the International Textile Machinery Exhibition, is held usually every three years; the location rotates between Milan, Paris and Hanover. It is by far the largest exhibition of its type. Organised by the CEMATEX-Comite Europeen des Constructeurs de Materiel Textiles, General Secretary, Bredewater 20, Postbus 1 90, NL-2700 AD Zoetermeer, Netherlands.

8.2.4 Automotive conferences

This selection of regular conferences feature automotive development and automotive interiors in general and are not specific to textiles. In the USA, the IFAI periodically organises conferences on automotive interior trim.

1. *Automotive and Transportation Interiors Expo*, organised in the USA by the magazine of the same name.
2. *Autoplas* conferences on use of plastics in the automobile (and also recycling), organised by Schotland Business Research Inc., 16 Duncan Lane, Skillman, New Jersey 08558–2313 USA.
3. *Autotech Conference*, held annually at the NEC, Birmingham organised by Centre Exhibitions jointly with the Institute of Mechanical Engineers, Automobile Division. NEC House, Birmingham B40 1NT.
4. *Comfort in the Automotive Industry*, organised approximately every two years by the Italian Associazione Technica Dell'Automobile (ATA) and Bologna University and held at Bologna. ATA, Strada Torino, 32/A, 10043 Orbassano (TO), Italy.
5. *Inter Auto*, organised by *Inside Automotives International* in conjunction with the Turret group. Events in Europe have been held at the RAI Centre, Amsterdam. *Inside Automotives International*, ANCAR Publications Inc., 21700 Northwestern Highway, Suite 565, Southfield MI 48075 USA.

6. *ISATA* (International Symposium on Automotive Technology and Automation), an annual event organised by Dusseldorfer Messegesellschaft mbH-NOWEA, Postfach 101006, D40001, Dusseldorf, Germany.

8.3 Journals

Many of the journals publish an annual resource file, where to buy guide, yearbook and review of progress.

8.3.1 Textile journals

1. *African Textiles* (for African and Arab markets), Alain Charles Publishing Ltd, 27 Wilfred Street, London SW1E 6FR.
2. *ATI* (America's Textiles Industries), Billian Publishing Inc., 2100 Powers Ferry Road, Atlanta GA 30339 USA.
3. *Canadian Textile Journal*, Textile Technology Centre, 3000 Boulle, St-Hyacinthe, Quebec, Canada J2S 1H9.
4. *International Fiber Journal*, International Media Group Inc., 1515 Mockingbird Lane, Suite 210, Charlotte NC 28209 USA.
5. *International Nonwovens Journal*, INDA, TAPPI and ANIC – see Section 8.4.1 for addresses.
6. *International Textile Bulletin*, *Textile Leader*, International Textile Services, Univer Hans, Kessler strasse 9, CH-8952, Schlieren, Zurich, Switzerland.
7. *JTN Monthly* (Japanese Textile News), Osaka Senken Ltd, 3-4-9, Bingomachi, Chuo-ku, Osaka 541-0051, Japan.
8. *Kettenwirk-Praxis* (German, with English translations), Karl Mayer Textilmaschinenfabrik GmbH, Postbox 1120, D-63166, Obertshausen, Germany.
9. *Knitting International*, World Textile Publications – see *Textile Horizons* below.
10. *Nonwoven Industry*, Redman Publications Inc., 70 Hilltop Road, 3rd Floor, Ramsey NJ 07446, USA.
11. *Nonwovens Report International*, World Textile Publications Ltd, see *Textile Horizons* below.
12. *Textile Asia*, Business Press Ltd, California Tower 11th Floor, 30–32 D'Aguilar Street, Hong Kong.
13. *Textile Horizons*, The Textile Institute and World Textile Publications Ltd, Perkin House, 1 Longlands Street, Bradford, West Yorkshire BD1 2TP.
14. *Textile Month*, World Textile Publications – see *Textile Horizons* above.

15. *Textiles World*, Maclean Hunter Publishing Co., 29N Wacker Drive, Chicago IL 60606 USA.

8.3.2 Technical textiles

1. *AATCC Review*, AATCC, see Section 8.4.1.
2. *Chemical Fibers International*, issued six times a year with annual Year Book, IBP International Business Publishers Gmbh, Mainzer Landstrasse 251, D-60326 Frankfurt am Main, Germany.
3. *Clothing and Textiles Research Journal*, International Textiles and Apparel Association Inc., PO Box 1360, Monument CO 80132–1360 USA.
4. *High Performance Textiles*, monthly newsletter from International Newsletters – see *Technical Textiles International* below.
5. *Industrial Products Review*, published by IFAI USA, see Section 8.4.1.
6. *Industrial Textiles*, published by Impact! on behalf of MUTA, see Section 8.4.1.
7. *International Textile Bulletin, Yarn and Fabric Forming*, International Textile Services, see Section 8.3.1.
8. *Journal of the Textile Machinery Society of Japan*, Osaka Science and Technology Center Building (Utsubo Park), 8-4 Utsubo-hommachi 1-chome, Nishi-ku, Osaka 550, Japan.
9. *New Journal of Industrial Fabrics*, Technomic Publishing Company Inc., 851 New Holland Avenue, Box 3535, Lancaster PA 17604 USA, and in Europe Missionsstrasse 44, CH-4055 Basel, Switzerland.
10. *Melliand Textilberichte* (also *Melliand English*), 60264, Frankfurt am Main, Germany.
11. *Technical Textiles International* (with annual Where to Buy Guide), International Newsletters, PO Box 133, Witney, Oxford OX8 6ZH.
12. *Technische Textilien* (German with English Translations), PO Box 10 0606, D-6000 Frankfurt am Main, Germany.
13. *Textile Research Journal*, TRJ Princeton, 601 Prospect Avenue PO Box 625, Princeton NJ 08542 USA.
14. *Textiles Usage Textiles* (TUT) (French with English summaries), Euredia SA – editorial collaboration with the French Textile Institute, 16 Rue Ballu F-75311, Paris Cedex 09 France.

8.3.3 Dyeing and finishing

1. *American Dyestuff, Reporter*, Promenade A, Suite 2, Harmon Cove Towers, Secaucus NJ 07094 USA.
2. *International Dyer*, World Textile Publications Ltd, – see *Textile Horizons* in Section 8.3.1.

3. *International Textile Bulletin, Dyeing, Finishing, Printing*, International Textile Services, see Section 8.3.1.
4. *Journal of the Society of Dyers and Colourists* (with annual Review of Progress in Colouration and annual Resource File), SOC, see Section 8.4.1. for address.
5. *Textile Chemist and Colorist*, AATCC, see Section 8.4.1.

8.3.4 Plastics and chemicals

1. *British Plastics and Rubber*, MCM Publishing Ltd, 37 Nelson Road, Caterham, Surrey CR3 5PP.
2. *European Adhesives and Sealants* (with annual Year Book), FMJ International Publications, Queensway House, 2 Queensway, Redhill, Surrey RH1 1QS.
3. *European Plastics News*, Emap Maclaren, Maclaren House, 19 Scarbrook Road, Croydon, Surrey CR9 1QH.
4. *Modern Plastics International* (with annual Buyers Guide), McGraw Hill, Modern Plastics International, PO Box 605, Hightstown NJ 08520 USA.
5. *Plastics and Rubber Weekly*, EMAP Business Communications, Ruislip, Middlesex HA4 9LT. The Journal of the Malaysian Rubber Research and Development Board, Tun Abdul Razak Centre, Brickendonbury, Hertford, SG13 8NL.
6. *Rubber World Magazine*, PO Box 5451, Akron OH 44334-0451 USA.
7. *Urethanes Technology*, Crain Communications Ltd, see *Urethanes Technology* in Section 8.2.2.

8.3.5 Composites

1. *Advanced Composites Bulletin*, monthly International Newsletters – see *Technical Textiles International* (Section 8.3.2).
2. *Composite Science and Technology*, Elsevier Science Ltd, The Boulevard, Langford Lane, Kidlington, Oxford OX5 1GB.
3. *Material Science*, Institute of Materials, see Section 8.4.3.
4. *Plastics, Rubber and Composites*, Institute of Materials, see Section 8.4.3.

8.3.6 The environment

1. *The ENDS Report*, Environmental Data Services Ltd, Finsbury Business Centre, 40 Bowling Green Lane, London EC1R 0NE.
2. *Energy and Environmental Management*, DEFRA, Ashdown House, 123 Victoria Street, London SW1E 6DE.

3. *Environment*, Heldref Publications, 1319 Eighteenth Street NW, Washington DC 20078-6117, USA.
4. *Environmental Manager*, Monitor Press, Suffolk House, Church Field Road, Sudbury, Suffolk CO10 2YA.
5. *Environment Business Magazine* (annual Directory), Information for Industry Ltd, 4 Valentine Place, London SE1 8YX.
6. *Environment Monthly*, Informa Publishers Ltd, 69–67 Paul Street, London EC2A 4LQ.
7. *Environment Times*, Beckhouse Media, 22 Warwick Street, Adlington, Lancashire PR7 4JQ.
8. *European Environment*, John Wiley and Sons, Baffins Lane, Chichester, West Sussex PO19 1UD.
9. *EUWID Recycling and Waste Management*, PO Box 1332, D-76586 Gernsbach, Germany.
10. *Material Recycling Week*, EMAP Maclaren Ltd, 19 Scarbrook Road, Croydon CR9 1QH.

Other environmental contacts include:

11. The Environment Agency, Rio House, Waterside Drive, Aztec West, Almondsbury, Bristol BS32 4UD. The Department of the Environment publish a large amount of free literature on environmental matters, energy conservation and other issues. Free advice is offered as well as information on technical matters, environmental legislation, conferences and seminars. They have a telephone helpline 0800 585794 (in the UK) and have a web site http://www.etsu.com/etbpp/
12. The EU also produce information on the environment – the general contact is Office for Official Publications of the European Communities, L-2985 Luxembourg.
13. The National Centre for Business and Ecology, Peel Building, University of Salford, Manchester M5 4WT.
14. The National Society for Clean Air and Environment Protection (NSCA), 136 North Street Brighton BN1 1RG. NSCA issue their *Pollution Handbook* every two or three years and also publish a series of well written and well presented information leaflets for the general public.
15. Warren Springs Laboratory (UK DTI Environment), Hertfordshire. Helpline 0800 585794.

Web sites:

16. Department of the Environment, *www.detr.gov.uk/*
17. The Environmental Agency, *www.environment-agency.gov.uk/*
18. US Environmental Protection Agency, *www.epa.gov/*

8.3.7 Automotive journals

1. *Automotive & Transportation Interiors* (with annual Where to Buy Guide), 6255 Barfield Road, Suite 200, Atlanta GA 30328-4300 USA.
2. *Automotive Engineering* (a publication associated with SAE of the USA), PO Box 5004, Pittsfield MA 01203-9990 USA.
3. *Automotive Industries AI*, 201 King of Prussia Road, Radnor PA, or 3011 W. Grand Boulevard, Ste. 2600 Detroit MI 48202 USA.
4. *Automotive Interiors International*, Turret-RAI plc., Armstrong House, 38 Market Square, Uxbridge, Middlesex UB8 1TG.
5. *Automotive News Europe*, Crain Communications Ltd, see *Urethanes Technology* in Section 8.2.2.
6. *Automotive World Publications*, Financial Times, Maple House, 149 Tottenham Court Road, London W1P 9LL.
7. *Inside Automotives International* (with annual Where to Buy Guide), ANCAR Publications Inc., 21700 Northwestern Highway, Suite 565, Southfield MI 48075 USA.

8.3.8 More specialist technical journals

1. *Converting Today*, Angel Business Communications Ltd, Kingsland House, 361 City Road, London EC1V 1PQ.
2. *Film, Foil and Paper Converter*, PO Box 12978, Overland Park KS 66282–2978 USA.
3. *Filtration and Separation*, Elsevier Advanced Technology, PO Box 150, Kidlington, Oxford OX5 1AS.
4. '*Fire*' (and Fire group of journals), DMG World Media (UK) Ltd, Queensway House, 2 Queensway, Redhill Surrey RH1 1QS.
5. *Geotechnical Fabrics Report*, official publication of the Geosynthetic Materials Division of IFAI USA, see Section 8.4.1.
6. *Journal of Coatings Technology*, 492 Norristown Road, Blue Bell PA 19422–2350 USA.
7. *Journal of the International Geosynthetics Society*, The Boulevard, Langford Lane, Kidlington, Oxford OX5 1GB.
8. *Journal of Science and Medicine in Sport*, MS Ester Guerzoni, Sports Medicine Australia, PO Box 897, Belconnen ACT 2616, Australia.
9. *World Sports Activewear* (international magazine for sports textiles and apparel), PO Box 6, 36 Crosby Road North, Liverpool L22 0QN.

8.3.9 General interest scientific journals

1. *Focus*, G + J of the UK, Portland House, Stag Place, London SW1E 5AU.

2. *Nature*, Porters South, 4 Crinan Street, London N1 9XW.
3. *New Scientist*, Reed Business Information Ltd, 151 Wardour Street, London WIV 4BN.
4. *Science*, American Association for the Advancement of Science, 1200 New York Avenue NW, Washington DC 20005, USA.
5. *Scientific American*, 415 Madison Avenue, New York, NY 10017-1111 USA.

8.3.10 Popular sports magazines

1. *On the Hill*, Maze Media (2000) Ltd, 89 East Hill, Colchester, Essex CO1 2QN.
2. *Outdoor Pursuits Magazine*, Tudorseed Limited, Unit 3, Ripon House, 35 Station Lane, Hornchurch, Essex RM12 6JL.
3. *Practical Boat Owner*, Westover House, West Quay Road, Poole, Dorset BH15 1JG.
4. *Practical Caravan*, Haymarket Magazines Ltd, 60 Waldegrave Road, Teddington, Middlesex TW11 8LG.
5. *The Rambler* (official magazine of the Ramblers' Association), 1–5 Wandsworth Road, London SW8 2XX.
6. *Yachting World*, Kings Reach Tower, Stamford Street, London SE1 9LS.

8.4 Technical and professional organisations and institutions

8.4.1 Textile organisations

1. AATCC (American Association of Textile Chemists and Colorists), PO Box 12215, Research Triangle Park NC 27709-2215, USA.
2. American Flock Association, c/o NTA Bolgen, 230 Congress Street, Boston MA 02110 USA.
3. American Textile Manufacturers Institute, 1130 Connecticut Avenue, NW-Suite 1200, Washington DC 20036-3954, USA.
4. ANIC (Asia Nonwoven Fabrics International Conference), Soto Kanda 6 Chome, Building 3Fl, 2–9 Chiyoda-ku, Tokyo 101, Japan.
5. Association of Nonwovens Fabric Industry, PO Box 1288, Cary NC 27512, USA.
6. British Textile Machinery Association, 20 Ralli Courts, West Riverside, Manchester M3 5FL.
7. BTTG (British Textile Technology Group), Shirley House, Wilmslow Road, Didsbury, Manchester M20 2RB.
8. Canadian Textile Institute, 66 Slater Street, Ste 1720, Ottawa, ON K1P 5H1, Canada.

9. CENTEBEL, National Textile Research Centre, Belgium Technologiepark, B-9052, Zwjinaarde, Belgium.
10. CIRFS (Comité International de la Rayon et des Fibres Synthetiques), Avenue E Van Nieuwenhuyse 4, B-1160 Brussels.
11. DCTA (Defence Clothing and Textiles Agency) (formerly SCRDE – Stores, Clothing, Research and Development Establishment – of the MOD), Flagstaff Road, Colchester, Essex CO2 7SS.
12. DWI (Deutsches Wollforschungsinstitut), Veltmanplatz 8, D-52062 Aachen, Germany (German Wool Institute – Organisers of the annual *Aachen Textile Conference*).
13. EDANA (the European Disposables and Nonwoven Association), 157 avenue Eugene Plasky, Bte 4-1030 Brussels, Belgium.
14. Ghent University, Dept of Textiles, Techologiepark 9, 9052 Gent Zwijnaarde, Belgium.
15. Hohenstein Institutes, Schloss Hohenstein, D-74357 Boennigheim, Germany.
16. IFAI (International Industrial Fabrics Association). Several divisions including a Transportation Division and European and Japanese Branches. IFAI USA, 1801 County Road B W, Roseville MN 55113-4061 USA. European Office: IFAI Europe, Marcel Thirylaan 204, B-1200 Brussels, Belgium. Japanese Office: IFAI Japan, 3-8-9 Nishidai, Itami, Hyogo 664, Japan.
17. INDA (Industrial Nonwovens and Disposable Association), 1700 Broadway 25th Floor New York, NY 10019 USA.
18. Institut Textile de France, Avenue Guy de Collongue-B.P. 60-69123 Ecully Cedex, France.
19. MUTA (Made-Up-Textiles Association) Ltd, 42 Heath St Tamworth, Staffordshire B79 7J.
20. The Nonwovens Network, BTTG WIRA House, West Park Ring Road, Leeds LS16 6QL.
21. North Carolina State University, Dept of Textiles, Box 8301, Raleigh NC 27695-8301, USA.
22. Osterreichisches Chemiefaser, Kolingasse 1, A-1090 Vienna, Austria.
23. Sachsisches Textil Forschungs Institut, Annaberger Strasse 240, D-09125, Chemnitz, Germany.
24. SATRA (The Shoe and Allied Trades Research Association), SATRA House, Rockingham Road, Kettering, Northants NN16 9JH.
25. SDC (Society of Dyers and Colourists) (publishes monthly journal and annual resource file), PO Box 244, Perkin House, 82 Gratton Road, Bradford, West Yorkshire BD1 2JB.
26. Swedish Institute for Fibre & Polymer Research (IFP), Box 104, SE-431 22 Molndal, Sweden.

27. Swiss Federal Laboratories For Materials Testing and Research, Lerchenfeldstrasse 5, CH-9014, St Gallen, Switzerland.
28. TAPPI (Technical Association of the Pulp and Paper Industry), PO Box 105113, Atlanta GA 30348, USA.
29. The Textile Institute, St James's Buildings, Fourth Floor, Oxford Street, Manchester M1 6EJ.
30. The Textile Technical Institute (AKZO Nobel), Kasinostrasse 19–21, D-42103 Wuppertal, Germany.
31. TNO, Institute of Industrial Technology (Netherlands), Centre for Textile Research, NL-2600 JA Delft, Netherlands.
32. UNITEX (Union of High Members of Belgium Textiles), Rivierstraat, 19/102, B-9080 Beervelde, Belgium.
33. Verband de Flockindustrie eV (Association of the Flock Industry), Tannenberger Strasse 66/62 D-72760 Reuttingen, Germany.
34. WIRA (Wool Industries Research Association) Technology (BTTG), West Park Ring Road, Leeds LS16 6QL.
35. Woolmark Company, Valley Drive, Ilkley, West Yorkshire LS29 8PB.
36. Wools of New Zealand, Design and Development, Little Lane, Ilkley, West Yorkshire LS29 8UG.

8.4.2 Some UK universities/institutions with textile departments

1. Bolton Institute, Deane Road, Bolton BL3 5AB.
2. Heriot-Watt University, Scottish Borders Campus (formerly Scottish College of Textiles), Netherdale, Galashiels, Selkirkshire TD1 3HF.
3. Huddersfield University, Queensgate, Huddersfield HD1 3DH
4. Leeds University, Leeds LS2 9JT.
5. Manchester Metropolitan University, John Dalton Building, Chester Street, Manchester M1 5GD. Hollins Facility (Garment Design & Development).
6. Nottingham Trent University, Burton Street, Nottingham NG1 4BU.
7. Royal College of Art (Postgraduate Art & Design/Textile Design), Kensington Gore, London SW7 2EU.
8. UMIST, PO Box 88, Manchester M60 1QD.

8.4.3 Plastics, rubber and composites

1. APME (Association of Plastics Manufacturers in Europe), Avenue E. Van Nieuwenhuyse 4, Box 3, B-1160 Brussels, Belgium.
2. BPF (British Plastics Federation), 6 Bath Place, Rivington Street, London EC2A 3JE.

3. British Rubber Manufacturers' Association Ltd, 90 Tottenham Court Road, London W1P 0BR.
4. The Center for Professional Advancement, Box 1052, 144 Tices Lane, East Brunswick, NJ 08816-1052, USA.
5. Institute of Materials, 1 Carlton House Terrace, London SW1Y 5DB.
6. The Malaysian Rubber Producers' Research Association, Tun Abdul Razak Laboratory, Brickendonbury, Hertford SG13 8NL.
7. Nottingham University Composites Club, School of Mechanical Materials, Manufacturing Engineering and Management, University of Nottingham, University Park, Nottingham NG7 2RD.
8. RAPRA Technology Ltd (Rubber Industries Research Association), Shawbury, Shrewsbury, Shropshire SY4 4NR.
9. SAMPE (Society of Advanced Materials and Process Engineers), International Business Office, 1161 Parkview Drive, Covina, California 91724-3748 USA.
10. SATRA Footwear Technology Centre, see Section 8.4.1.
11. The School of Polymer Technology, University of North London, Holloway Road, London N7 8DB (run courses in rubber technology and compounding).
12. The Society of Plastics Engineers, 14 Fairfield Drive, PO Box 403, Brookfield CT 06804-0403, USA.
13. VDI-Gesellschaft Kunstofftechnik (VDI-K), Graf-Recke-Strass 84, 40239 Dusseldorf, Germany (Institute of German Engineers – Technical Plastics).

8.4.4 Joining technology/other research organisations

1. FIRA (Furniture Industries Research Association), Maxwell Road, Stevenage, Herts SG1 2EW.
2. Laboratory of the Government Chemist, Queens Road, Teddington, Middlesex TW11 0LY.
3. Loughborough University, Institute of Surface Science and Technology, Ashby Road, Loughborough, Leics LE11 3TU.
4. National Physics Laboratory, Teddington, Middlesex TW11 0LW.
5. Oxford Brookes University, Joining Technology Research Centre, Oxford OX3 0BP.
6. SAE (Society of Automobile Engineers), 400 Commonwealth Drive, Warrendale, PA 15096-001, USA.
7. TWI (The Welding Institute), Abington Hall, Abington, Cambridge CB1 6AL. N.B. 'Welding' encompasses all joining techniques – including adhesives.

8. Warrington Fire Research Centre Ltd, Holmesfield Road, Warrington WA1 2DS.

8.4.5 Testing and standards

1. AATCC (American Association of Textile Chemists and Colorists), see Section 8.4.1.
2. ANSI (American Standards Institute), 11 West 42nd Street, 13 Floor, New York, NY 10036-8002, USA.
3. Association Francaise de Normalisation, F-92049 Paris, La Defense Cedex, France.
4. ASTM (American Society for Testing and Materials), 1916 Race Street, Philadelphia, PA 19103-1187, USA.
5. BSI (British Standards Institution), 389 Chiswick High Road, London W4 4AL.
6. CEN (The European Committee For Standardisation), Central Secretariat, Rue de Stassart 36, B-1050 Brussels, Belgium.
7. Detroit Testing Laboratory Inc, 7111 E Eleven Mile, Warren, MI 48092, USA.
8. DIN (Deutsches Institut fur Normung), Burggrafenstrasse 6, D-10787 Berlin, Germany.
9. Ente Nazionale Italiano di Unificazione, Via Battistotti Sassi 11/b 1-20133 Milano, Italy.
10. International Organisation for Standardisation, Central Secretariat, 1 Rue de la Varenbe Case Postale 56, CH-1211, Geneva 20, Switzerland.
11. Japanese Industrial Standards Committee, Department of Standards, Industrial Science and Technology Agency, Ministry of International Trade and Industry, 1-3-1, Kasumigaseki, Chiyoda-ku Tokyo 100.
12. Korean National Institute of Technology and Quality, 1599 Kwanyang-dong, Dongan-ku, Anyang City Kyonggi-Do 430-060, Republic of Korea.
13. Secretary of the International Oeko-Tex Association, Oko-Tex Association, Postfact 585, CH-8027, Zurich.
14. Swiss Federal Laboratories for Materials Testing and Research, Lerchenfeldstrasse 5, CH-9014, St Gallen, Switzerland.
15. United States Testing Inc., 1415 Park Ave., Hoboken, NJ 07030, USA.

8.4.6 Recycling organisations

1. Bureau of International Recycling, 24 Avenue Franklin Roosevelt, 1050 Brussels, Belgium. Web site *www.bir.org*
2. Textile Recycling Association, PO Box 124, Huntingdon, Cambridgeshire PE18 7DP (Forum on Recycled Textiles).

8.5 Market information on technical textiles industry

1. David Rigby Associates (DRA), Peter House, St Peter's Square, Manchester M1 5AQ. *www.davidrigbyassociates.com*
2. Datamonitor Plc., 106 Baker Street, London W1M 1LA. *www.marketprofiles@datamonitor.com*
3. The Economist Intelligence Unit, PO Box 200, Harold Hill, Romford, Essex RM3 8UX.
4. Euromonitor, 60-61 Britton Street, London EC1M 5NA.
5. Freedonia Group Inc, 767 Beta Drive, Cleveland OH 44143-2326, USA.
6. IFAI (Industrial Fabrics Association International), 1801 County Road, BW Roseville MN 55113-4061, USA (Members).
7. Intercontuft (Automotive Fabrics), 3080 Tervuren, Moorelboslaan 15, Belguim.
8. Textile Outlook International, 10 Beach Lane, Wilmslow, Manchester M1 5AQ.

8.6 General further reading

8.6.1 Coating and laminating of textiles

There are very few textbooks devoted to this subject, apart from the following. Much of the available literature is in the form of articles and papers, manufacturers' technical information leaflets and manuals and proceedings of the *International Coating and Laminating Conferences* organised by Technomic (11 consecutive years 1990–2001), AATCC in the USA, BTTG (Shirley Institute) in the UK and Unitex/Centebel in Europe.

1. Holker JR. *Bonded Fabrics*, Merrow Monograph MM/TT/14 Merrow, Watford 1975.
2. Lennox Kerr P (Editor), *Flexible Textile Composites*, Textile Trades Press/WRC Smith, Manchester/Atlanta 1973.
3. Van Parys M, *Coating*, Eurotex (EEC Comet Program), Guimaraes 1994.
4. Wypych J, *Polymer Modified Textiles*, New York, John Wiley 1988.

8.6.2 Textile reference

1. Anstey H and Weston T, *The Anstey Weston Guide to Textile Terms*, Woodhead Publications, Cambridge 1997.
2. *European Index of Yarns and Fibres*, Third Edition, World Textile Publications, Bradford 1999.
3. Gordon Cook J, *Handbook of Textile Fibres*, Vol. 1 *Natural Fibres* and

Vol. 2 *Man-made Fibres*, Both Fifth Edition, Merrow, Shilden Co Durham 1984.
4. *Index to Textile Auxiliaries*, 17th Edition, World Textile Publications, Bradford 2000.
5. McIntyre JE and Daniels PN (Editors), *Textile Terms and Definitions*, Tenth Edition, The Textile Institute, Manchester 1995.
6. Tortora PG and Collier BJ, *Understanding Textiles*, Prentice Hall, New York 1997.

8.6.3 General rubber and plastic reference

1. Gachter R and Muller H (Editors), *Plastics Additives Handbook*, Fourth Edition, Hanser, Munich 1993.
2. Karsa DR, Goode JM and Donnelly PJ, *Surfactants Application Directory*, Blackie, Glasgow 1991.
3. Katz HS and Milewski V (Editors), *Handbook of Fillers for Plastics*, Van Nostrand, New York 1987.
4. Pritchard G (Editor), *Plastics Additives An A–Z Reference*, Chapman and Hall, London 1998.
5. Pruett KM, *Chemical Resistance Guide for Elastomers*, Compass, Le Mesa CA, USA 1994.
6. Pruett KM, *Chemical Resistance Guide for Elastomers 11*, Rubber World Publishers, PO Box 5451 Akron OH 44334-0451, USA 2000.
7. Roff WJ, Scott JB and Pacitti J, *Fibres, Films and Plastics (A Handbook of Common Plastics)*, Butterworths, London 1971.
8. *Rubber Blue Book* – as above.
9. *Rubber Red Book 2000* (buyers guide for rubber industry), Rubber World Magazine, Lippincott & Peto Publications, PO Box 5451, Akron OH 44334-0451, USA.
10. Rubin II, *Handbook of Plastics Materials*, Interscience, New York 1990.

8.7 Glossary of unfamiliar terms and abbreviations

These are for information only in the context of technical textiles and coating and lamination. Many of the terms will be familiar to those in the plastics, rubber and adhesive industries but unheard of by the average textile technologist. They are not official definitions, and the list is not intended to be exhaustive. A *Textile Terms and Definitions* publication is available from the Textile Institute, and in addition there are the various ASTM Terminology documents: D 123 relating to textiles, D 3990–93 (1998) relating to fabric defects, D 4391–93a relating to burning behaviour of textiles, D 907 relating to adhesives, D 883–98 relating to plastics, D1566–99a abbreviated terms relating to plastics, D 1566a relating to rubber, D 1517–98 relating to leather and F 1494–94 relating to protective clothing.

AATCC – Association of American Textile Chemists and Colorists
ABS – acrylonitrile-butadiene-styrene plastic. Not to be confused with automatic braking system
ALREM – Association of Load Restraint Equipment Manufacturers
APME – Association of Plastics Manufacturers in Europe
ASTM – American Society for Testing and Materials
ATH – aluminium trihydrate (FR agent)
BATNEEC – best available techniques not entailing excessive cost (for prevention of release of harmful substances to the environment)
BCF – bulked continuous filament
Benchmarking – a procedure which compares competitive products to one's own and then seeks to establish the 'best practice'
BOD – biological oxygen demand
BPF – British Plastics Federation
BRITE EURAM – Basic Research for Industrial Technologies in Europe, European Research in Advanced materials. (EU sponsored research)
BSI – British Standards Institution, hence BS = British Standard
BS AU – British Standard Automobile Series
BTTG – British Textile Technology Group
CAD – computer assisted design
CAM – computer assisted manufacturing
CEFIC – European Chemical Industry Council (Federation)
CEN – Comite European de Normalisation
CFC – chlorofluorocarbons, chemicals once used in refrigeration, foam making, etc. Identified as damaging to the environment and now being phased out worldwide
CHIP – Chemical hazard information and packaging regulations
CIELAB – Commission International du Eclairage (LAB refers to colour differences calculated)
CIRFS – Comite International de la Rayonne et des Fibres Synthetiques (Brussels)
COD – chemical oxygen demand, the amount of oxygen needed to purify effluent
COSHH – Control of Substances Hazardous to Health
CR – chloroprene (e.g. Neoprene, DuPont)
CRAG – Composite Research Advisory Group (UK Ministry of Defence)
Crosslinking – *see* curing
CRE – constant rate of extension
CRL – constant rate of load
CSM – chlorosulphonated polyethylene (e.g. Hypalon – DuPont)

CTR – constant rate of traverse
Curing – crosslinking of a compound or resin
DBDO – decabromodiphenyl oxide (FR agent)
DCTA – Defence Clothing and Textile Agency (UK)
DEFRA – Department of the Environment, Food and Rural Affairs (UK)
DIN – Deutsches Institut fur Normung (German Standards Institute)
Dioxins – a generic term applied to groups of chemicals including polychlorinated dibenzo-*p*-dioxins (PCDDs) and chlorinated dibenzofurans. They are sometimes created as by-products in certain combustion processes and some are extremely toxic
DMF – dimethylformamide
EDANA – European Disposables and Nonwovens Association
EfW – energy from waste
ELV – end-of-life vehicle. Vehicle that has come to the end of its useful life – sometimes referred to as EOL – end-of-life vehicle
EMAS – European Eco-Management and Audit Scheme
EMS – Environmental Management System
Energy recovery – the useful energy produced by using waste material as fuel
EOL – *see* ELV
EPDM – ethylene propylene diene monomer rubber
EPM – ethylene propylene monomer rubber
EVA(c) – ethylene vinyl acetate
FAAS – flame atomic absorption spectroscopy
FAKRA – Fachnormenausschuss Krafflahzeuge – German body for automotive standards – probably best known in textiles' industry for accelerated light stability test, DIN 75202
FES – flame emission spectrophotometry
FISITA – International Federation of Automotive Engineering Societies
FKM – fluoro-elastomer (Viton, DuPont)
FMEA – failure mode and effects analysis – a systematic procedure to assess the potential failure of a component or process and its effects and, to establish ways to prevent re-occurrence
FPM – hexafluoropropylene vinylidene fluoride (Viton – DuPont)
FR – flame retardant properties
FRP – fibre reinforced plastic
FTIR – Fourier transform IR (analysis)
GC – gas chromatography, an analytical method
GLC – gas liquid chromatography, an analytical method
GRP – glass reinforced plastic
GRU – glass reinforced urethane
GRV – gross vehicle weight
HDPE – high density polyethylene

HMPE – high modulus polyethylene
HT – high tenacity – a term applied to yarns of above average tensile strength, i.e. generally in excess of 7 g/dtex
IFAI – Industrial Fabrics Association International
IMMFC – International Man-Made Fibres Congress (Dornbirn, Austria)
IMO – International Maritime Association
INDA – Industrial Nonwoven and Disposables Association (USA)
IPC – integrated pollution control
IPPC – integrated pollution prevention and control
IR – infra red
ISO – International Standards Organisation
ISOPA – European Isocyanate Producers' Association (Polyurethane Industry) – an organisation within the European Chemical Industry Federation (CEFIC)
ITMA – International Textile Machinery Association (Exhibition every four years in Europe)
JAR – Joint Aviation Regulations
JIT – just in time
LCA – life cycle analysis – a method of assessing a product or process for impact on the environment which takes into consideration every factor, from the raw materials, through production, distribution, use and final disposal. Also referred to as 'cradle to grave' analysis
LCP – liquid crystal polymer – used in instrument panel displays
LDPE – low density polyethylene
LLDPE – linear low density polyethylene
LOI – limiting oxgen index, FR test used mainly by researchers
LS – low shrinkage
MEK – methyl ethyl ketone
MF – melamine formaldehyde
MFI – melt flow index – viscosity measurement for molten polymers and adhesives
MOD – Ministry of Defence
MPW – mixed plastics waste
MUTA – Made up Textiles Association
MVP – moisture vapour permeability (of perspiration)
MVTR – moisture vapour transfer (of perspiration)
NAMAS – National Measurement Accreditation Service – by the NPL (UK)
NBC – nuclear, biological and chemical protection
NBR – acrylonitrile butadiene rubber
NBS – National Bureau of Standards (US)
NFPA – National Fire Protection Agency (USA)

NPL – National Physics Laboratory (UK)
NR – natural rubber
NSCA – National Society for Clean Air and Environmental Protection (UK)
NVH – noise, vibration and harshness
OEL – occupational exposure limit (of chemicals)
OEM – original equipment manufacturer of automobiles, i.e. Ford, BMW, etc
PA – polyamide, known also as nylon – designated with numbers which relate to the starting materials from which they are made, e.g. nylon 6, nylon 66, nylon 46, etc
PAN – polyacrylonitrile, usually refers to acrylic fibre, e.g. Courtelle
PBDE – polybrominated diphenyl ethers
PBI – polybenzimidazole, a material with very high FR properties
PBTP – polybutylene terephtalate, a polymer related to PET, with more stretch in yarn form and a lower melting point
PCB – polychlorinated biphenyls
PCDD – see dioxins
PE – polyethylene
PET – polyethylene terepthalate, known commonly as polyester
PO – polyolefin, a general chemical term used to describe plastics made from polyethylenes and polypropylenes
PP – polypropylene
PPE – personal protective equipment
PRAVDA – Project Altfahrzeugverwertung Deutscher Automobilhersteller (German automobile manufacturers used vehicle recycling project)
Prepreg – an assembly of fibres impregnated with resin which will be formed into a composite by curing
PSAB – pressure sensitive adhesive backed
PSB – polystyrene – butadiene rubber
PTFE – polytetrafluoroethylene
PU – polyurethane, interior trim can be made from polyurethane polyester foam or from polyurethane polyether foam. In the trade they are referred to as 'polyester foam' or 'polyether foam', and even simply as 'ether' and 'ester'. Not to be confused with polyester
PUR, (PU) – polyurethane
PURRC – Polyurethane Recycle and Recovery Council (of the Society of the Plastics Industry – USA)
PVA(c) – polyvinyl acetate
PVA(l) – polyvinyl alcohol
PVC – polyvinylchloride

PVDC – polyvinylidene chloride
PVDF – polyvinylidine fluoride
QS 9000 – quality system based on IS0 9000 specially designed for the auto industry
Rad XL PP – radiation crosslinked polypropylene (foam)
RAPRA – Rubber and Plastics Industry Research Association (UK)
RF – radio frequency
SAE – Society of Automotive Engineers (USA)
SAMPE – Society of Advanced Materials and Process Engineers (composites)
SBR – styrene butadiene rubber
SCRDE – Stores, Clothing, Research and Development Establishment
SDC – Society of Dyers and Colourists (UK)
SEPA – Scottish Environmental Protection Agency
SERA – Socialist Environmental and Resources Association
SI – Systeme International des Unites
SME – small medium enterprise
SPC – statistical process control
SUV – sports utility vehicle
Tack – the property of an adhesive to form a measurable bond, immediately it is brought into contact with a surface under low pressure. Sometimes referred to as 'grab' or 'green tack'
Taguchi Methods – statistical techniques for prototyping and optimising product and process design
TDI – toluene di-isocyanate
TI – Textile Institute
TLV – threshold limit value
TPE – thermoplastic elastomer
TPO – thermoplastic polyolefin
TPU – thermoplastic polyurethane
TQM – total quality management
UF – urea-formaldehyde
UL – Underwriter's Laboratory (e.g. for FR testing)
UNCED – United Nations Conference on the Environment
UV – ultra violet
VDI – Verein Deutscher Ingenieure – Association of German Engineers
Viscosity – the perceived resistance of liquids to flow ('thickness')
VOC – volatile organic compound
Vulcanisation – a crosslinking process which improves the properties of rubber materials
WIRA – Wool Industries Research Association

Index